Fabio Sartori Miguel Ángel Sicilia
Nikos Manouselis (Eds.)

Metadata and Semantic Research

Third International Conference, MTSR 2009
Milan, Italy, October 1-2, 2009
Proceedings

Springer

Volume Editors

Fabio Sartori
Department of Computer Science,
Systems and Communication (DISCo)
University of Milan-Bicocca
Milan, Italy
E-mail: sartori@disco.unimib.it

Miguel Ángel Sicilia
Department of Computer Science
University of Alcalá
Alcalá de Henares, Madrid, Spain
E-mail: msicilia@uah.es

Nikos Manouselis
GRNET S.A., Athens, Greece
E-mail: nikosm@grnet.gr

Library of Congress Control Number: 2009934596

CR Subject Classification (1998): H.2.8, H.3, I.2.6, E.1, I.7, J.5

ISSN 1865-0929
ISBN-10 3-642-04589-8 Springer Berlin Heidelberg New York
ISBN-13 978-3-642-04589-9 Springer Berlin Heidelberg New York

springer.com

Printed in Germany

Typesetting: Camera-ready by author, data conversion by Scientific Publishing Services, Chennai, India
Printed on acid-free paper SPIN: 12753581 06/3180 5 4 3 2 1 0

Communications
in Computer and Information Science 46

Preface

This volume collects the papers selected for presentation at the Third International Conference on Metadata and Semantic Research (MTSR 2009), held in Milan at the University of Milano–Bicocca (October 1-2, 2009).

Metadata and semantic research is today a growing complex set of conceptual, theoretical, methodological, and technological frameworks, offering innovative computational solutions in the design and development of computer–based systems. From this perspective, researchers working in this area must tackle a broad range of issues on methods, results, and solutions coming from different classic areas of this discipline. The conference has been designed as a forum allowing researchers to present and discuss specialized results as general contributions to the field.

In order to give a novel perspective in which both theoretical and application aspects of metadata research contribute in the growth of the area, this book mirrors the structure of the conference, grouping the papers into three main categories: (1) Theoretical Research: Results and Proposals; (2) Applications: Case Studies and Proposals; (3) Special Track: Metadata and Semantics for Agriculture, Food and Environment. The book contains 31 full papers (10 for the first category, 10 for the second and 12 for the third), selected from a preliminary initial set of about 70 submissions.

Many people contributed to the success of the conference and the creation of this volume, from the initial idea to its implementation. Our first acknowledgement is to the members of the Steering Commitee, George Bokos and David Raitt. We would also like to thank all Program Committee members and reviewers for their collaboration. Special thanks to Carlo Batini, on behalf of the Department of Computer Science, Systems and Communication of the University of Milan–Bicocca, who kindly hosted our conference.

A particular acknowledgement to Stefania Bandini, the director of the Research Center on Complex Systems and Artificial Intelligence (CSAI) of the University of Milano–Bicocca, and to the "official" Local Organizing Committee (Andrea Bonomi, Daniela Micucci, Eliana Magnolo, Lorenza Manenti, Roberto Pedroli) for their fundamental work.

Finally, our gratitude to the authors of papers presented at the conference and included in this volume of proceedings: your precious collaboration was the main reason for the success of MTSR 2009!

October 2009

F. Sartori
N. Manouselis
M.Á. Sicilia

Organization

MTSR 2009 was organized by the Complex Systems and Artificial Intelligence Research Centre (CSAI) of the Department of Computer Science, Systems and Communication (DISCo) of the University of Milano–Bicocca, in cooperation with the University of Alcalá de Henares (Spain) and the Ionian University (Greece).

Congress Organization

Congress Chair	Fabio Sartori, Miguel Ángel Sicilia and Nikos Manouselis
Local Organization	Andrea Bonomi, Daniela Micucci, Eliana Magnolo, Lorenza Manenti and Roberto Pedroli

Steering Committee

Giorgios Bokos	Ionian University (Greece)
David Raitt	ESA–ESTEC (The Netherlands)
Miguel Ángel Sicilia	University of Alcalá (Spain)
Fabio Sartori	University of Milano-Bicocca (Italy)

Program Committee

Rajendra Akerkar	Technomathematics Res. Foundation (India)
Ioannis Athanasiadis	Dalle Molle Institute for Artificial Intelligence (Switzerland)
Tomaz Bartol	University of Ljubljana (Slovenia)
Carlo Batini	University of Milano-Bicocca (Italy)
Paolo Bouquet	University of Trento (Italy)
Gerhard Budin	University of Vienna (Austria)
Tobias Bürger	University of Innsbruck (Austria)
Caterina Caracciolo	Food and Agriculture Organization of the United Nations (Italy)
Artem Chebotko	University of Texas - Pan American (USA)
Stavros Christodoulakis	Technical University of Crete (Greece)
Oscar Corcho	Universidad Politéchnica de Madrid (Spain)
Constantina Costopoulou	Agricultural University of Athens (Greece)
Sally Jo Cunningham	Waikato University (New Zealand)
Emilia Currás	Universidad Autónoma de Madrid (Spain)
Darina Dicheva	Winston-Salem State University (USA)
Asuman Dogac	University of Ankara (Turkey)

Koraljka Golub	University of Bath (UK)
Asunción Gómez-Pérez	Universidad Politéchnica de Madrid (Spain)
Stefan Gradmann	University of Berlin (Germany)
Jane Greenberg	University of N. Carolina at Chapel Hill (USA)
Claudio Gutierrez	University of Chile (Chile)
Francisca Hernández	Fundación Marcelino Botín (Spain)
Diane Hillmann	Cornell University (USA)
Eero Hyvönen	Helsinki University of Technology (Finland)
Johannes Keizer	Food and Agriculture Organization of the United Nations (Italy)
Pankaj Jaiswal	Oregon State University (USA)
Pete Johnston	Eduserv Foundation (UK)
Dimitris Kanellopoulos	University of Patras (Greece)
Christian Kop	University of Klangenfurt (Austria)
Nikos Manouselis	Agricultural University of Athens (Greece)
Brahim Medjahed	University of Michigan (USA)
Eva Méndez	Carlos III University (Spain)
Daniela Micucci	University of Milano-Bicocca (Italy)
Akira Miyazawa	National Institute of Informatics (Japan)
Ambjörn Naeve	Royal Institute of Technology (Sweden)
William Moen	University of North Texas (USA)
Xavier Ochoa	Centro de Tecnologías de Información Guayaquil (Ecuador)
Petter Olsen	Nofima (Norway)
Matteo Palmonari	University of Milano-Bicocca (Italy)
Laura Papaleo	University of Genova (Italy)
Panayiota Polydoratou	London City University (UK)
Marios Poulos	Ionian University (Greece)
Tadinada. V. Prabhakar	Indian Institute of Technology Kanpur (India)
Edie Rasmussen	University of British Columbia (Canada)
Andrea Emilio Rizzoli	Dalle Molle Institute for Artificial Intelligence (Switzerland)
Stefanie Rühle	University of Göttingen (Germany)
Gauri Salocke	Food and Agriculture Organization of the United Nations (Italy)
Salvador Sánchez-Alonso	University of Alcalá (Spain)
Inigo San Gil	Long Term Ecological Research Network (USA)
Giovanni Semeraro	University of Bari (Italy)
Javier Solorio Lagunas	University of Colima (Mexico)
Elena Simperl	University of Innsbruck (Austria)
Praditta Sirapan	National Science and Technology Development Agency (Thailand)
Aida Slavic	UDC Consortium (The Netherlands)
Shigeo Sushimoto	University of Tsukuba (Japan)
Hussein Suleman	University of Cape Town (South Africa)
David Taniar	Monash University (Australia)

Emma Tonkin	University of Bath (UK)
Joseph Tennis	University of Washington (USA)
Giovanni Tummarello	National University of Ireland (Ireland)
Gottfried Vossen	University of Muenster (Germany)
Andrew Wilson	National Archives of Australia (Australia)
Telmo Zarraonandia	Universidad Carlos III de Madrid (Spain)
Thomas Zschocke	United Nations University (Germany)

Referees

Gunes Aluc
Rajendra Akerkar
Paolo Bouquet
Gerhard Budin
Tobias Bürger
Artem Chebotko
Stavros Christodoulakis
Constantina Costopoulou
Emilia Currás
Damian Gessler
Koraljka Golub
Stefan Gradmann
Claudio Gutierrez
Diane Hillmann
Eero Hyvönen
Dimitris Kanellopoulos
Christian Kop
Nikos Manouselis
Brahim Medjahed
Akira Miyazawa
William Moen
Sergio Munoz-Venegas
Ambiörn Naeve
Cagdas Ocalan
Xavier Ochoa
Matteo Palmonari
Laura Papaleo
Axel Polleres
Marios Poulos
Tadinada. V. Prabhakar
Edie Rasmussen
Andrea Emilio Rizzoli
Stefanie Rühle
Gauri Salokhe
Inigo San Gil
Elena Simperl
Praditta Sirapan
Javier Solorio Lagunas
Shigeo Sugimoto
David Taniar
Joseph Tennis
Emma Tonkin
Gottfried Vossen
Andrew Wilson
Telmo Zarraonandia

Sponsoring Institution

Department of Computer Science, Systems and Communication (DISCo), University of Milano–Bicocca, Milan.

Table of Contents

Theoretical Research: Results and Proposals

Applications: Case Studies and Proposals

Special Track: Metadata and Semantics for Agriculture, Food and Environment

VMAP: A Dublin Core Application Profile for Musical Resources

Carlos A. Iglesias[1], Mercedes Garijo[2], Daniel Molina[1], and Paloma de Juan[2]

[1] Germinus XXI (Grupo Gesfor)*
[2] Depto. Ingeniería de Sistemas Telemáticos,
Universidad Politécnica de Madrid**

Abstract. This paper details a Dublin Core Application Profile defined for cataloguing musical resources described within the European eContentPlus project Variazioni. The metadata model is based on FRBR and has been formalised with DC-Text and implemented in an available web portal where users and music institutions can catalogue their musical assets in a collaborative way.

1 Introduction

According to Lynch [1], the new context for bibliographic control in the new millennium will certainly reorder priorities for investment in bibliographic control practices and will change the way that cataloguing information is used. He identifies three general approaches to retrieve potentially relevant information: through bibliographic surrogates that represent an intellectual analysis and description of aspects and attributes of a work; through content-based techniques that compare queries to parts of the actual works themselves (or to computationally-derived surrogates for the works); and through social processes that exploit the opinions and actions of communities that author, read, and evaluate works, and the information seeker's view of those communities of people involved.

This paper details a Dublin Core Application Profile (DCAP) defined for cataloguing musical resources described within the European eContentPlus project Variazioni. Variazioni project aims at enriching musical contents provided by musical institutions and end-users by combining the three processes: manual cataloguing based on well-defined metadata by musical institutions and end users; automatic cataloguing based on audio analysis, and social tagging. This paper is focused on the definition of the metadata schema for manual cataloguing.

The article is structured as follows. First, section 2 reviews the limitations of existing metadata standards and projects for cataloguing musical assets. Then,

* This research has been co-funded by the European Community under the programme eContentPlus. The authors are solely responsible for this article and it does not represent the opinion of the European Community. The European Community is not responsible for any use that might be made of information contained within it.

** This work has been partially funded by the Spanish Ministry of Industry, Trade and Tourism under the Avanza Programme in the project Musiteca.

F. Sartori, M.Á. Sicilia, and N. Manouselis (Eds.): MTSR 2009, CCIS 46, pp. 1–12, 2009.

section 3 presents the model of the Variazioni Musical Dublin Core Application Profile (VMAP). Finally, section 4 draws some conclusions and future work.

2 Limitation of Metadata Standards for Musical Assets

After reviewing relevant metadata standards in different domains related with music assets (Libraries, Museums, Education, Audivisual and Music), the first conclusion is that any of the reviewed standards deal with the cataloguing of music resources with enough detail for fitting user requirements in terms of search facilities and collocations. In addition, there are important limitations in traditional cataloguing systems for music resources.

Traditional library cataloguing records, based on AACR2R [2] cataloguing rules and MARC [3] bibliographic and authority standards, have provided a solid foundation for the required descriptive metadata elements for searching and retrieving works of music and are used by music cataloguing agencies worldwide [4]. Nevertheless, they present limitations for the music domain [5,4]. First, they lack of adequate structural and administrative metadata for expressing the internal structure of the musical objects. In addition, they consider only the role of author, while other roles such as peformer or composer which are relevant for grouping results for end users are not considered. Other examples include limitations with just one title (track title, CD container title, alternative title, etc.) and with just one date (date of performance, composition, record creation, etc.), or the lack of an object oriented metadata model which provides facilities for including key entities such as composition, performer or composer.

FRBR (Functional Requirements for Bibliographic Records) [6] has accomplished a shift in the cataloguing area, putting emphasis on a conceptual model which is focused on the Work rather than on the Manifestation. FRBR has been applied previously in the musical domain, and new library standards, such as RDA [7] or IAP [8] are based on FRBR. Our conclusion is that FRBR is a good starting point for defining and modelling Variazioni metadata. This conclusion could be considered in a wider scope. According to Gartner [9], *"given the complexity of metadata requirements, it is perhaps not surprising that no single standard has yet emerged which addresses them all. Nonetheless, the emergence of the standards detailed in this report, all of which are based on the Functional Requirements for Bibliographical Records (FRBR) conceptual model, and the interoperability allowed by their common language, does allow for a coherent metadata landscape to be constructed on a sector-wide basis."*

METS [10] and MPEG-21 [11] are two standards that attempt to provide overall frameworks within which descriptive, administrative and structural metadata have emerged from different communities [9]. While METS comes from the library community (the MARC standards office), MPEG-21 comes from the multimedia community. The project Variazioni counts with experts in MPEG-21, and the resulting metadata will be available in MPEG-21.

Regarding the standards developed in the museum community, they deal with aspects not relevant for Variazioni (physical location or provenance of the items) and, in addition, there is an adaptation of FRBR, so-called FRBRoo, which provides an effort in modelling CIDOC CRM [12] based on FRBR entities. The usage of controlled vocabularies will be discussed in the presentation of the Variazioni metadata model definition.

Another interesting specification is IAP (Image Application Profile) [13,8], which defines an application profile based on DC, FRBR, VRA [14], CDWA [15] and MIX [16]. The basic FRBR model is modified. For sake of clarity, FRBR Work is renamed as Image, in order to distinguish the common use of work in image cataloguing (the image, physical thing, manifestation) from the abstract Work defined in FRBR. In addition, expression is omitted in this profile, since it is not considered useful for cataloguing images.

In the musical domain, the most relevant projects are Variations [17,5,18,19], Music Australia [20,21] and Harmos [22].

Harmos was an eContent project which preceded Variazioni and was focused on cataloguing master classes. The educational taxonomy for cataloguing master classes [22] has been included in this model.

Variations and MusicAustria have used FRBR as their basis. According to Riley [23], FRBR is the most influential model and represents a great deal of potential for search systems based on high-quality, structured metadata. Several FRBR-like systems have been implemented for the music domain, such as Variations2 [24] from Indiana University, the British Library Sound Archive [24], Opera Archives [25] and MusicAustralia [21].

The application of FRBR [26] in the next version of AACR2, called RDA, has provided some findings, such as 78% of works have only one manifestation, which could question its benefits. Nevertheless, "works in fields like music, that appear in multiple versions over time, will benefit the most from systems that implement FRBR principles" [26]. Buchanan [27] has shown how users and librarians can benefit from the exploitation of FRBR relationships, as well as the inclusion of new services, such as alerts.

The music community has been very active in discussing modelling issues [28]. One of the problems that has been pointed out is how to deal with Music and Lyrics in vocal music or which is the work (music or lyrics or both). A second problem is how FRBR model fits in non Western music tradition, where there is not a previous Work.

Variations is focused on a digital music library, while our project addresses a wide range of music contents, which has led to distinguish different kinds of musical contents and its physical embodiment in productions.

The Music ontology [29] is also based on FRBR, but, as well as MusicBrainz[30], it is also focused on digital music libraries, which are not the only target of our project. The Music ontology has been considered very interesting and a mapping between our model and this ontology could be defined as future work. Some of the concepts of the Music ontology have been considered too complex for this model. For example, a Composition is an Event which produces the Musical Work. Variazioni Metadata model can contribute to this ontology by refining some of their concepts.

3 Metadata Modelling for Musical Resources Using FRBR

In this section it is discussed how the FRBR conceptual model can be applied for cataloguing different types of musical contents, such as master classes videos or digitalized scores. In order to understand better the relationship with FRBR, a first identification of FRBR entities per musical content type has been carried out as shown in table 1.

Table 1. Identification of FRBR entities

Variazioni Contents	FRBR 1st Group Entities			
	Work	Expression	Manifestation	Item
Master class	Master Class	Class Event	Production	Media File
Score	Composition	Editorial Event	Printing	Media File
Concert	Composition	Concert event	Production	Media File
Image*	Image itself	"Event"	Production	Media File
Studio Recording	Composition	Production Event	Production	Media File
Libretto	Composition, Lyrics	Editorial Event	Production	Media File

From the previous table, the following issues can be pointed out.

- Expression and Work entities are not easy to identify in some cases, such as Master Classes or Conferences. This happens because the intellectual or artistic activity (Work) emerges while the activity (Expression) is being carried out. A similar issue has been previously reported for Western Music or Jazz improvisation in FRBRList [28] or MusicAustralia [20].
- According to FRBR, an Expression is the realization of one and only one Work [6]. This can cause some problems while cataloguing if the final digital file contains several Expressions (for example, a video recording with several performances or a digitalised score book with several scores, or a CD in only one track) and there is not a segmentation tool available in the system. The main Work entity in the music domain is Composition. Nevertheless, in some musical contents, such as Master Classes or Conferences, the Composition is not the intellectual / artistic activity of the Master class / Conference, but it is commonly used to exemplify a concept. They are used as *subjects*.
- Managing image and 'event material'. The image content is problematic. For example, let us consider a concert, where there are a video recording, an audio recording and photos of the event. One natural alternative is considering that all of them are 'Manifestations' of the same Expression (the Concert) but recorded in different media (image, video or sound). The main problem is that the photo may not be easily linked to the performance of one particular Work, but to the general event. A similar case happens when cataloguing related material such as the announcement poster of the Concert. According

to [31], these augmentations (illustrations, notes, glosses, etc.) of the Expression should be considered separate Expressions of their own separate works, but this makes cataloguing more laborious.

- In digital libraries, the distinction between Manifestation and Item is not so relevant, since there is only one copy of the work (the digital media). FRBR cannot be considered as a data model, but a conceptual schema. FRBR does not even require implementing the four entities of the first FRBR Group [31].
- While FRBR follows a top-down approach for cataloguing, cataloguing process follows a bottom-up approach. Users or librarians catalogue an Item, not a Work. Users should have an easy interface in order to catalogue their media files, without being aware of the FRBR model. Expertise in implementing FRBR in standard databases [20] has shown its utility for end users to find relationships between items, which were hidden before its implementation. Nevertheless, these experiences have shown that since FRBR provides several alternatives during the cataloguing process, this can make the process difficult to understand. Some examples of these difficulties are deciding whether music and lyrics should be catalogued as different items, the definition of relationships between expressions (i.e. an interpretation (e1) based on a libretto (e2) of a work (o1)), and the cataloguing of expressions based on improvisation, such as jazz music and folk traditions.
- Cataloguing is commonly carried out in an iterative way in musical institutions. Depending on the available resources, a media file can be catalogued with very few metadata and catalogued more exhaustively when there are more resources. This includes the identification of the Composition (Work), which can be delayed.

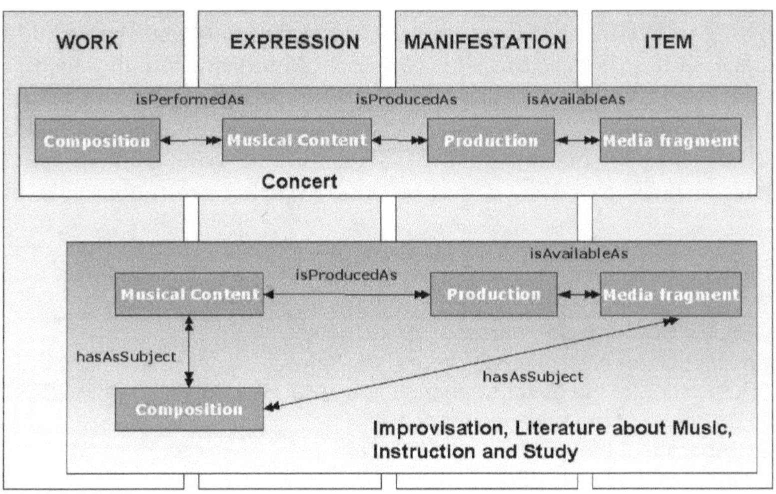

Fig. 1. VMAP and FRBR entities

Based on the limitations identified above, an adaptation of FRBR for musical resources is here proposed.

This model, Variazioni Musical Application Profile (VMAP) is based on the existing FRBR entities, although these entities has been renamed and redefined as shown in figure 1, in order to overcome the identified limitations. In particular:

- *Work* has been limited to *Composition.* A Composition is an original piece of music.
- *Expression* has been redefined as *Musical Content.* A Musical Content (Musical Content Type) is a classification scheme of digital items which defines the nature and descriptive metadata of the digital item. Some of the musical content types identified are Master Class, Conference, Libretto, Musical Score, etc.
- *Manifestation* has been renamed as *Production.* A Production maintains all the metadata related to the physical edition of a Musical Content, as well as the structural metadata when the manifestation is composed of more than one Media Fragment. The structural metadata can include the order of different Media Fragments or the starting and end points of one media file with different fragments (pages, seconds, frames, etc.). Two subtypes of Production have been identified, document and audiovisual, for defining specific metadata.
- *Item* has been renamed as *Media Fragment.* A Media Fragment is a media file or a fragment of it, and maintains all the relevant metadata of the media file, including its title and licence.

In order to clarify these elements, an example of how the same items are catalogued according to standard FRBR and Variazioni Music Application Profile is shown in tables 3 and 3 respectively.

From this example, the main differences of the model can be outlined. First of all, according to FRBR, and Expression has one and only one Work, and this has involved the shift in focus from the resource (Manifestation) in the traditional cataloguing world to the Work in FRBR. Our proposal consists of modifying the cardinality of the relationship hasWork between Work and Expression, from 1-1 in FRBR to M-M (many-to-many). This allows solving some of the previous issues pointed out: (a) , since Compositions (Works) are not mandatory for a

Table 2. FRBR modelling. Legend: W: Work, E: Expression, M: Manifestation, I: Item.

W1. J. S. Bach's Six suites for unaccompanied cello
E1. Transcription for classic guitar by Stanley Yates
M1. Publication of the guitar transcription by Mel Bay Publisher in 1988
I1. Exemplar of the book in library 1.
I2. Separata of the guitar edition in library 1.
E2. Performances by Janos Starker recorded in 1963 and 1965
M1. Recordings released on 33 1/3 rpm sound discs in 1965 by Mercury
M2. Recordings re-released on CD in 1991 by Mercury

Musical Content (Expression); and (b), since one Musical Content (Expression) may have more than one associated Compositions (Works).

Another interesting change is the usage of the relationship *hasSubject*, in particular for linking any element of the model with *Composition*. FRBR only considers this relationship for *Works*. In our case, for example, for Master classes, several Compositions could be the subject (or example) of a master class. In the example previously presented, a Composition can be assigned as subject of a *Media Fragment*, avoiding the need for creating a new *Expression*. This is depicted in figure 1, which points out two different kinds of semantic relationships between *Composition* and *Musical Content*: *isRealizedAs* and *hasSubject*. In terms of searchability, we have not found the need to distinguish between

Table 3. VMAP modelling. Legend: C: Composition, MC:Music Content, P: Production, MF: Music Fragment.

- MC1. Score. Transcription for classic guitar by Santley Yates
  - C1: J. S. Bach's Six suites for unaccompanied cello
  - P1: Book edition
    - MF1: Media file of the book (page range if book includes more compositions)
  - P2 Separata of the guitar edition
    - MF2: Media file of the separata
- MC2. Studio Recording. Performances by Janos Starker recorded in 1963 and 1965
  - C1: J. S. Bach's Six suites for unaccompanied cello
  - P3: Recordings released on 33 1/3 rpm sound discs in 1965 by Mercury.
    - MF3: Suite 1 media file (and details of the fragment, full or time range)
      - C2: J. S. Bach Suite 1 for unaccompanied cello [is-part-of C1]
    - MF4: Suite 2 media file (and details of the fragment, full or time range)
      - C3: J. S. Bach Suite 2 for unaccompanied cello [is-part-of C1]
    - ...
    - MF8: Suite 6 media file (and details of the fragment, full or time range)
      - C7: J. S. Bach Suite 1 for unaccompanied cello [is-part-of C1]
  - P4: Recordings re-released on CD in 1991 by Mercury
    - MF9: Single media file of the suites

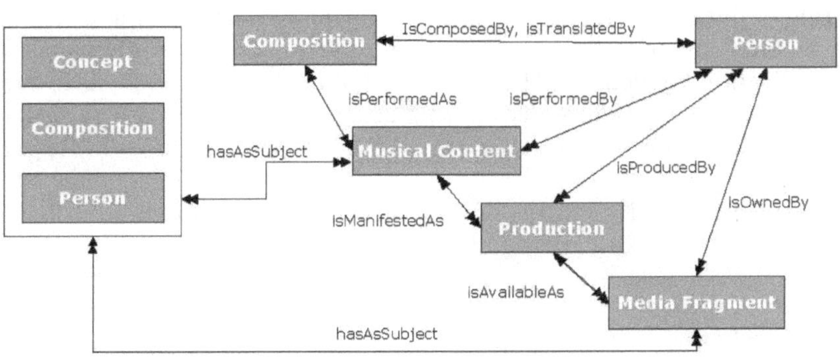

Fig. 2. VMAP model

both in the implementation of the model. Furthermore, it is possible to define the subject of a media fragment, allowing a direct asignation.

The relationships between the entities of the VMAP model are shown in figure 2. It is important to point out the cardinality between Composition and Musical Content. A Musical Content can have more than one Composition (as a realization or as a subject). In addition, Media Fragments can have one or more Compositions as a subject or realization. In order to simplify the cataloguing process, no distinction is made between the relationships *isPerformedAs* and *hasAsSubject*, as shown previously in figure 1.

VMAP considers *MusicalContent* as the central entity. The entity *Composition* can be identified when there are resources to do this catalogation, but it is not required, in contrast with FRBR, where the entity *Work* should be identified in first place. According to the requirements of the content providers in the project, the following musical contents have been identified:

Score: Musical score, it can be a scanned handwritten, autograph or printed score or a computerized score.

Musical document: Document related with the musical domain, such as image, libretto, literature on music, reference works, correspondence between musicians, methods, press releases or concert programs.

Class: Audiovisual content about some pedagogical activity, such as master class or regular class.

Conference: Audiovisual content recording of a conference about music.

Musical performance: Audiovisual content with the recording of a musical performance, such as concert, competition or musical event.

For each musical content type, specific metadata has been identifed as shown in figure 4 and is formalized in [32].

This model has been formalized as a Dublin Core Application Profile for describing digital musical contents hold by musical institutions or end users. The concept of Application Profile has emerged within the Dublin Core Metadata Initiative. Dublin Core Application Profiles (DCAP) can be defined [33] as schemas consisting of data elements drawn from one or more name spaces, combined together by implementers, and optimized for a particular local application Dublin Core Application Guidelines [34] and the Singapore guidelines [35] have been followed to document attributes drawn from other name spaces and the declaration of new elements under the name space VMAP (*Variazioni Musical Application Profile*).

The resulting DCAP is available at [32] and has been formalized using DC-Text [36], where all the entities and properties have been formalized, as well as 26 classification schemes. VMAP reuses several controlled vocabularies, such as Getty Thesaurus for Geographical names [37], the thesaurus of Musical Instruments [38] as well as metadata schemas such as FOAF [39]. In addition, a mapping to Simple Dublin Core has been defined in order to provide OAI-PMH interoperability.

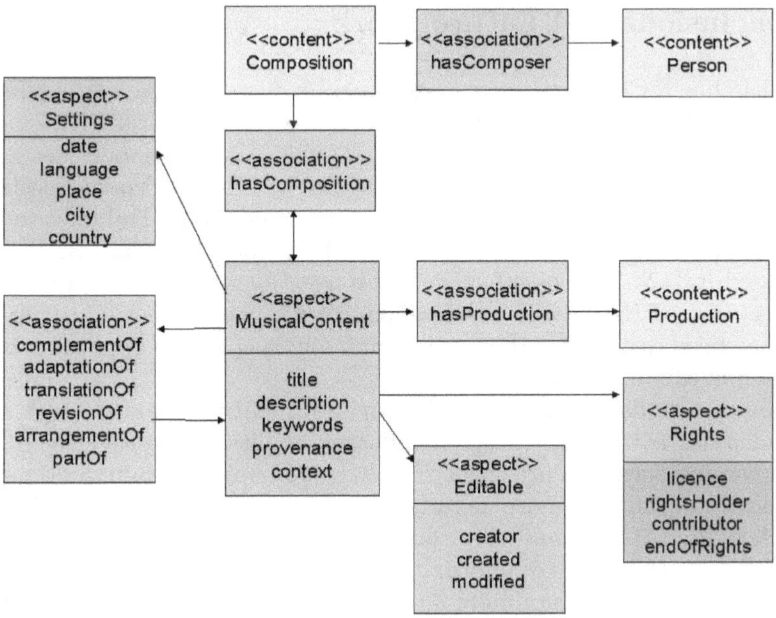

Fig. 3. Musical Content Entity

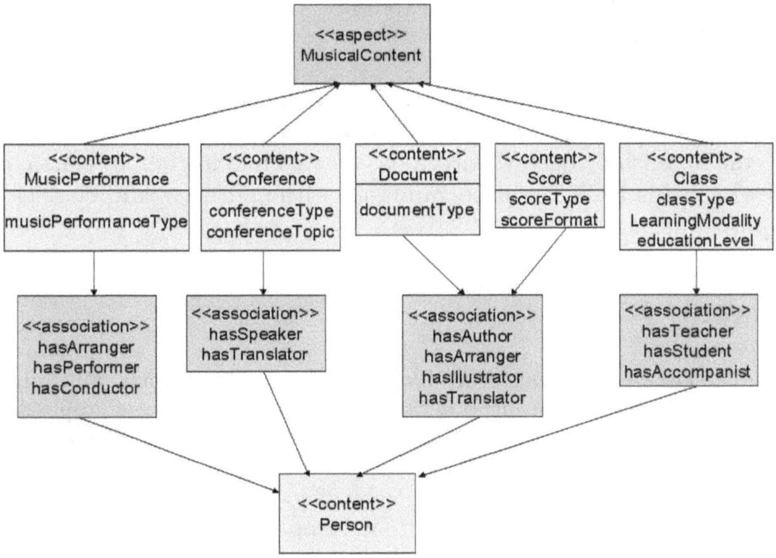

Fig. 4. Subtypes of Musical Content Entity

The model has been implemented following the aspect oriented content model of Alfresco [40] (figure 3) and is publicly available at the Variazioni Content Enrichment Model (CEP) [41].

4 Conclusions and Future Work

This article has reviewed the practices in cataloguing musical contents and has proposed a Dublin Core Application Profile for Musical Digital Contents.

The usage of FRBR as a basis for the musical metadata mode has provided a suitable model, although the need to redefine the entities could question the suitability of this model and its interoperability, since several Dublin Core Application Profile have followed the same approach in redefining core FRBR entities, as discussed in [42]. This could suggest the need for defining a more extensible model based on FRBR, which would improve the interoperability. Although VMAP has been defined for the musical domain, it could be applicable in other arts, which is left to future work.

Defining a metadata schema which provides quality metadata is not an easy task. It enables a trade-off between different quality properties, such as consistency, completeness, accuracy, shareability, economic feasibility and usability. One could think that the more metadata a schema has, the more quality it has. Nevertheless, one of the common pitfalls pointed out in [43] for providing shareable metadata is the excess of information. Economic and human factors are also important in the definition of metadata, in order to provide understandable and effective metadata. Furthermore, natural resistance to change is a challenge for adopting a new metadata model.

In order to define quality metadata, the process defined in [44] has been followed for defining Variazioni Metadata Model, identifying external and internal requirements. The implementation of the project has provided feedback about the decisions taken in this definition.

Currently partners of Variazioni are working in the dissemination of both the VMAP model and the portal. Feedback from users will guide its evolution. In addition, the Spanish project Musiteca has started a living lab for classic musical contents where this DCAP will be validated with end users and classical music content providers.

References

1. Lynch, C.: The new context for bibliographic control in the new millennium
2. Hunter, E.J.: AACR 2: an introduction to the second edition of Anglo-American cataloguing rules. C. Bingley, London; Linnet Books, Hamden, Conn. (1979)
3. Library of Congress: MARC (MAchine Readable Catalogue) Web Site
4. Hemmassi, H.: Why not MARC? In: Proceedings of the 3rd International Conference on Music Information Retrieval, ISMIR, pp. 242–248 (2002)
5. Minibbayeva, N., Dunn, J.W.: A digital library data model for music. In: Proceedings of the Second ACM/IEEE Joint Conference on Digital Libraries, pp. 154–155 (2002)
6. IFLA: FRBR functional requirements for bibliographic records. Technical report, International Federation of Library Associations and Institutions, IFLA (1998)
7. RDA: Resource Description and Access, Joint Steering Committee for Development of RDA, `http://www.collectionscanada.gc.ca/jsc/rda.html`

8. JISC: Images Application Profile, http://www.ukoln.ac.uk/repositories/digirep/index/Images_Application_Profile
9. Gartner, R.: Metadata for digital libraries: state of the art and future directions. JISC Technology and Standards Watch (2008)
10. Digital Library Federation: Metadata encoding and transmission standard: Primer and reference manual (September 2007)
11. Wan, X.: Mpeg-21 rights expression language: enabling interoperable digital rights management. Multimedia IEEE 11(4), 84–87 (2004)
12. CIDOC Web Site. International Council of Museums, http://www.cidoc.ics.forth.gr
13. Eadie, M.: Towards an application profile for images. Ariadne Electronic Magazine (55) (April 2008)
14. VRA (Virtual Resources Association), International Association of Image Media Professionals, http://www.vraweb.org
15. Paul Getty Trust, J.: CDWA (Categories for Description of Works of Art) (2006), http://www.getty.edu/research/conductiong_research/standards/cdwa/index.html
16. NISO: MIX. NISO Metadata Standard for Images in XML Schema. technical metadata for still images standard
17. Dunn, J.W., Mayer, C.A.: Variations: A digital music library system at indiana university. In: Proceedings of the 4th ACM Conference on Digital Libraries, Berkeley, CA, pp. 12–19 (1999)
18. Riley, J., Hunter, C., Colvard, C., Berry, A.: Definition of a frbr-based metadata model for the indiana university variations3 project. Technical report, University of Indiana (2007)
19. Riley, J., Mullin, C., Colvard, C., Berry, A.: Definition of a frbr-based metadata model for the indiana university variations3 project. phase 2: Frbr group 2 and 3 entities and frad. Technical report, University of Indiana (2008)
20. Ayres, M.L.: Case studies in implementing functional requirements for bibliographic records [frbr]: Auslist and musicaustralia. Australian Library Journal 54(1) (2004)
21. MusicAustralia: Musicaustralia web site, http://www.musicaustralia.org
22. Iglesias, C.A., Sánchez, M., Guibert, M.J., Guibert, A., Gómez, E.: A multilingual a multilingual web-based educational system for professional musicians. In: Current Developments in Technology-Assisted Education, Proceedings of Fourth International Conference on Multimedia and Information and Communication Technologies in Education, Sevilla, Spain (2006)
23. Riley, J., Mayer, C.A.: Ask a librarian: The role of librarians in the music information retrieval community. In: Proceedings of the International Conference on Music Information Retrieval, ISMIR (2006)
24. Mark Notess, J.R., Hemmasi, H.: From abstract to virtual entities: Implementation of a work-based searching in a multimedia digital library. In: Heery, R., Lyon, L. (eds.) ECDL 2004. LNCS, vol. 3232, pp. 157–167. Springer, Heidelberg (2004)
25. Pompilio, A., Bianconi, L.: RADAMES: A New Management Approach to Opera: repertory, archives and related documents. In: Proceedings of the First International Conference on Automated Production of Cross Media Content for Multichannel Distribution, AXMEDIS 2005 (2005)
26. Riley, J.: The essence of information technology for library decision-makers. In: FRBR, TechEssence. info (May 2006)

27. Buchanan, G.: Frbr: Enriching and integrating digital libraries. In: JCDL 2006. Proceedings of the 6th ACM/IEEE-CS Joint Conference on Digital Libraries, pp. 260–269 (2006)
28. le Boeuf, P.: Frbr mailing list, list address frbr@bnf.fr
29. Giasson, F., Raimond, Y.: Music ontology specification
30. Friedich, M., Kaye, R.: Musicbrainz metadata initiative 2.1. XML metadata format (2008)
31. IFLA Working Group on the Expression Entity: FRBR Chapter 3: Entities, Proposed changes to the FRBR Text
32. Iglesias, C.A., Molina, D.: Variazioni musical application profile (2009), http://www.variazioniproject.org/vmap
33. Heery, a., Patel, M.: Application profiles: mixing and matching metadata schemas. Ariadne Electronic Journal (25) (September 2000)
34. Baker, T., Makx Dekkers, T.F., Heery, R.: Dublin core application profile guidelines
35. Nilsson, M., Baker, T., Johnston, P.: The singapore framework for dublin core application profiles
36. Johnston, P.: Expressing Dublin Core metadata using the DC-Text Format (2007), http://www.dublincore.org/documents/dc-text/
37. Paul Getty Trust, J.: Arts and Artchitecture Thesaurus (AAT), http://www.getty.edu/research/conducting_research/vocabularies/aat/
38. Jenkins, K.: Thesaurus of Musical Instruments (2003), http://www.alteriseculo.com/instruments/
39. Bricklye, D., Miller, L.: Foaf vocabulary specification 0.91, namespace document 2, openid edn.
40. Alfresco: Alfresco web page, http://www.alfresco.com
41. Variazioni Project: Variazioni content enrichment portal, http://cep.variazioniproject.org
42. Chaudhri, T.: Assessing frbr in dublin core application profiles. Ariadne Electronic Magazie (58) (January 2009)
43. Sarah, L.S., Riley, J., Milewicz, L.: Moving towards shareable metadata. First Monday 11(8) (August 2006)
44. Marieke Guy, A.P., Day, M.: Improving the quality of metadata in eprint archives. Ariadne (38) (January 2004)

Usage-Oriented Topic Maps Building Approach

Nebrasse Ellouze[1,2], Nadira Lammari[1], Elisabeth Métais[1], and Mohamed Ben Ahmed[2]

[1] Laboratoire Cedric, CNAM
292 rue Saint Martin, 75141 Paris cedex 3, France
{metais,lammari}@cnam.fr
[2] Ecole Nationale des Sciences de l'Informatique, Laboratoire RIADI
Université de la Manouba, 1010 La Manouba
{nebrasse.ellouze,mohamed.benahmed}@riadi.rnu.tn

Abstract. In this paper, we present a collaborative and incremental construction approach of multilingual Topic Maps based on enrichment and merging techniques. In recent years, several Topic Map building approaches have been proposed endowed with different characteristics. Generally, they are dedicated to particular data types like text, semi-structured data, relational data, etc. We note also that most of these approaches take as input monolingual documents to build the Topic Map. The problem is that the large majority of resources available today are written in various languages, and these resources could be relevant even to non-native speakers. Thus, our work is driven towards a collaborative and incremental method for Topic Map construction from textual documents available in different languages. To enrich the Topic Map, we take as input a domain thesaurus and we propose also to explore the Topic Map usage which means available potential questions related to the source documents.

Keywords: Topic Map (TM), thesaurus, incremental building process, collaborative build, multilingual documents, Topic Map enrichment, multilingual documents.

1 Introduction

Topic Map standard [1] is started in the 90's from work on managing indexes to computer documentation and was worked on from several years before being an ISO standard in 2000 and a semantic web standard in 2001 with XTM (XML Topic Maps) [2]. In this context, Topic Maps are used as semantic structures to organize contents of documents provided from different information sources and different languages in order to enhance navigation and help users find relevant information in theses resources. Since, nowadays people want to obtain and to access information not only in their native language, but also information provided in foreign languages.

We present in this paper an approach to build a Topic Map based on an incremental building process, taking into account multilingual textual documents and Topic Map usage by considering available potential questions related to source documents.

Many Topic Maps building approaches can be found in the literature. Generally, they are dedicated to particular data types like text, semi-structured data, relational data, etc. We note that most of these approaches take as input monolingual documents

F. Sartori, M.Á. Sicilia, and N. Manouselis (Eds.): MTSR 2009, CCIS 46, pp. 13–23, 2009.

to build the Topic Map. The problem is that the large majority of resources available today are written in various languages, and these resources are relevant even to non-native speakers. The work we present in this paper describes a semi-automatic method for building Topic Maps from multilingual documents. The resulting global Topic Map should give the possibility to navigate multilingual information that should allow users to access information in the transparent way as they do in their mother language.

The paper will be structured as follows: In section 2, we give an overview of our multilingual Topic Maps construction approach. Section 3 and 4 details the different steps of the approach. Section 5 discusses some related works proposed for building and generating Topic Maps. At least, we conclude and give some perspectives for this work.

2 An Overview of Our Multilingual Topic Map Construction Approach

A Topic Map is a semantic structure which allows organizing and representing knowledge from information resources (documents, databases, videos, etc). The main concept of Topic Maps is topic which represents the subject being referred to. A topic may have a base name and variant names. A Topic Map can contain many topics that can be categorized into topic types. A topic may be linked to one or more information sources that are deemed to be relevant to the topic. Such links are called occurrences of the topic. An association is a link element, showing relationships between topics. The role played by a topic in an association is one of the topic characteristics.

Topic Maps don't have the same purpose as ontologies. In fact, an ontology is defined as a formal conceptualisation of the real world shared by a community to enable knowledge sharing and reuse independently from its usage whereas Topic Maps aims at organizing a content of documents according to its usage. Furthermore, the notion of concept in ontologies doesn't have the same meaning as the notion of topic in Topic Maps: in a Topic Map everything could be represented as a topic (a person, an object, a concept, a theme, etc).

Having a content composed of multilingual textual documents, our approach aims at supplying an organization of this content in the form of a Topic Map which would

Table 1. Relationships abbreviations in a multilingual thesaurus

English	**Meaning**	**French**	**German**
USE	*Use term instead*	EM *Employer*	BS *Benutze*
UF	*Used for*	EP *Employer pour*	BF *Benutzt für*
BT	*Broader term*	TG *Terme générique*	OB *Oberbegriff*
NT	*Narrower term*	TS *Terme spécifique*	UB *Unterbegriff*
RT	*Related term*	TA *Terme associé*	VB *Verwandter Begriff*
SN	*Scope note*	NE *Note explicative*	D *Definition*

allow, given a user query in the native language, to get back all the available documents containing the response to the query. It takes into account the document multilingual aspect by integrating it in the produced Topic Map specification. This is realized thanks to the contribution of domain experts and the use of a multilingual domain thesaurus. The latter is above all a thesaurus that groups terms of a specific domain with their definitions and their relationships. In a multilingual thesaurus, both terms and relationships are represented in more than one language. The following table describes the relationships of a thesaurus as presented in the ISO standard for multilingual thesaurus (ISO-5964-1985) [3]:

A thesaurus can also be viewed as a form of ontology since its grammar is not formalized and since its type of relationships are not diversified as well as those of an ontology. Moreover, some ontology building approaches propose to explore an existing thesaurus as a starting point to create the ontology. In fact, hierarchical relations in the thesaurus, namely BT and NT, can be considered as corresponding to the "is-a" relation in an ontology. RT relations are defined between two terms that are generally used together in the same context. All SN terms can be treated as properties. The preferential relation (UF/USE) can be converted to synonymy in the ontology.

Another specificity of our approach is that a Topic map is built in an incremental way. In fact, it produces a Topic Map that has been evolved, during the building process, gradually, with the construction of the content at every introduction of a new document in the content. Its main idea is to construct a Topic Map TM_i corresponding to a set of documents D= $\{d_1, d_2 \ldots d_i\}$ by enriching the topic map TM_{i-1} associated to the set of document D-$\{d_i\}$. This enrichment of TM_i is realized by integrating the topic map associated to a document d_i into TM_{i-1}. Each of the process of building a topic map for a document d_i and its integration into the global topic map uses the domain thesaurus and interacts with the domain expert community. Its general algorithm is the following:

Algorithm 1. The general algorithm of our approach

Inputs: *A set of documents to be organized into a topic map.*
Output: *A global topic map*

Action 1. *Build the root of the global Topic Map (Topic Map 0). We mean by root the topic which has the domain name in different languages.*
Action 2. *Select a community of domain experts to build the global Topic Map.*
For *each document i in the multilingual document set* ***do:***
- ***Action 3.*** *Extract a list of topics from document i and organize them to build Topic Map i.*
- ***Action 4.*** *Validate the Topic Map i by all domain experts and evaluate the produced topics.*
- ***Action 5.*** *Enrich the global Topic Map (Topic Map i-1) with the Topic Map i created from document i.*

End

Details of each action 3, 4 and 5 are given in the two next sections of this paper.

3 The Topic Map Building and Validation Steps

3.1 The Topic Map Building Step

This step is referred to the action 3 in our general algorithm. It consists on the extraction from a document d_i a set of topics and their organization into a topic map that is to say on the construction of a topic map associated to a document d_i.

This phase requires a kickoff phase where the group of domain expert mastering the language in which is written d_i gives (1) a query sample attached to d_i and (2) for each query of the sample the response to the query. This response is a portion of d_i that could be a data or a segment of d_i. That supposes that domain experts' groups were beforehand constituted (an expert can be in several groups) and that the documents were beforehand sorted out according to the language.

Algorithm 2. Topic Map building algorithm from one document

Action 3. *Extract a list of topics from document i and organize them to build Topic Map i:*

- ***Action 3. 1.*** *Ask each selected expert to elaborate a list of questions related to the treated document i and annotate the associated fragment of the document that allow finding the answer to this question. The answer could be a data or a fragment from the document.*
- ***Action 3. 2.*** *For each data considered as an answer to a question, ask the expert to supply an associated note (meaning) and then consider it as a topic having as name the attributed meaning.*
- ***Action 3. 3.*** *Attribute a list of keywords for all the other answers. Each keyword is a concept in the thesaurus and constitutes a topic which has a name in each language (if there is one).*
- ***Action 3. 4.*** *Organize all the obtained topics using the thesaurus relationships and the occurrence links.*

To identify the list of keywords that are candidates to be topics, we take into account the list of document fragments proposed by the experts as answers of their questions and the thesaurus. It is an extraction task of thesaurus concepts that appear in the fragments. For this purpose we reuse one of the existing techniques. In fact, many approaches have been proposed to solve this problem. The majority of them have been developed through the creation and enrichment of ontologies from textual data [4][5][6][7][8]. They are distinguished by the type of the techniques used: statistical techniques, technical or syntactic techniques for data mining. Most of them have been implemented and propose tools to extract concepts and relations from textual documents.

All the determined topics are then organized using thesaurus relationships. Relations between terms present in the thesaurus (BT and NT) are directly exploited to add relations between topics and obtain a first version of the Topic Map i associated to document i. We also exploit the relationships USE and UF to merge topics or to add a name to a topic.

Finally, each produced topic is linked to the studied document by an occurrence link. This latter is annotated by the language of the related document. This annotation can be expressed by the concept of facet already existing in the Topic Map model. In fact, the TM standard provides the notion of facets. A facet is a set of attributes characterizing an occurrence link. It could be applied to view a topic in different ways. In our case, since we are in a multilingual context, this attributes could be used later, by the system, as a selecting criteria: the user gives the query in his native language and the system should return first all related documents in the user language and then documents in other languages.

After creating Topic Map i, a pruning process have to be applied to keep only relevant topics and eventually add new topics and associations. This process is referred to the Topic Map validation step by domain experts.

3.2 The Topic Map Validation Step

Our approach is based essentially on collaborative build. In fact, all domain experts are involved in the validation process. At this stage, they validate the resulting Topic Map i, derived from the previous step. The outcome of this stage is a new version of the Topic Map containing knowledge acquired from all domain experts.

This step is referred to the action 4 of our general algorithm (algorithm 1). Its execution will ensure a reasonable broad coverage of the document for the studied domain. It consists in adding new topics, deleting proposed topics, modifying proposed topics by affecting a new name to a topic, renaming a topic, grouping topics, etc. It is also an enrichment step since it involves experts'intervention to add semantic associations between topics.

After the expert validation, each topic is weighted by its frequency in experts' questions. This evaluation could serve a pruning process of the topic map. In fact, to facilitate the management of the evolution of a Topic Map it seems to us judicious to collect knowledge about the pertinence of the topics. This knowledge is stored as meta-properties associated to topics. The frequency of a topic in expert questions could be an example of this type of knowledge to store.

The most popular measure for weighting terms is *term frequency–inverse document frequency* measure [9], denoted $tfidf_{t,d}$. In our case it will be $tfiqf_{t,q}$ and defined in equation (1). This measure is used throughout the Topic Map creation process to assign a weight to each topic according to its frequency in the questions elaborated by domain experts and related to the source document.

$$tfiqf_{t,q} = \frac{freq_{t,q}}{|q|} * log\left(\frac{|Q|}{qf_t}\right) \qquad (1)$$

$tfiqf_{t,q}$ is tfiqf measure of term t for question q
$freq_{t,q}$ is the number of occurrences of term t in question q
$|q|$ is the number of all terms in question q
$|Q|$ is the number of questions in the experts' questions collection
qf_t is the number of questions that term t occurs in.

4 The Topic Map Integration Step

As we said previously, the main design considerations of our approach is the incremental build. Thus, our approach is based on an iterative process (see Fig.1). Each iteration includes two steps. The first one consists of building a Topic Map (Topic Map i) from one document (document i) according to the algorithm presented in the previous section (algorithm 2). The second step consists of enriching the global Topic Map with Topic Map i created from document i. The enrichment step is referred to action 5 in the global Topic Map building algorithm (algorithm 1).

The enrichment step of our approach is based on integration techniques. Most of them propose a four step integration process [10]: a pre-merging step where the sources are standardized into a unified model, a comparison step to identify links between concepts belonging to different sources (this corresponds to a matching process), an integration step that merge the sources according to identified links and to available merging rules and finally a re-organization step of the resulting target.

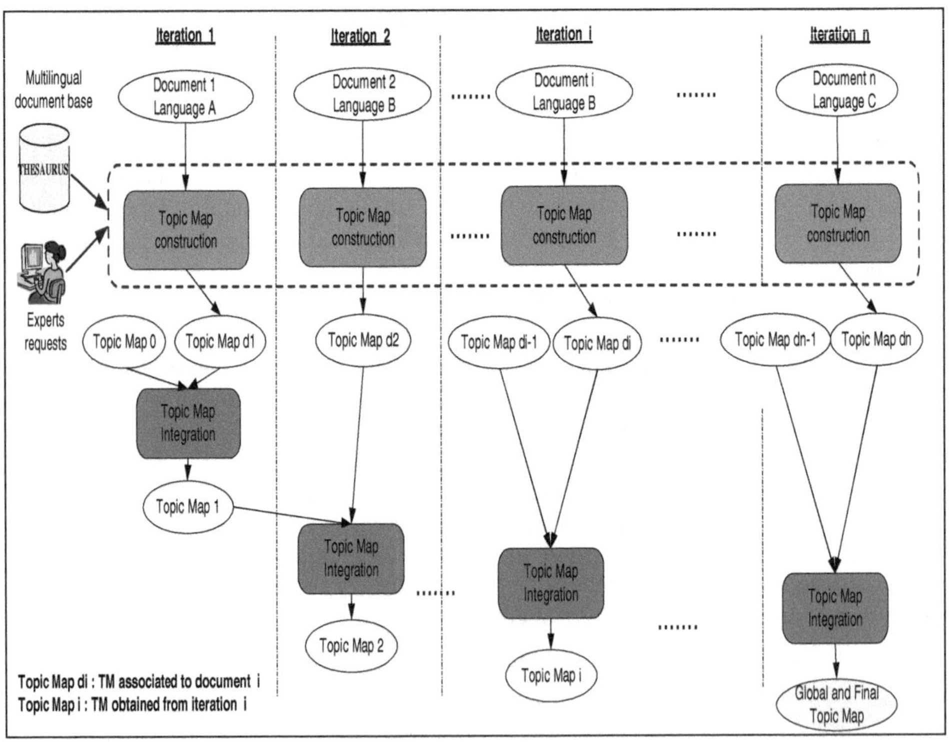

Fig. 1. The multilingual Topic Map incremental building

There are many and various research works in the field of integration schemas [11][12][13][14][15][16][17][18][19][20]. These approaches can be distinguished according to the types of schemas treated (Entity/Relationships schemas, object-oriented schemas, ontologies, XML documents, etc...), to the matching techniques used or to the integration techniques adopted. Very few approaches supply a complete integration process and few of them offer a free tool. For that, we choose to use and adapt the integration process proposed in [18].

Topic Map Reference Model [21] defines a generic merging function based on the equivalence rules to determine if two or more topics can be merged, these rules can be briefly stated as follows:

1. When two Topic Maps are merged, any topics that the application determines to have the same subject are merged and any duplicate associations are removed.
2. When two topics are merged, the result is a single topic whose characteristics (names, association role, and occurrences) are the union of the characteristics of the original topics, with duplicates removed.

Since the global Topic Map of the step i and the topic map associated to a document i+1 are produced using the same process, they are described in the same language. So the first step of a usual integration process, described above, is ignored in our integration approach. The latter encompasses three steps. The first one consists in matching the two Topic Maps. This matching allows to detect similarities between concepts of the two Topic Maps and to generate two Topic Maps that take into account these similarities. The second step merges the resulting Topic Maps into an integrated one. The last one is a validation step.

In the following, we present our integration algorithm. It takes as input the Topic Map i created from document i and the global Topic Map. The output is a new Topic Map.

Algorithm 3. Topic Maps integration algorithm

Action 5. *Enrich the global Topic Map with Topic Map i:*
- ***Action 5. 1.*** *Match the global Topic Map with Topic Map i*
- ***Action 5. 2.*** *Merge the Topic Maps resulting from action 5.1*
- ***Action 5. 3.*** *Validate the integrated Topic Map*

The matching of two Topic Maps consists of matching topics followed by matching associations. Matching associations must take into account the topics that it relies and the name attributed by the experts during the validation step (action 4 of algorithm 1). This process is still in progress. In our future works, we will discuss in more detail steps and issues related to the matching process.

The merging of two Topic Maps is about merging hierarchies of topics while taking into account the other existing topic relationships except the occurrence links. We consider the fact that two hierarchies belonging respectively to TM1 and TM2 can be merged if and only if they share topics. The process is composed of three phases. The first phase is a pre-merging phase. It selects pairs of hierarchies to merge. The second phase is the merging phase itself. It is applied to each group of hierarchy determined in the pre-merging phase. It is obvious that if the two hierarchies are equivalent there is no need to apply the phase to this group. To construct from two non equivalent hierarchies an integrated hierarchy, the process uses two schema transformation techniques described in [22]. The first one consists on translating a hierarchy into a Boolean function. The second one is the inverse of the first one. It is executed in the process after merging the two Boolean functions associated to the two hierarchies. The third phase of the merging process consists of grouping integrated hierarchies taking into account roles played by topics in the associations represented as links between topics in different hierarchies.

Once the integrated Topic Map is obtained, it is supplied to the experts for validation. The validation process is the same as the one described in the action 4 of algorithm 1.

5 Related Works

In this section, we discuss some works related to Topic Maps construction. We can observe a growing interest in the use of Topic Maps for modeling and sharing knowledge, and consequently several approaches have been proposed to build, manage and maintain Topic Maps. Based on the state of the art [23] on Topic Maps construction approaches, we notice that, very few Topic Maps have been developed from scratch, since, in most cases; construction approaches are based on existing data sources and use a priori knowledge (such as domain ontologies, thesaurus) to build and populate the Topic Maps. These approaches take as input different data types: structured documents, databases, unstructured documents and semi-structured data, and propose different techniques to generate Topic Maps from these sources. Some of these approaches take as input XML documents and propose to apply automated processes to leverage these documents [24][25][26]; other approaches propose to map directly RDF [27] metadata to Topic Maps [28][29]. Some other works propose to use learning techniques and Natural Language Processing (NLP) techniques to extract topics and associations from textual documents [30][31]. Learning methods can be applied with different automation levels: manual, semi-automatic or automatic. Some research works are dedicated to cooperative TM building involving different actors in the construction process [32][33][34][35]. Another research area deals with merging Topic Maps [36]. It concerns Topic Map applications using multiple Topic Maps. Each one may emanate from a different source, generated by a different technique or written in a different syntax.

Based on the state of the art, we note that:

- Most Topic Map building approaches are a combination of auto-generation, manual enrichment and merging techniques.
- In all proposed approaches, automatic construction is not enough developed. In fact, all these approaches require the participation of the user during the building process. Indeed, TM creation can be very costly and can quickly become a bottleneck in any large-scale application if recourse is not made to automatic methods. Problems of maintenance and coherence may arise when the TM is applied to heterogeneous and multilingual information sources, as manual construction and maintenance can not keep pace with any incoming documents.
- The multilingual aspect is not handled in all existing approaches except Kasler's approach [37] which takes into account English and Hungarian texts. The problem

is that the large majority of resources available today are written in various languages, and these resources are relevant even to non-native speakers. In this case, it is difficult for users to find relevant information in documents using their native language.

- The main lacks for all the approaches are that there are not integrated methods and techniques, that combines heterogeneous knowledge sources with existing knowledge structures such as ontologies or thesaurus to accelerate the building process.
- The majority of these approaches are dedicated to specific application domains.
- At least, we note that most of TM construction approaches do not propose techniques for the evaluation of the resulting Topic Map.

In our approach, we propose to combine four main aspects: use a priori knowledge which is the multilingual domain thesaurus, involve a community of domain experts to participate in the building process, use enrichment techniques based on merging process to deal with multilingual issues, build a topic map according to the potential usage of the content it will represent and finally explore the experts requests related to source documents to assign a weight to each topic according to its frequency in the questions. Thus, our goal is to provide a new method for Topic Map construction allowing the integration of information resources available in different languages.

6 Conclusion and Future Work

In this paper, we present a usage-oriented approach to build a Topic Map from multilingual documents. The proposed approach aims at providing a system, based on the Topic Map model that, given a user query in a particular language, will return relevant documents in any of the languages available in the multilingual document base. The main design considerations of our approach are the incremental and collaborative build of the global Topic Map with domain experts, we use merging techniques to enrich the Topic Map and we take into account a multilingual domain thesaurus as starting point and a list questions already elaborated by domain experts related to the source documents.

In the future, we will study methods and techniques to further improve the Topic Map building and enrichment processes. We will discuss, in more details, the multilingual merging aspect and related issues through the merging step already defined in our general approach. Finally, our work will be driven towards multilingual access using the Topic Map in a way that users can access information in the transparent way as they do in their mother language.

References

1. ISO/IEC: 13250. Topic Maps: Information technology-document description and markup languages (2000), `http://www.y12.doe.gov/sgml/sc34/document/0129.pdf`
2. Pepper, S., Moore, G.: XML Topic Maps (XTM) 1.0, Topic Maps. Org. (2001), `http://www.topicmaps.org/xtm/`

3. ISO 5964:1985 Documentation - Guidelines for the establishment and development of multilingual thesauri, `http://www.iso.org/iso/en/ISOOnline.frontpage`
4. Faatz, A., Steinmetz, R.: Ontology enrichment with texts from the WWW. In: The Semantic Web Mining Conference WS 2002 (2002)
5. Parekh, V., Gwo, J.-P., Finin, T.: Mining Domain Specific Texts and Glossaries to Evaluate and Enrich Domain Ontologies. In: International Conference of Information and Knowledge Engineering (2004)
6. Bendaoud, R., Rouane Hacene, M., Toussaint, Y., Delecroix, B., Napoli, A.: Construction d'une ontologie à partir d'un corpus de textes avec l'ACF, IC (2007)
7. Maedche, A., Staab, S.: Mining ontologies from text. In: Dieng, R., Corby, O. (eds.) EKAW 2000. LNCS (LNAI), vol. 1937, pp. 189–202. Springer, Heidelberg (2000)
8. Velardi, P., Missikof, M., Fabriani, P.: Using text processing techniques to automatically enrich a domain ontology. In: Proceedings of ACM- FOIS (2001)
9. Salton, G., Buckley, C.: Term-weighing approaches in automatic text retrieval. Information Processing & Management 24(5), 513–523 (1988)
10. Parent, C., Spaccapietra, S.: Issues and Approaches of Database Integration. Communications of the ACM 41(5) (1998)
11. Rahm, E., Bernstein, P.A.: A survey of approaches to automatic schema matching. Journal VLDB 10, 334–335 (2001)
12. Rasgado, A.D.C., Guzman, A.A.: A language an Algorithm for Automatic Merging of Ontologies. In: Proc. Conf. on Computing (CIC), pp. 180–185 (2006)
13. Stumme, G., Maedche, A.: FCA-MERGE: Botton-up merging of ontologies. In: Proc. of IJCAI, pp. 225–234 (2001)
14. Lukovic, I., Ristic, S., Mogin, P., Pavicevic, J.: Database schema integration process- A methodology and aspects of its applying. Journal of Mathematics Novi Sad (NSJM) 36(1), 115–140 (2006)
15. Nguyen, B., Varlamis, I., Haldiki, M., Vazirgianis, M.: Construction de classes de documents web, JFT'2003. Journées francophones de la Toile, Tours, France (Juin 2003)
16. Kong, H., Hwang, M., Kim, P.: Efficient Merging for heteregeneous doamin ontologies based on Wordnet. Journal JACIII 10(5), 733–737 (2006)
17. Ding, H., Solvberg, I.: A schema Integration Framework over Super-Peer based network. In: Proc. International Conference IEEE on Service Computing, SCC 2004 (2004)
18. Lammari, N., Essanaa Besbes, S.: Méthode de rétro-ingenierie pour l'analyse des sites Web, Rapport de recherche, laboratoire Cedric, CNAM, paris, France (2008)
19. Benazet, E., Comyn-Wattiau, I., Guehl, H.: A method for object insertion in a generalization hierarchy. Rapport de recherche ESSEC-CR-DR - 96-068, ESSEC (1996)
20. Noy, N.F., Musen, M.A.: Prompt: Algorithm and tool for automated ontology merging and alignment. In: Proceedings of the Seventeenth National Conference on Artificial Intelligence and Twelfth Conference on Innovative Applications of Artificial Intelligence. AAAI Press/The MIT Press (2000)
21. Durusau, P., Newcomb, S., Barta, R.: Topic Maps – Reference Model ISO/IEC JTC1/SC34, Version 6.0 (2006), `http://www.isotopicmaps.org/tmrm`
22. Lammari, N., Métais, E.: Building and Maintaining Ontologies: a set of Algorithms. Data and Knowledge Engineering Journal 48(2), 155–176 (2004)
23. Ellouze, N., Métais, E., Ben Ahmed, M.: State of the Art on Topic Maps Building Approaches. In: Kutsche, R.-D., Milanovic, N. (eds.) MBSDI 2008, Model Based Software and Integration Systems. CCIS, vol. 8, pp. 102–112. Springer, Heidelberg (2008)
24. Reynolds, J., Kimber, W.E.: Topic Map Authoring With Reusable Ontologies and Automated Knowledge Mining. In: XML 2002 Proceedings by deepX (2002)

25. Lin, X., Qin, J.: Building a Topic Map Repository (2002), http://www.knowledgetechnologies.net/proceedings/presentations/lin/xialin.pdf
26. Librelotto, G.R., Ramalho, J.C., Henriques, P.R.: TM-Builder: An Ontology Builder based on XML Topic Maps. Clei electronic journal 7(2), paper 4 (2004)
27. Ora, L., Swick, R.: Resource Description Framework (RDF) Model and Syntax Specification, W3C Recommendation (1999), http://www.w3.org/TR/REC-rdf-syntax/
28. Pepper, S.: Topic Map Erotica RDF and Topic Maps "in flagrante" (2002), http://www.ontopia.net/topicmaps/materials/MapMaker_files/frame.htm
29. Pepper, S.: Methods for the Automatic Construction of Topic Maps (2002), http://www.ontopia.net/topicmaps/materials/autogen-pres.pdf
30. LeGrand, B., Soto, M.: Topic Maps et navigation intelligente sur le Web Sémantique, AS CNRS Web Sémantique, CNRS Ivry-sur-Seine (October 2002)
31. Folch, H., Habert, H.: Articulating conceptual spaces using the Topic Map standard. In: Proceedings XML 2002, Baltimore (December 8-13, 2002)
32. Ahmed, K.: TMShare – Topic Map Fragment Exchange in a Peer-To-Peer Application (2003), http://www.idealliance.org/papers/dx_xmle03/papers/02-03-03/02-03-03.pdf
33. Lavik, S., Nordeng, T.W., Meloy, J.R.: BrainBank Learning - building personal topic maps as a strategy for learning. In: XML, Washington (2004)
34. Zaher, L.H., Cahier, J.-P., Zacklad, M.: The Agoræ / Hypertopic approach. In: International Workshop IKHS - Indexing and Knowledge in Human Sciences, SdC, Nantes (2006)
35. Dicheva, D., Dichev, C.: TM4L: Creating and Browsing Educational Topic Maps. British Journal of Educational Technology - BJET 37(3), 391–404 (2006)
36. Ouziri, M.: Semantic integration of Web-based learning resources: A Topic Maps-based approach. In: Proceedings of the Sixth International Conference on Advanced Learning Technologies (ICALT 2006) IEEE, Los Alamitos (2006), 0-7695-2632-2/06 $20.00 ©
37. Kasler, L., Venczel, Z., Varga, L.Z.: Framework for Semi Automatically Generating Topic Maps. TIR-06. In: Proceedings of the 3rd international workshop on text-based information retrieval, Riva del Grada, pp. 24–30 (2006)

ManagemOnt: A Semantic Approach to Software Engineering Management Process

Baris Ulu and Banu Diri

Yildiz Technical University, Computer Engineering Department
Electric-Electronics Faculty, 34349, Besiktas-Istanbul, Turkey
baris47@gmail.com, banu@ce.yildiz.edu.tr

Abstract. Software engineering processes, today, tend to have a gap between the assets because of non-manageable experiences in the domain which causes the organizations to fail in process improvement activities and software engineering practices in terms of time and cost. The data maintained in current software engineering process models, such as project and resource plans, documents, metrics, etc. is syntactic and out of interpretation. The lack of interpretation results in redundant data for an asset of software engineering process. It is well-known for years that each asset in software engineering domain generates an output which is an input for another asset in the domain in a logically related manner. This approach to software engineering process assets reveals knowledge-based software engineering process modeling via inference and reuse of domain experiences. It is proposed to model semantic software engineering processes and their assets by means of ontologies to achieve the inference and reuse of domain knowledge in a way different from syntactic approach. In order to trigger semantic software engineering processes, project planning activity is prototyped from software engineering management process since this activity almost comprises the mentioned process data because of its position in software engineering processes and practices.

Keywords: Ontological software engineering management, semantic software engineering management.

1 Introduction

The practices in software engineering domain have so many outputs in each asset such as project plans, resource plans, documents, metrics, etc. These data result in garbage because of redundant data after a while. The reason for this redundant is the lack of intelligent query and recording of experiences, where experience stands for knowledge in the domain. Thereby, each asset is modeled from scratch since the knowledge in a related asset is lost after the activities have been completed just before a new activity is due to start. This upcoming gap among the assets causes potential failures in process improvements and direct impact on software engineering practices in terms of time and cost.

F. Sartori, M.Á. Sicilia, and N. Manouselis (Eds.): MTSR 2009, CCIS 46, pp. 24–34, 2009.

Each asset in software engineering domain needs an input which is generated by another asset as output to keep the process to continue. This relation between the software engineering assets is valid as far as these relations are well defined and applied for each activity to be performed. It is proposed in this study to put forward the formal logical representation of software engineering assets to be able to reuse the knowledge. This reuse idea is put into practice with the semantic modeling of software engineering assets. The knowledge in the domain is stored by ontologies to keep the logical relations alive and valid for each software engineering activity in order to prevent the users to reinvent the wheel. Regarding the main idea of the study, the sample activity prototype developed in the ongoing article indicates that project plans can be created automatically by reusing the stored knowledge in the proposed ontology.

The ongoing article is structured in a way that second section generates a brief motivation for the discussion to define what the infrastructure is for the proposed model. Third section explains the proposed model, named as ManagemOnt, for semantic software engineering processes with the mentioned prototype of a sample activity in software engineering management process both in architectural and end-user perspective.

2 Motivation

The semantic approach [1][2][3][4][5], an evolving collection of knowledge, built to allow anyone on any network to add what they know and find answers to their questions. It is built on RDF [6][7][8][9][10], that uses URI [20] that defines the meta-data [11][12][13][14][15] to represent data. These collections of data that are logically related and formed by RDF are stored by ontologies [16][17][18][19][20], a common communication language based on the meta-data models of distributed data. Besides the terminological meaning, ontology is a collection of data for a group of agents. Information is stored in ontologies for the aim of inference [4][8]. Inference is the provisioning of information via using different collections of data. Many of the instances are not needed to be stored because of inference. The logically related collections of data and sets of inference rules have to be accessed by agents to gain reasoning on Semantic Web.

Software engineering processes [21][22][23][24][25] are modeled for modularization, distribution, reuse and integration of software components based on Entity/Relationship model [32] to gain the conceptual modeling [33][34][35]. Many tools based on this model influenced software engineering domain till object-oriented methodology has come up. Object-oriented methodology tried to express the behavior modeling by defining the operation definitions in entities and their attributes via UML [36][37]. UML represented the interrelationships among the entities in a domain. These entities were syntactic where it encapsulates the naming, classification and structural description of entities with their associations. These associations are expressed by operations just with their names, parameters and result types, such as the name of a document, the milestone in a project plan or a standard cost for a resource in a project, etc. This is known as the traditional conceptual modeling.

Current software engineering process modeling techniques tend to focus on the high-level activities and products involved in software engineering domain. These models are too abstract to be of much practical help other than illustrating the logical dependencies among the entities. Table 1 lists these entities transformed as software engineering knowledge assets [26][27].

Table 1. Software engineering knowledge assets

Primitive Processes	Supporting Processes	Organizational Processes
Acquisition Process	Documentation Process	Management Process
Supply Process	Configuration Management Process	Infrastructure Process
Development Process	Quality Assurance Process	Improvement Process
Operation Process	Verification Process	Training Process
Development Process	Validation Process	
Maintenance Process	Joint Review Process	
	Audit Process	
	Problem Resolution Process	

However, the traditional conceptual modeling is convenient for a known or limited group of software engineering objects where the definitions do not refer to anything out of the scope of the mentioned object models. This causes the gaps to happen between logically related assets in software engineering domain existing in different group of objects. On the other hand, the traditional models result in inefficient usage of time, resources and budget since parsing in huge and garbage documents, planning from scratch and re-inventing the wheel is in question. This is basically because most of the experiences tend to disappear with human.

It seems a requisite to re-model the mentioned models in a semantic way to gain re-use of experiences and further conditions, constraints, connections and meanings in order to overcome the mentioned kind of gaps. The architecture of semantic software engineering processes should be able to find the answer to question "how" instead of "what" where the former carries towards semantic and the latter to syntactic model. The obtained answers to the former one result as strengthening the communication among software engineering processes and reusing the experiences as well.

The motivation of the ongoing article is based on the above idea which is to apply semantic approach to software engineering processes from the viewpoint of software engineering management process.

3 Semantic Software Engineering Processes

Software engineering management process is positioned in the center of the remaining software engineering knowledge assets due to its conditions, constraints,

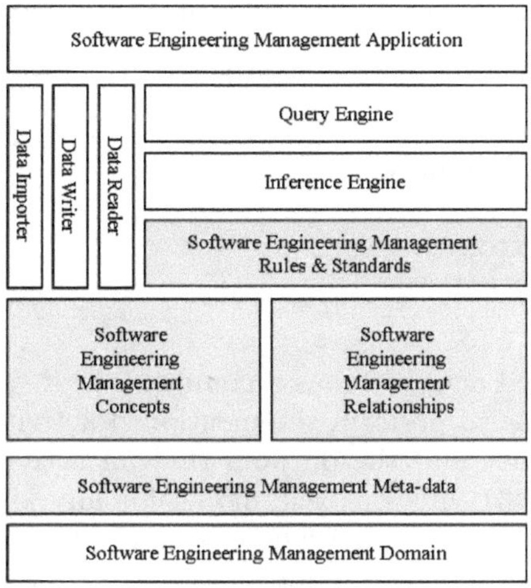

Fig. 1. Semantic software engineering management model

connections and meanings. For this reason, the rest of the paper focuses on software engineering management process as a suggestion on how to model a semantic software engineering process. The proposed model has the following high-level architecture as indicated in Fig 1. The layered architecture in Fig 1 is based on the domain covered with software engineering management process domain which includes the process data defined in [26][27]. The second layer on top of process domain presents the process meta-data with the concepts, relationships, rules and standards for the software engineering management domain. The mentioned meta-data includes the relationships among the process assets in the domain with the defined concepts and relationships. The concepts related to one other are based on the domain rules and standards. The third layer includes the decision logic for the underlying domain via help of inference and query engines in order to form the knowledge from the data defined in lower layer. This layer is supported with the reader, writer and importer utilities for the top most layer that includes the software engineering management application for the end-users.

Semantic software engineering management will result in;

- Re-using the experience in requirements,
- Re-using documentation and cost in successfully closed projects,
- Re-using the project plans prepared,
- Re-using resources, tasks and agents involved in the projects with correct or near-correct timelines,
- The adoption of the quality assurance methods to newly coming projects to reach appropriate maturity.

3.1 Software Engineering Management Meta-model

The software engineering management process includes the following list of activities:

- Initiation and scope definition,
- Planning,
- Execution and control,
- Review and Evaluation,
- Closure.

Each of the above listed activities has a common agent, where agent stands for the resource responsible to perform the mentioned activity, the inputs that are generated for an activity and the outputs that an activity generates, and the activity itself. SPEM [31] gives a base model for any kind of Agent-Activity-Resource model as a generic model where it can easily be applied to semantic software engineering management model as the basis of the software engineering management meta-model. This atomic model for the meta-model of the domain is depicted in Fig 2 with the mentioned relationships.

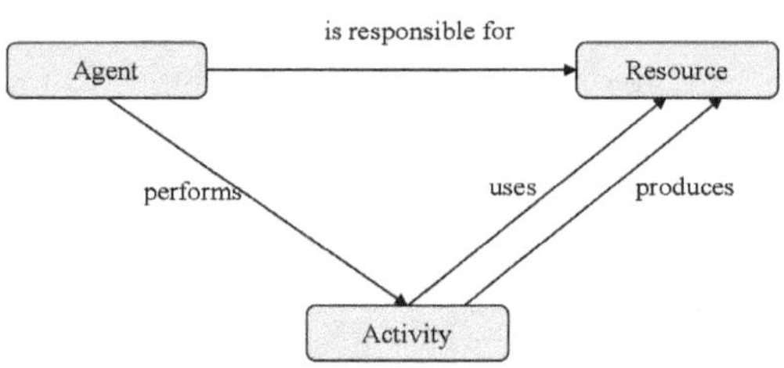

Fig. 2. Atomic model for ManagemOnt meta-model

Fig 3 depicts the application of the atomic model to semantic software engineering management meta-model to generate the meta-data for the domain. It represents the planning activity of the software engineering management process partially with basic Agent-Resource-Activity model.

There exist two different software engineering processes in Fig 3, software engineering management and software supply processes. These software engineering processes are connected to each other via atomic model which was depicted in Fig 2 in opposite to the current process models mentioned in the beginning of Section 3. The atomic model for the software supply process includes the software product manager, requirement specifications and roadmap, and analysis for the agent, resource and activity concepts respectively. Similarly, software engineering management process includes software engineering manager, project plan and planning. Each process is logically related to other although these processes have their own definitions and object models which do not refer to anything out of their scope. In Fig 3, these object models have been modeled as concepts with their inputs and outputs as connections to other concepts.

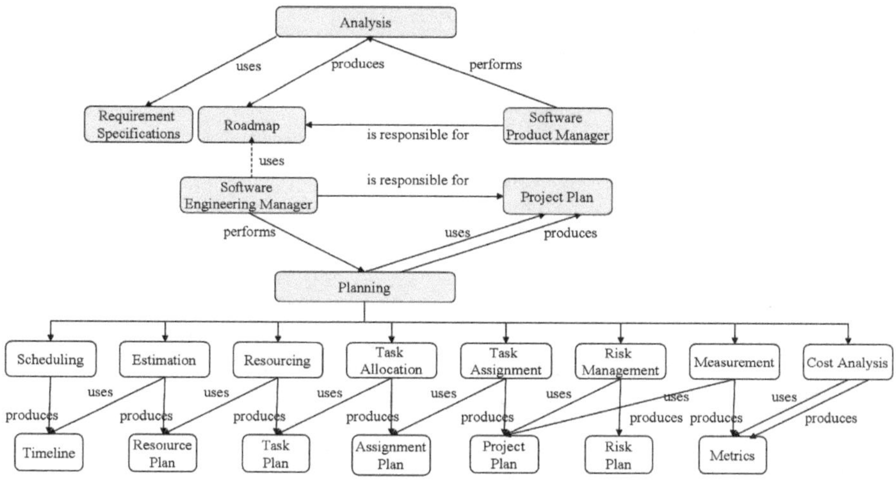

Fig. 3. Semantic software engineering management meta-model for planning activity

Semantic software engineering management meta-model (or meta-data as depicted in Fig 3) is built on the atomic model (as depicted in Fig 2). This meta-model represents the concepts and relationships on top of software engineering management domain (as depicted in Fig 1), and is called as *ManagemOnt* [28][29][30], where it stands for software engineering management ontology.

3.2 Software Engineering Management Ontology

Ontologies are used to establish a common understanding of a domain by making the shared conceptualization explicit in a machine-accessible manner. Ontology represents the domain knowledge by expressing the concepts, relationships and their meanings to expose the knowledge to applications for the aim of query, inference and reuse. The domain knowledge in software engineering stands for experiences such as project plans, metrics, cost analysis, etc. and the relationships among these. The promise of ManagemOnt is to keep the software engineering management domain knowledge live by encapsulating it in an ontology or group of ontologies. The following usage areas are identified for ManagemOnt without human intervention in the way towards ontology implementation in the very first stage:

- Determination of timeline for a project,
- Determination of task plan for a project,
- Determination of resource plans for a project,
- Determination of assignment plan for a project,
- Decision of estimations on timeline, resource, assignment and task plans based on previous cost analysis,
- Manual or automatic reporting on a project status to involve ManagemOnt in other activities of software engineering management process such as execution and control, and review and evaluation.

The above usage areas lead ManagemOnt to be modeled by performing the below set of tasks respectively in the way towards ontology implementation in the second stage:

- Construction of the software engineering management process concept hierarchy, where stands for taxonomy,
- Identification of the needed attributes for a software engineering management process concept,
- Construction of the connections among the software engineering management process concepts,
- Identification of the software engineering management process rules and standards, expressed in IEEE [26][27], to use for constraint checking,
- Creation of the instances throughout the ontology for the aim of data import, query and inference to gain software engineering management process domain knowledge.

ManagemOnt is modeled in Protegé v3.3.1 to obtain the construction of the following;

- ManagemOnt concept hierarchy (taxonomy definition),
- Identification and integration of the needed attributes for the ManagemOnt concepts,
- Construction of the logical connections among the ManagemOnt and,
- Identification of the ManagemOnt constraint checking.

In the final step of the tasks in second stage of ManagemOnt modeling, some instances are created throughout the model. The output as the RDF graph of the above represented ManagemOnt model is created as in the following Fig 4 with RDF and RDFS as well. The graph in Fig 4 basically includes the concepts, concept attributes and relationships among these concepts.

The proposed ontology, ManagemOnt, is interfaced with an application through Jena [38][39][40] application programming interface to support data insertion, navigation, query and inference activities. The software architecture for ManagemOnt and the application on top of the proposed ontology for end-user interaction is depicted in Fig 5.

ManagemOnt software architecture is composed of three functionalities, such as data import, query execute and inferencing, and six components to support these functionalities as listed below;

- Jena API: This component is a software development kit provided by HP's Jena including ontology writer, ontology reader and RDF Reasoner packages. The Jena API will have the ability to directly perform onto ManagemOnt.
- Data Importer: This component is responsible for feeding data to ManagemOnt through ontology writer Jena API in one of the following ways;
  - Traditional project plans that have been extracted into a standardized format such as CSV, XLS, etc. which is supported by most of the project planning tools, or

  • Manual data inserted via semantic software management client from scratch.

- Query Executer: This component is responsible for querying data from ManagemOnt through ontology reader Jena API via semantic software management client. Query executer is in coordination with inference performer to provide the exact or near-exact results to the client.
- Inference Performer: This component is responsible for the inference of knowledge from data queried by query executer. The inference performer is using RDF Reasoner by default which is bundled with Jena API.

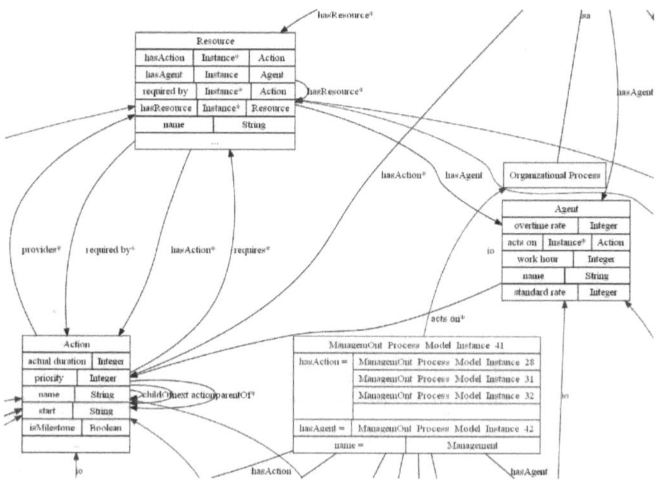

Fig. 4. RDF graph of ManagemOnt (partial)

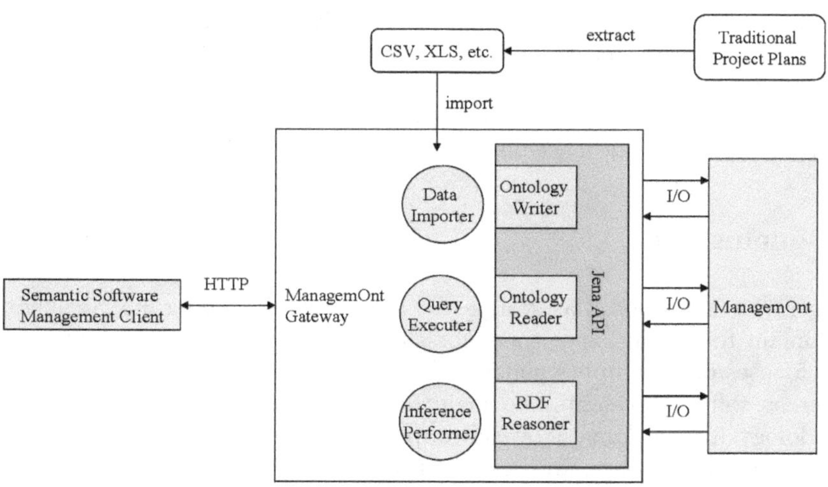

Fig. 5. Software architecture of ManagemOnt

- ManagemOnt Gateway: This component is responsible for transforming the incoming requests from semantic software management client to Jena API calls via data importer, query executer and inference performer. ManagemOnt Gateway also returns responses from Jena API calls to HTTP commands for the client to gather the operation results. This is a Java application with an embedded web application server and naturally can be executed on each type of computer regardless of the operating system.
- Semantic Software Management Client: This component is a thin client with a user-friendly graphical user interface which comprehends HTTP communication interface in and out. This thin client includes fields for entering data in the needs of ManagemOnt concepts and relationships, and forming queries with constraints to gain the knowledge from ManagemOnt. This client is also a Java client and naturally can be executed on each type of computer regardless of the operating system.

4 Future Work

Research in this study, named as ManagemOnt which has come up as a new idea in [28], has started with the investigation of current software engineering process models. The gaps happened in software engineering practices have been put forward because of the deficiencies of current software engineering process models. The layered architecture for the ontological management process is presented. The meta-model, concepts and relationships for this approach are implemented as software engineering management ontology in Proteg. The software engineering management ontology has resulted in RDF graph and RDF/RDFS documents.

Currently, ManagemOnt software architecture is finalized with ManagemOnt Gateway and Semantic Software Management Client with defined interfaces based on already prepared use cases and requirements. Furthermore, the presented software architecture will be implemented for data import, query and inference operations for end-user. The promise of ManagemOnt is to be able to automatically prepare a project plan in a domain with inferenced timeline, task, resource, assignment and cost plans.

5 Conclusions

ManagemOnt is proposed as storage for software engineering management process domain to obtain experiences for the aim of reusability on practices. The model focuses on the inputs generated by a concept and outputs generated for another as well. It presents the logically related assets in a concept-oriented methodology in the domain by the implementation of inference on the assets and connections to reach knowledge on data. As a result of this re-model on the domain, the promise of ManagemOnt is the rapid development of software engineering management activities in terms of time and cost saving.

Additionally, the new model aims the reuse of domain experiences in software engineering which are mostly lost after an activity is completed. Besides, the proposed model demonstrates the logical relationship of software engineering management process to others in order to overcome the gap that is to happen because of syntactic approach of current process models.

This idea is also valid for other activities in other software engineering assets likewise.

References

1. Allen, J.: Making a Semantic Web (2001), `http://www.netcrucible.com/semantic.html`
2. Barners-Lee, T.: Uniform Resource Identifiers (URI): Generic Syntax, W3C Publications (1998), `http://www.ietf.org/rfc/rfc2396.txt`
3. Berners-Lee, T.: What The Semantic Web can Represent, W3C Publications on Semantic Web (1998), `http://www.w3.org/DesignIssues/RDFnot.html`
4. Berners-Lee, T.: Semantic Web on XML, W3C Publications on Semantic Web (2000), `http://www.w3.org/2000/Talks/1206-xml2k-tbl/slide1-0.html`
5. Berners-Lee, T., et al.: The Semantic Web, The Scientific American Publications on Semantic Web (2001), `http://www.scientificamerican.com/article.cfm?id=the-semantic-web`
6. Brickley, D., Guha, R.V.: Resource Description Framework (RDF) Schema Specification, W3C Publications on Semantic Web (1999), `http://www.w3.org/TR/2000/CR-rdf-schema-20000327/`
7. Miller, et al.: Resource Description Framework (RDF) v1.173, W3C Publications on Semantic Web (2006), `http://www.w3.org/RDF/`
8. Palmer, S.B.: The Semantic Web: Taking Form, Semantic Web Agreement Group Publications (2001), `http://infomesh.net/2001/06/swform/`
9. Swick, R.: Resource Description Framework (RDF) Model and Syntax Specification, W3C Publications on Semantic Web (1999), `http://www.w3.org/TR/1999/REC-rdf-syntax-19990222/`
10. Powers, S.: Practical RDF. O'Reilly, US (2003)
11. Ceri, S., Pelagatti, G.: Distributed Databases. McGraw Hill, US (1984)
12. Hay, D.C.: Data Model Patterns. Dorset House Publishing, US (1996)
13. Tannenbaum, A.: Metadata Solutions. Addison Wesley, US (2002)
14. Tozer, G.: Metadata Management. Artech House, US (1999)
15. Sumpter, R.M.: Data Management. Lawrence Livermore National Lab. (1994)
16. Noy, F.N., McGuiness, D.L.: Ontology Development 101: A Guide to Creating Your First Ontology. Stanford Uni. Publications (2004)
17. Spyns, P., Meersman, R., Jarrar, M.: Data Modelling versus Ontology Engineering. STARLab, Belgium (2002)
18. Kalinichenko, L., Missikoff, M., Schiapelli, F., Skvortsov, N.: Ontological Modelling. In: Proceedings of the 5th Russian Conference on Digital Libraries RCDL 2003, Russia, (2003)
19. Gardner, S.P.: Ontologies and Semantic Data Integration. DDT 10(14) (2005)
20. Cruz, I.F., Xiao, H.: The Role of Ontologies in Data Integration. ADVIS Lab., US (2005)
21. Humphrey, W.S.: Managing the Software Process, SEI Series, US (1998) ISBN: 0-201-18095-2

22. Pressman, R.S.: Software Engineering: A Practitioner's Approach, 5th edn. McGraw Hill, US (2000)
23. Pfleger, S.L.: Software Engineering: Theory and Practice, 2nd edn. Prentice Hall, US (2001)
24. Behrofoz, A., Hudson, F.J.: Software Engineering Fundamentals. Oxford Press, UK (1996)
25. Abran, A., Moore, J.W., Bourque, P., Dupuis, R., Tripp, L.L.: SWEBOK: Software Engineering Body of Knowledge Trial Version. IEEE Computer Society, US (2001)
26. ISO/IEC 12207, Software Lifecycle Processes - Implementation Considerations, IEEE/EIA Guide 12207.2-1997, US (1998)
27. ISO/IEC 12207, Software Lifecycle Processes, IEEE/EIA Guide 12207.0-1996, US (1998)
28. Ulu, B., Diri, B.: Software Process Ontology. In: International MultiConference of Engineers and Computer Scientists 2007, IMECS 2007, 21-23 Mart 2007, Hong Kong. Lecture Notes in Engineering and Computer Science, pp. sf. 1110–1115 (2007), ISBN: 978-988-98671-4-0
29. Ulu, B., Diri, B.: Software Management Ontology. In: 3rd National Software Engineering Conference, UYMS 2007, September 27-30, pp. 103–108. UYMS 2007 Publishings (2007) ISBN: 978-9944-89-337-4
30. Ulu, B., Diri, B.: Software Engineering Management Process Ontology. In: 10th Academical Science Conference, AB 2008, Çanakkale, January 30 -February 018. AB 2008 Publishings (2008) ISBN: 978-97581-00-73-6
31. OMG, The Software Process Engineering Metamodel (SPEM), OMG Documents: ad/2001-03-08, US (2001)
32. Ramakrishnan, R., Gehrke, J.: Database Management Systems, 2nd edn. McGraw Hill, US (2000)
33. Jarrar, M., Demey, J., Meersman, R.: On Using Conceptual Data Modeling for Ontology Engineering. In: Spaccapietra, S., March, S., Aberer, K. (eds.) Journal on Data Semantics I. LNCS, vol. 2800, pp. 185–207. Springer, Heidelberg (2003)
34. Calero, C., Rulz, F., Plattini, M.: Ontologies in Software Engineering and Software Technology. Springer, US (2006)
35. Patel-Schneider, P.F., Horrocks, I.: Position Paper: Comparison of Two Modelling Paradigms in the Semantic Web. In: WWW 2006, US (2006)
36. Cranefield, S., Purvis, M.: UML as an Ontology Modelling Language. In: IJCAI 1999 Proceedings, Nz (1999)
37. Booch, G., Jacobson, I., Rumbaugh, J.: UML Distilled, 2nd edn. Addison-Wesley, Reading (2000)
38. McBride, B.: Jena: A Semantic Web Toolkit. IEEE Internet Computing Series. Hawlet-Packard Labs, UK (2002)
39. Rajagopal, H.: JENA: A Java API for Ontology Management, IBM Corporation, Colorado Software Summit (2005)
40. Magkanaraki, A., Karvounarakis, G., Christophides, V., Plexousakis, D.: Ontology Storage and Querying, Foundation for Research and Technology Hellas Institute of Computer Science Information Systems Lab., Technical Report No: 308 (2002)

Clarifying the Semantics of Relationships between Learning Objects

M. Elena Rodríguez[1], Jordi Conesa[1], and Miguel Ángel Sicilia[2]

[1] Estudis d'Informàtica, Multimèdia i Telecomunicació,
Rambla del Poblenou 156, E-08018 Barcelona
{mrodriguezgo,jconesac}@uoc.edu

[2] Departamento de Ciencias de la Computación, Universidad de Alcalá,
Edf. Politécnico, Ctra. Barcelona km 33.6, E-28871 Alcalá de Henares
msicilia@uah.es

Abstract. In this paper we discuss about the ambiguities and deficiencies of the Learning Object Metadata (LOM) standard to specify relationships between learning objects (LOs), specially those relationships that relate LOs instances. We also study the impact of relationships in the internal organizational structure of LOs. As main contribution, we develop a taxonomy of possible relationships between LOs that has been created by refining the LOM standard relationships with other meaningful relationships from a common sense ontology.

1 Introduction

The use of metadata as a means of describing knowledge about an item without requiring the examination of the item itself [6] has received a lot of interest in different fields of computer science. The metadata value derives from saving human time and effort, given that metadata summarize the items of interest that are being described. Therefore metadata must be effective at distinguishing between relevant and irrelevant items of interest.

We are interested in the description of learning resources (also known as learning objects (LOs)) according to the metadata proposed by the IEEE LOM standard [10]. The purpose of LOM is to facilitate the search, evaluation, acquisition and reuse of LOs that learners, instructors and automated software processes need. LOM is well suited for LO cataloguing and string based searches but it lacks of the semantic expressiveness necessary for allowing semantic searches. For this reason applications using LOs metadata are evolving their metadata representations by adding semantic structures [1].

Our mid-term general objective is to improve LOM semantic expressiveness. This requires dealing with structural issues as identifying classes of interest and relationships between these classes, as well as the specification of their integrity constraints and the appropriate derivation rules. In a similar way, the assigned values to metadata should be concepts of an ontology instead of values belonging to a controlled list of terms without explicit semantics. Our proposal derives the

F. Sartori, M.Á. Sicilia, and N. Manouselis (Eds.): MTSR 2009, CCIS 46, pp. 35–47, 2009.

ontology of the possible values of LOM metadata from common sense ontologies, in concrete from OpenCyc [13]. Even though we could have chosen other common sense ontologies with similar results, such as SUMO, we have chosen OpenCyc because its query, inference and extra capabilities facilitate the identification and extraction of relevant information. We have chosen common sense ontologies for two reasons. First, a common sense ontology provides general, domain, task and application knowledge and it is usually larger than others helping to deal with the several domains included in LOM. Secondly, this kind of ontology provides reliable background and top level concepts which facilitate to establish the appropriate correspondences between LOM metadata (and their values) and the ontology. As main problem, we need to overcome the usability problems that a common sense ontology (as it is the case of OpenCyc) imposes [2].

In this paper we focus on improving the semantics of explicit and implicit relationships that LOM proposes, specially those relationships that relate LO instances. The paper is organized as follows: Sect.2 revises the ambiguities and deficiencies of the LOM standard in specifying relationships between LOs. Section 3 reviews related work. Sections 4 and 5 develop our proposal. While Sect.4 studies the impact of relationships in the internal organizational structure of LOs, Sect.5 presents a taxonomy of possible relationships between LO instances. Finally, Sect.6 concludes the paper and provides an outlook to future work.

2 Critical Review of Relationships in LOM

LOM allows specifying binary relationships between LOs through metadata belonging to the *Relation* category. Given that a LO could be related with different LOs, multiple metadata instances of *Relation* category could be defined in the LO metadata record. For every existing relationship, we should specify: 1) the target LO (through the compound metadata 7.2 *Resource*) related with the LO that is being described; and 2) the nature of the relationship between both LOs (through the metadata 7.1 *Kind*).

LOM also provides several kinds of relationships between LOs organized as a recommended list of appropriate values (vocabulary, in terms of LOM). These relationships are based on the relationships proposed by Dublin Core. In addition, each relationship has associated its inverse relationship. Table 1 shows all these relationships, their meaning and an example per relationship.

Previous relationships relate LOs instances. None of them allows establishing relationships at class level, as a generalization relationship would permit. This is because LOM does not explicitly include the notion of LO class ([16], [14]).

In addition, any assessment about LOM relationships would be incomplete without the study of some other metadata proposed in LOM whose values clearly depend (directly or indirectly) on specified relationships. This is the case of the metadata 1.7 *Structure* and its related metadata 1.8 *Aggregation level* both belonging to the *General* category. The metadata 1.7 *Structure* allows to define the underlying organizational structure of a LO. By means of this metadata we can distinguish between atomic and compound LOs. On the other hand, the

Table 1. Relationships provided by the IEEE LOM standard

Relationship	**Meaning**	**Example**
Is part of (has part)	A LO is a physical or logical part of another LO	An exercise solution *is part of* its corresponding self assessment exercise
Is version of (has version)	A LO is an historical state or edition of another LO by the same creator	A new, updated edition of a lesson *is version of* the original lesson
Is format of (has format)	A LO has been derived from another by a reproduction or reformatting technology	Lesson.pdf *is format of* Lesson.doc
References (is referenced by)	A LO cites, acknowledges, disputes etc. about another LO	A lesson explaining the relational data model *references* C. J. Date's text book "An Introduction to Database Systems"
Is based on (is basis for)	A LO is a production, derivation, adaptation or interpretation of another LO	An edition of "El Quijote" for children *is based on* the original work written by Cervantes
Requires (is required by)	A LO requires another LO for its functioning, delivery, or content and cannot be used without the related LO being present	Example.java *requires* the Java Development Kit

metadata 1.8 *Aggregation level* expresses the functional granularity of a LO, and its value depends on the value assigned to the metadata 1.7 *Structure.*

Table 2 shows the organizational structures proposed by LOM through the metadata 1.7 *Structure* as well as their definitions. The following problems (P) overview the ambiguity of this metadata and its vocabulary.

P1. For non atomic LO, the relevant relationships (and their integrity constraints) that allow to create composite LOs remain hidden.

P2. The organizational structure of a LO is determined by the relationships that relate its component LOs. The ambiguity here is about what kind of relationships should be taken into account. If we take into account any kind of relationship then the syntactic structure of the LO is addressed, while if only distribution and ordered relationships are evaluated then structure also deals with the LO semantic structure. Some things can be inferred from the semantics, such as the expected precedences or dependencies between the components or the value of the metadata 1.7 *Structure.*

P3. The definition associated to networked structure is absolutely ambiguous and thus non-operative. One may think that "networked" is the structure used when the components of a LO do not fit with any other structure (linear, hierarchical etc.), but this is not its meaning according to the standard.

In regard to P1, and taking into account the relationships proposed by LOM, there is only one relationship that can be used to specify the components of a LO: the "is part of" relationship. This generic relationship corresponds to the

Table 2. Internal organizational structures in the IEEE LOM standard

Organizational structure	Meaning
Atomic	An indivisible LO
Collection	Set of LOs with no specified relationships between them
Networked	Set of LOs with relationships that are unspecified
Hierarchical	Set of LOs whose relationships can be represented by a tree structure
Linear	Set of LOs that are fully ordered

part-whole relation. Applying this relationship to LOs, it means that one LO plays the role of a part in another LO (the composite LO) who plays the role of the whole. The "is part of" relationship must be antisymmetric and transitive across all its participants (see [12] for a complete discussion).

In the case of P2, a deeper reading of LOM suggests that we need to determine the relationships that allow organizing the set of LOs that compose a compound LO from a rhetorical point of view, i.e. the relationships for expressing a coherent distribution and arrangement of the LOs that form the composite LO (examples of this kind of relationship can be found in [5]). None of the proposed relationships in LOM seems to be appropriated to structure the components of a LO. Relationships as "is version of", "requires", "is format of", "references" and "is based on" constitute dependency or referential relationships [15]. A generic relationship for organizing constituents LOs of a composite LO would be a rhetorical precedence relationship as, for example, it is the case of the "previous" (and its inverse "next") relationship proposed by LOM in the definition of the linear structure. However, this relationship is surprisingly missing in the relationships proposed in the *Relation* category.

Finally, in regard to P3, networked structure allows that the relationships between component LOs could be unspecified, and it is not clear what unspecified means here (without the metadata 7.1 *Kind*, partially specified relationships ...).

Since rhetorical relationships are not defined in LOM, the only manner to distinguish between the internal organizations of composite LOs will be by means of the value associated to the metadata 1.7 *Structure*. And even when this metadata is defined, it is not possible to know the rhetorical relationships between component LOs. Figure 1 provides some metadata record excerpts of different composite LOs (in the form of lessons). Each composite LO has a different internal organization (graphically represented by means of arrows) which cannot be expressed in the metadata records due to the lack of rhetorical relationships in LOM.

It is important to note that the use of precedence relationships as well as the use of metadata as 1.7 *Structure* and 1.8 *Aggregation level* is controversial. In fact, these metadata are frequently under-used. As main reason, in [7] authors argue that such metadata reflect the nature of a LO understood as a software artifact instead of a piece of learning content. Furthermore, there are other specifications that allow defining complex aspects related to the organization of LOs in a learning experience. Therefore, some other authors [18] suggest dealing with all these issues as a matter of instructional design. We humbly believe that a consensual position should be reached. LOM allows the description of LOs with different complexity levels and different internal organizational structures which partially reflect the intention of LOs authors. To improve the retrieval, the reuse and the restructuring of LOs, their internal structure, that means the relationships between their components, should be known in advance. In a similar way, the use of small LOs as basic building blocks promotes the creation of composite LOs. This is not contradictory with the fact that issues as the sequencing of contents and activities properly contextualized in a learning experience should not be reflected by means of LOM, but with some other specifications (e.g. IMS Learning Design, [9]).

Summing up, in general terms, relationships and related metadata in LOM, suffer from ambiguous definitions. Moreover, some relevant relationships are missing, some others remain implicit scattered over several metadata (out of the scope of the *Relation* category), and some others are expressed in a very generic form. As in [15], we agree that this situation may mislead LOs users, and it might be one of the main reasons why this kind of metadata is not included in LOs metadata records [7].

3 Related Work

The approaches for the enhancement of LOM can be classified in two main categories:

1. Adding new elements to LOM: extending existing vocabularies as well as the inclusion of new metadata (and their vocabularies).
2. Developing ontologies. In this case, instead of using vocabularies for certain metadata, the values associated to the metadata and/or the metadata are expressed in ontological terms.

In [17] an extension of LOM is presented. The overall objective is to capture the context and the pedagogical instructional role of LOs as a mechanism to increase LOs reusability, adding new metadata in the *Educational* category. These new metadata are *learning style*(which captures users' learning preferences in the form of a vocabulary) and *history* (which expresses the usage of the LO in different learning contexts). The main problem is that these metadata partially overlap with existing LOM metadata belonging to the *Education*, *Annotation* and *Classification* categories, compromising interoperability. In order to intensify reusability and improve LOs search mechanisms, they propose a new vocabulary for the LOM metadata 7.1 *Kind* of the *Relation* category. The relationships adopted have their origin in two sources: 1) relationships for structuring text documents and 2) semantic relationships, i.e. relationships that can be used in content selection, extracted from different thesauri.

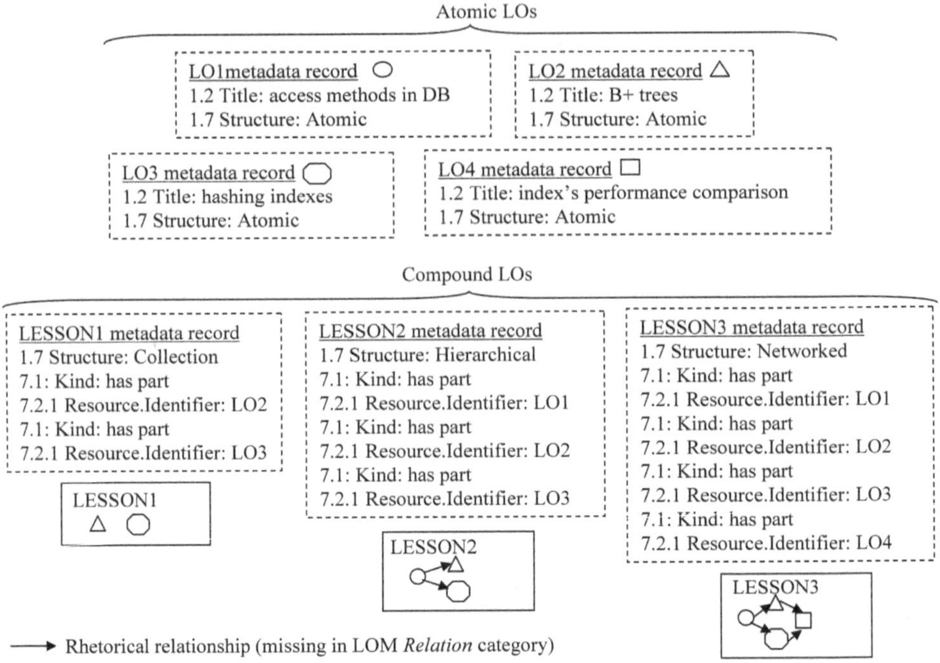

Fig. 1. Fragments of LOs metadata related to the structure and relationship metadata

The idea of organizing relationships (or any metadata) in the form of a vocabulary is quite problematic; it is a criticism that can be globally extended to LOM. In particular, the empirical analysis done in [8] reveals that a vocabulary does not lead to a consistent description of LOs. Differences occur, given that the LOs description partially depends on persons. Therefore the value assigned to a metadata is based on the interpretation that people do about the available values. In this sense, ontologies can be a means to promote consistent (and semi-automated) description of LOs.

An evaluation on the semantics of aggregation and specialization of LOs and the constraints these relationships impose is presented in [15]. The analysis is done by establishing analogies with the object-oriented paradigm (OOP). They propose approximate mappings between LOM relationships and OOP relationships and further refinements would be required. For example, for defining LOs classes, as first step they propose the creation of a generic LO (in the sense that its content is not directly usable) describing common properties; this generic LO constitutes the LO class. As second step, they relate LOs instances with the generic LO by means of "is based on" relationship. To deal with constraints imposed by generalization they propose using LO contracts. Given that LOM was conceived for describing LO instances, and generalization is a class level relationship, we believe that it should be treated out of the scope of LOM.

In [11] authors remark the importance of relationships not only for promoting reusability but also for enhancing learning effectiveness. They propose an ontology of instructional objects which considers generalization and referential ("is for") relationships. Unfortunately, it is unclear how these relations are accommodated in LOM *Relation* category.

A similar situation can be found in [5]. In this work the author presents the relationships that can be applied to the description of knowledge domains (illustrated in the multimedia domain) as well as relationships that can be used to represent dependencies between atomic LOs, separating concepts from content. Although [4] mentions that each LO can have a relationship to one or more topics of the multimedia domain, how all these relationships are included in LOs metadata records, is not discussed.

Finally in [14] we discussed the importance of explicitly consider the specialization of LOs according to different criteria. The work presents a specialization which distinguishes between conceptual LOs (abstract LOs consequence of a creative work) and existing LOs (the final available digital/paper LOs, probably in different formats). The Copy Conceptual LOs class is a subclass of conceptual LOs that collects those LOs that are derived as versions of the conceptual LOs. This specialization allows, amongst others, to deal with multilingual learning environments. Some relevant relationships between LO instances were identified, as "translation of" (connecting conceptual and copy conceptual LOs) and "instance of" (connecting conceptual LOs and existing LOs). By mapping all previous elements to common sense ontologies (specifically OpenCyc), most metadata can be derived by using the proposed relationships. Discussion about how integrating or mapping these relationships in LOM was omitted.

4 Impact of Relationships in the Metadata 1.7 Structure

While the Sect. 2 presents the ambiguity problems (P) of the metadata 1.7 *Structure*, this section addresses their solutions (S).

Firstly, networked structure definition is absolutely ambiguous (P3) because it allows that relationships between component LOs could be unspecified. In the case where no relationships are specified, for example, a networked LO also satisfies the collection structure definition. On behalf of LO reusing, it seems reasonable to require that these relationships, as well as their semantics, were known. Hence, we redefine (S3) the networked LO structure as the structure where the component LOs follow an acyclic directed graph structure (not necessarily connected).

Secondly, from the standard it is possible to deduce that sometimes ordered relationships between LOs (P2) can form composite LO. These relationships established between the LOs that form a composite LO may be seen as strict ordered relationships (S2). Strict ordered relationships verify irreflexivity, antisymmetric and transitivity properties. None of the relationships proposed by LOM in the *Relation* category is appropriate. Therefore, at least one generic relationship that allows establishing rhetorical ordering (i.e. content arrangement

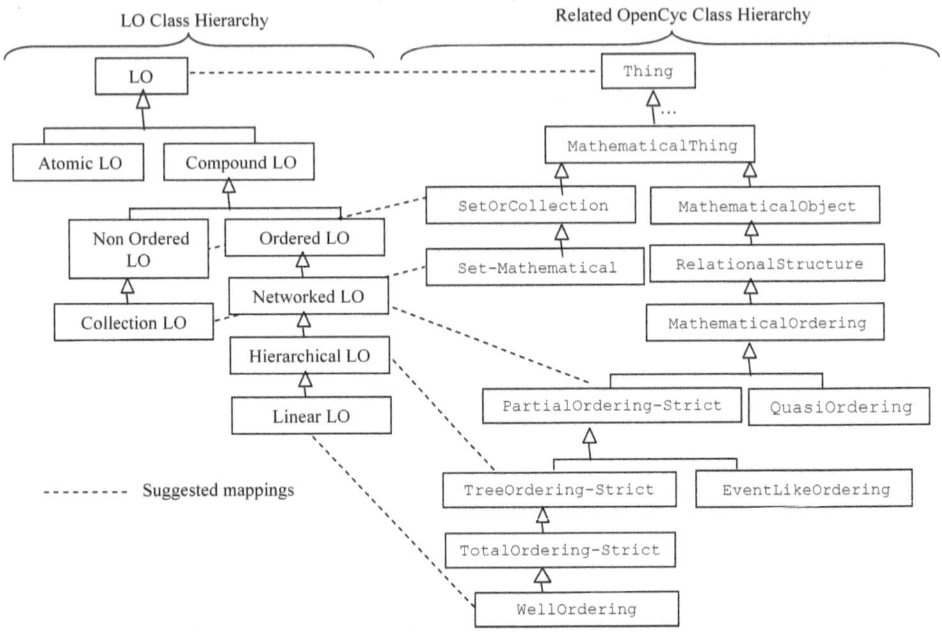

Fig. 2. LO specialization according to their structure and their mapping to OpenCyc

in a document) between component LOs must be added. We suggest to add (S2) the "previous" (and its inverse "next") relationship included by LOM in the definition of the metadata 1.7 *Structure*. This generic relationship is equivalent to relationship "follows" (jointly with its inverse relationship "precedes") from [5]. Finally, just remind that by means of "is part of" relationship (S1) we can specify the component LOs that give raise to a compound LO, thus solving P1.

In consequence, we can conclude that 1) LOM collection structure fits the definition of mathematical set; 2) LOM hierarchical structure follows a tree structure, i.e. we have a root LO which does not depend on any other LO. The remaining LOs depend only on one LO. 3) Networked and hierarchical structures impose strict partial ordered relationships between the component LOs. 4) Finally, LOM linear structure imposes strict total ordered relationships between all component LOs. From previous definitions, we can also conclude that some organizational structures are closely related. Networked is the most generic organizational structure. Hierarchical structure is a particular case of networked structure where each component LO has only one ancestor (if any). In turn, linear structure is a specific case of hierarchical structure where each component LO has only one successor.

Once we have disambiguated the semantics of each possible organizational structure for composite LOs as well as the relevant relationships, a specialization of the LOs according to their structure can be established as depicted in Fig.2. The classification of a LO according to its structure can be automatically

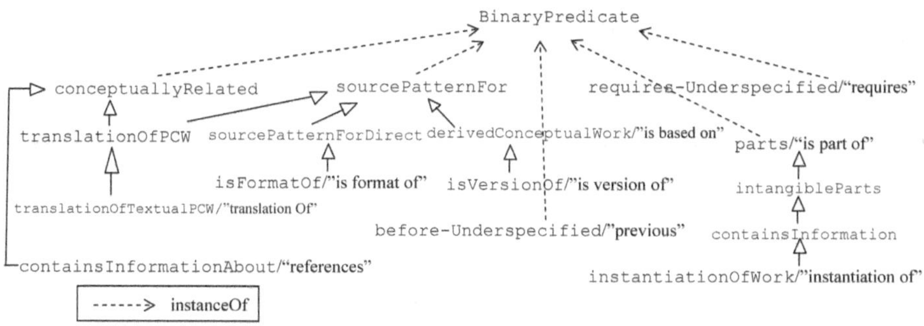

Fig. 3. Top level fragment of the proposed taxonomy of relationships

inferred. In order to do so for a given LO, the set of its component LOs have to be determined by examining the defined "has part" relationships in the composite LO. Thereafter, by exploring "next" relationships between component LOs, the specific kind of internal organizational structure can be calculated. Furthermore, many other LOM metadata for the composite LO can be partially derived from its component LOs: *General* metadata like *Language* (as the union of all languages of component LOs), *Technical* metadata like *Size*, *Duration* and *Format* (as the sum/union of *Size/Duration/Format* of component LOs), *Requirement* (made up of all requirements from component LOs), *Life cycle* metadata like *Status* (the *Status* of a compound LO takes the lowest *Status* of its component LOs), *Educational* metadata as *Interactivity type* (the composite LO will be "mixed" if it incorporates at least one "active" component LO and one "expositive" component LO), etcetera.

Summarizing, our proposal not only solves the ambiguity problems of LOM (exemplified in Fig.1) related to the metadata 1.7 *Structure*, but also takes advantage of the semantic structures that an ontological approach provides.

5 Modeling Relationships

Ontologies may be a means to minimize different interpretations in the LOs realm as [1], [6], [8] and [14] discuss. In addition, their use can help to (semi)-automate different processes. Some examples of these processes are: 1) the indexing of LOs; 2) the search of LOs in repositories; 3) the composition and sequencing of LOs and 4) learning personalization. Furthermore, having a meaningful set of relationships applicable to LOs based on formal semantics, as ontologies provide, clearly improves previous processes (see Sect.3).

Hereby we present a taxonomy of relationships (extracted from OpenCyc) that relate LOs instances (see Fig.3). In order to ensure interoperability with implementations that use the LOM vocabulary, all relationships proposed in the *Relation* category have been included in our taxonomy. The taxonomy has also been extended with OpenCyc relationships relevant for the LO realm. These extensions improve semantic expressiveness, producing better interoperability

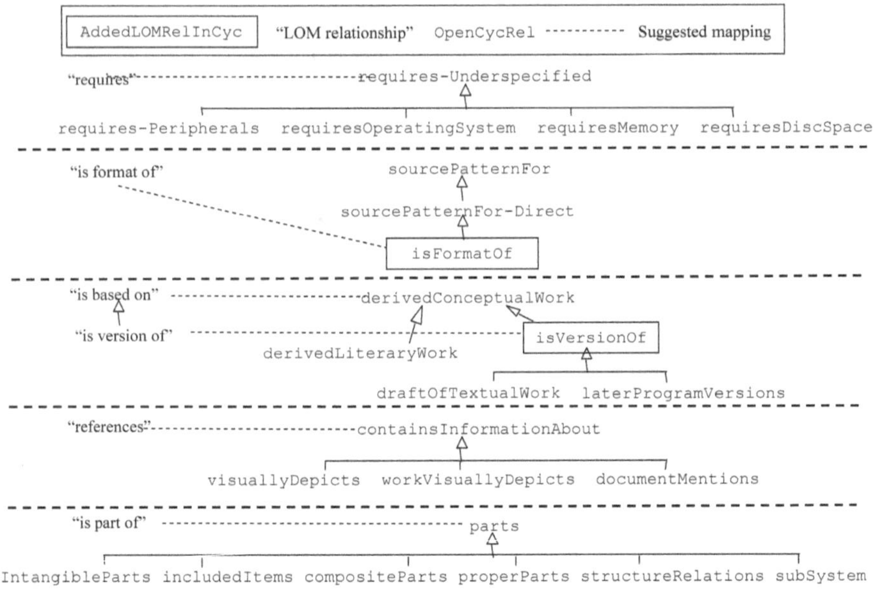

Fig. 4. Mapping between LOM and OpenCyc relationships and relevant first level subtypes of OpenCyc relationships

levels in LOs related services [16]. More information about the mapping between LOM and OpenCyc relationships is presented in Fig. 4.

We have followed a systematic way of retrieving the relationships applicable to LOs. For each meaningful relationship, we created an OpenCyc query that returns the relationships that follow its semantics. The results of the query are sent to a pruning algorithm (see [3] for a deeper discussion) that creates a fragment of OpenCyc that contains only the relevant relationships, their participants' types, and the necessary information to maintain the inheritance between them.

For example, the relationships of OpenCyc that satisfy the semantics of the "is part of" relationship of LOM are the binary relationships that are transitive, antisymmetric and specialize from the OpenCyc predicate `parts`. There are 117 relationships applicable to different knowledge domains in OpenCyc that satisfy the previous query. Figure 4 shows some relevant first level subtypes of the `parts` relationship (the OpenCyc counterpart of LOM relationship "is part of") as `compositeParts` which refers to composition [12].

In a similar way, LOM relationships "references", "is based on" and "requires" have been mapped to `containsInformationAbout`, `derivedConceptualWork` and `Requires-Underspecified` respectively. Our taxonomy also makes explicit the fact that "is version of" is a refinement of "is based on" relationship. Attending to LOM, "is version of" relates a LO with another LO (a new edition) realized by the original LO's creator. "Is version of" does not have an equivalent relationship in OpenCyc, although more specific relationships have been found (as `draftOfTextualWork` or `laterProgramVersions`). So, "is version of" has been added and aligned inside the OpenCyc knowledge base. This is also the case of "is format of" related to the generic relationship of OpenCyc `sourcePatternFor-Direct`.

The rhetorical relationship "previous" (omitted in LOM) has been mapped to `before-Underspecified` in OpenCyc (see Fig.3). Finally, new relationships, unrelated to relationships proposed by LOM in the *Relation* category have been added (also shown in Fig.3). This is the case of "translation of" and "instantiation of", mapped to `translationOfTextualPCW` and `instantiationOfWork`, respectively. The first one allows dealing with multilingual learning environments whereas the second one permits to relate conceptual LOs and their final, existing LOs (see [14] for a more detailed discussion).

Apart of the taxonomy, we defined a set of derivation rules to infer automatically the structure of a LO and some other elements of LOM related to the LO structure. Afterwards we present a derivation rule, written in OpenCyc, that shows how to derive, by means of the use of established relationships between LOs instances, the value for the metadata 1.7 *Structure*. Specifically, the rule infers that the composite LOs, whose components are not related by the rhetorical relationship `before-Underspecified`, have a collection structure. The `lomStructure` predicate (which represents the metadata 1.7 *Structure*) and `Collection` have been previously defined in the knowledge base of OpenCyc but not presented here for space constraints.

```
(#$implies
    (#$and
       (#$parts ?LO ?PART1)
       (#$parts ?LO ?PART2)
       (#$not (#$before-Underspecified ?PART1 ?PART2)))
    (#$lomStructure ?LO #$Collection))
```

6 Conclusions

This research contributes to the improvement of the semantics in the annotation of LOs in several ways. First, it studies the ambiguities and deficiencies of LOM to specify relationships between LOs. Second, it analyses why other approaches that extend LOM to enhance the representation of relationships between LOs improve but do not solve most of the semantic problems detected. Third, it studies the impact of relationships in the internal organizational structure of LOs. Fourth, it presents a taxonomy of possible relationships of LOs that has been created by refining LOM relationships with other meaningful relationships from a common sense ontology. Using the presented taxonomy, or any extension of it, would partially allow inferring automatically the value of other metadata as for example, the 1.7 *Structure*, 4.2 *Size*, 4.7 *Duration*, 4.4 *Requirements*, etc.

Future research will study the ambiguities of other LOM metadata and will propose an ontology, for each ambiguous element, that represents the possible values it can take and the associated derivation rules. As a result, we would obtain a repository of knowledge composed of several ontologies, including the taxonomy presented in this paper. The main objective would be to support the user in the annotation of LOs, automate the annotation of some metadata and promote an unambiguous representation of LO metadata values.

Acknowledgements. This work has been partly supported by the Ministerio de Ciencia y Tecnología under project TIN2006-15107-C02-01/02, the NET2LEARN research group (IN3-UOC) and the eLearn Center (UOC).

References

1. Al-Khalifa, H.S., Davis, H.C.: The Evolution of Metadata from Standards to Semantics in E-Learning Applications. In: 17th ACM Conference on Hypertext and Hypermedia Systems (2006)
2. Conesa, J., Storey, V.C., Sugumaran, V.: Usability of Upper Level Ontologies: The Case of Research Cyc. Accepted Manuscript to appear in Data & Knowledge Engineering (2009)
3. Conesa, J., Olivé, A.: A Method for Prunning Ontologies in the Development of Conceptual Schemas of Information Systems. Journal of Data Semantics 5, 64–90 (2006)
4. El Saddik, A., Fischer, S., Steinmetz, R.: Reusability and Adaptability of Interactive Resources in Computing. ACM Journal of Educational Resources in Computing 1(1), Article 4 (2001)
5. Fischer, S.: Course and Exercise Sequencing Using Metadata in Adaptative Hypermedia Learning Systems. ACM Journal of Educational Resources in Computing 1(1), Article 5 (2001)
6. Haase, K.: Context for Semantic Metadata. In: 12th Annual ACM International Conference on Multimedia, pp. 204–211. ACM, New York (2004)
7. ISO/IEC JTC1 SC36: Final Report on the International LOM Survey (2004)
8. Kabel, S., de Hoog, R., Wielinga, B., Anjewierden, A.: Indexing Learning Objects: Vocabularies and Empirical Investigation of Consistency. Journal of Educational Multimedia and Hypermedia 13(4), 405–425 (2004)
9. Koper, R., Oliver, B.: Representing the Learning Design of Unit of learning. Educational Technology and Society 7(3), 97–111 (2004)
10. LTCS WG12: IEEE Learning Technology Standards Committee. Draft Standard for Learning Object Metadata. Technical Report 1484.12.1, IEEE Inc. (2002)
11. Lu, J.E., Hsieb, C.: A Relation Metadata Extension for SCORM Content Aggregation Model. Computer Standards and Interfaces (2008)
12. Olivé, A.: Conceptual Modeling of Information Systems. Springer, Heidelberg (2007)
13. OpenCyc, `http://www.cyc.com/cyc/opencyc`
14. Rodríguez, M.E., Conesa, J., García-Barriocanal, E., Sicilia, M.A.: Conceptual Interpretation of LOM and its Mapping to Common Sense Ontologies. In: International Conference on Semantics Systems, pp. 126–133. Journal of Universal Computer Science, Graz (2008)
15. Sánchez, S., Sicilia, M.A.: On the Semantics of Aggregation and Generalization in Learning Object Contracts. In: 4th IEEE International Conference on Advanced Learning Technologies, pp. 425–429. IEEE Press, Joensuu (2004)

16. Sicilia, M.A.: Metadata, Semantics and Ontology: Providing Meaning to Information Resources. International Journal of Metadata, Semantics and Ontologies 1(1), 83–86 (2006)
17. Yahya, Y., Yussoff, M.: Towards a Comprehensive Learning Object Metadata: Incorporation of Context to Stipulate Meaningful Learning and Enhance Learning Object Reusability. Interdisciplinary Journal of E-Learning and Learning Objects 4 (2008)
18. Zouaq, A., Nkambou, R., Frasson, C.: An Integrated Approach for Automatic Aggregation of Learning Knowledge Objects. Interdisciplinary Journal of Knowledge and Learning Objects 3 (2007)

A Framework for Automatizing and Optimizing the Selection of Indexing Algorithms

Mihaela Brut, Sébastien Laborie, Ana-Maria Manzat, and Florence Sèdes

Institut de Recherche en Informatique de Toulouse (IRIT)
118 Route de Narbonne 31062 Toulouse Cedex 9, France
{brut,laborie,manzat,sedes}@irit.fr

Abstract. Inside an information system, the indexation process facilitates the retrieval of specific contents. However, this process is known as time and resource consuming. Simultaneously, the diversity of multimedia indexing algorithms is growing steeply which makes harder to select the best ones for particular user needs. In this article, we propose a generic framework which determines the most suitable indexing algorithms according to user queries, hence optimizing the indexation process. In this framework, the multimedia features are used to define multimedia metadata, user queries as well as indexing algorithm descriptions. The main idea is that, apart from retrieving contents, user queries could be also used to identify a relevant set of algorithms which detect the requested features. The application of our proposed framework is illustrated through the case of an RDF-based information system. In this case, our approach could be further optimized by a broader integration of Semantic Web technologies.

1 Introduction

Various domains (such as news, TV, resource collections for commercial or consumer applications, collaborative work, video surveillance...) develop information systems for managing and retrieving multimedia contents. In order to retrieve these contents, a multimedia indexing phase is priory required. This one mainly extracts some multimedia features for producing multimedia metadata and stores these descriptions into a metadata collection on which user queries will be applied.

Currently, information system developers select by themselves the appropriate indexing algorithms to be included into the indexation engine. However, the diversity of these indexing algorithms is permanently increasing. Furthermore, as information systems should handle extensive multimedia collections, it is not possible to execute all algorithms because the indexation process is time and resource consuming (e.g., CPU). Consequently, a solution for automatizing and optimizing the selection of indexing algorithms must be developed.

Our article provides such solution by selecting the indexing algorithms which are the most suitable for particular user needs. For that purpose, indexing algorithms are described by the features they extract and user queries are used to determine a relevant set of algorithms which extract all requested features.

F. Sartori, M.Á. Sicilia, and N. Manouselis (Eds.): MTSR 2009, CCIS 46, pp. 48–59, 2009.

Alongside with the time and computing resource economies, our solution has the following advantages:

- It determines a relevant set of indexing algorithms which could extract as many as possible multimedia features requested by user queries. Thus, developers are informed about the most suitable algorithms set corresponding to a specific query set.
- It is based on a general uniform modeling of user queries and indexing algorithm descriptions. This modeling uses the multimedia features in order to describe and relate the queries and the algorithms. Hence, our approach can be applied to different representation and query languages.
- It considers low-level multimedia features as well as high-level ones, acquired through indexing algorithms that make use of Semantic Web technologies.
- It can be adopted in the development phase of an information system as well as during the concrete system usage.

The remainder of the paper is structured as follows. Section 2 illustrates the diversity and heterogeneity of existing indexing algorithms, and presents some systems which integrate such algorithms in order to manage and retrieve various multimedia contents. Because existing systems do not provide a solution for a selective indexing process according to user queries, we propose such solution in Section 3. The proposed solution is generic and could be applied to multiple data representations (unstructured or structured descriptions). Section 4 illustrates the application of our approach for RDF descriptions as well as the supplementary benefits acquired in this case due to Semantic Web approach. Section 5 presents a brief conclusion and some perspectives.

2 Related Work

In general, an information system in charge with managing and retrieving multimedia contents is composed of [15,6]:

- A **multimedia collection** which contains several multimedia contents. These contents refer to media items, such as texts, images, videos or audios.
- A **metadata collection** which contains information about the media characteristics (e.g., size, name, file type) and their contents. [8] presents some specific metadata for multimedia contents. These metadata may be encoded in several standards, such as EXIF [12], Dublin Core [19], MXF [9], etc.
- An **indexation engine** which includes several indexing algorithms to be applied on the multimedia collection in order to enrich the metadata collection.

A fair amount of research has been conducted on developing indexing algorithms. In the following, we propose to illustrate their diversity and heterogeneity for each media type.

For textual documents, some indexing techniques, e.g., [14], are inspired by classic Information Retrieval [21], or by Web Information Retrieval, exploiting the hypertext features, such as page hyperlinks [5] and HTML general tags [1].

The progress from a term-based to a concept-based document indexation was possible due to the latent semantic indexing technique [3] or to some knowledge representation models and methods that are typical to artificial intelligence domain (such as neural networks, semantic networks, bayesian networks) [18].

Concerning images, content semantic indexation processes usually analyze object related information (e.g., how many objects are in this image?, is object X present?, which objects are in the image?). They exploit various a priori knowledges on the observed scenes and a general model of the world. To achieve this, current methods are based on feature extraction, clustering/segmentation, object descriptor extraction, object recognition [7]. Pattern recognition techniques [23], such as boosting or cascade of classifiers execution, have been also applied for image semantic indexation.

Audio analysis is accomplished in some main directions [11]: segmentation (splitting an audio signal into intervals for determining the sound semantic or composition), classification (categorizing audio segments according to predefined semantic classes, such as speech, music, silences, background noise) and retrieval by content (using similarity measures in order to retrieve audio items that syntactically or semantically match the query).

In the area of content-based video indexing and retrieval, many research efforts have been conducted on automatic techniques for extracting metadata which describe the content of large video data archives, such as [10]. Some of these metadata have been adopted in video standards, e.g., in MPEG-7 [17]. These systems typically use metadata elements such as content type, shot-boundaries, audio keywords, textual keywords or caption text keywords. Besides, there is also some research, e.g., [22], in determining the class of video scene objects (e.g., human, vehicle, type of vehicle, animal) and detecting activities (e.g., carrying a bag, raising arms in the air, the manner of running or walking).

A single indexation engine could not include all these algorithms because it would overload the indexation process with useless analysis for concrete user needs. Thus, the development phase of an indexation engine requires the selection of specific indexing algorithms. Many projects integrate such specific algorithms in their indexation engines in order to manage multimedia contents.

The K-Space project[1] is focused on semantic inferences for semi-automatic annotation and retrieval of multimedia contents. It integrates three research clusters: content-based multimedia analysis, knowledge extraction, and semantic representation and management of multimedia. The VITALAS project[2] (Video & image Indexing and reTrievAl in the LArge Scale) develops solutions for cross-media indexing and retrieval by developing intelligent access techniques to multimedia professional archives. The CANDELA project[3] (Content Analysis and Network DELivery Architectures) performed video content analysis in combination with networked delivery and storage functionalities. The ISERE project[4]

[1] http://kspace.qmul.net:8080/kspace/

[2] http://vitalas.ercim.org

[3] http://www.hitech-projects.com/euprojects/candela

[4] http://www.mica.edu.vn/Isere/

(Inter-media Semantic Extraction and Reasoning) aimed to study a unified multimedia model which enables to enhance the identification of semantic contents. The MUSCLE network of excellence[5] (Multimedia Understanding through Semantics, Computation and LEarning) deals with multimedia data mining and machine learning technologies in order to provide a set of showcases for object recognition, content analysis, automatic character indexing, content-based copy detection, unusual behavior detection, movie summarization, human detection, speech recognition, etc.

However, none of these projects propose a solution to select the suitable indexing algorithms according to user needs. Moreover, their indexation engines consider a fixed set of algorithms. In the following section, we propose a generic framework which determines, according to user queries, a relevant set of indexing algorithms to be included into a multimedia retrieval system.

3 A Generic Framework for Selecting Indexing Algorithms

When a user query asks for some multimedia contents, the system builds the results by searching into the metadata collection. Our proposition considers that the same query could be also used in order to identify a set of indexing algorithms which detect the requested multimedia features. In this section, we present a method for determining such list of algorithms (§3.1). Moreover, we show that our approach could be used in the development phase of an information system as well as during the concrete system usage (§3.2).

3.1 Determining Indexing Algorithms According to a Query

In order to be as much generic as possible, our proposition is based on a uniform modeling of the query, the multimedia metadata and the indexing algorithm descriptions. This uniformity is acquired mainly by considering multimedia features in all models. More precisely, a query is viewed as a list of features to be retrieved, a multimedia metadata contains a list of features presented in a multimedia content and an indexing algorithm identifies a list of features.

In this context, answering a query consists in finding multimedia contents by locating in the metadata collection the requested features. Similarly, we propose to determine the set of indexing algorithms necessary for answering a query, more precisely a relevant set of algorithms in order to optimize the indexation process.

Consider a query Q, a list of indexing algorithm L_A and a feature f, such that $f \in Q$. Algorithm 1 selects in L_A a relevant set of indexing algorithms which extracts the feature f and as many as possible features specified by Q.

Initially, the list of indexing algorithms L_A contains all available algorithms in the system. The feature f is selected in Q according to the maximum number

[5] `http://www.muscle-noe.org`

Algorithm 1. indexingAlgorithmsSelection

Input: A user query Q, a list of indexing algorithms L_A and a feature f, such that $f \in Q$.
Output: A list of indexing algorithms.
Data: A list of indexing algorithms L which gather all results returned by the recursive call of *indexingAlgorithmsSelection*.

```
L_A ← the indexing algorithms of L_A which extract f;
if L_A ≠ ∅ then
    mark f in Q;
    L ← ∅;
    foreach f_i unmarked in Q do
        L ← L ∪ indexingAlgorithmsSelection(Q, L_A, f_i);
    if L = ∅ then
        e ← selectOneAlgorithm(L_A);
        return {e};
    else
        return L;
else
    return ∅;
```

of indexing algorithms in L_A which extracts f. Actually, during the recursive call of Algorithm 1 this list L_A will be refined into a new list L.

During the execution of Algorithm 1, we propose to mark features that could be identified by some indexing algorithms included in L_A. Hence, at a time only features which have not already been identified by indexing algorithms are considered. This branch and bound technique is useful to prune the backtracking search tree of Algorithm 1. Moreover, when several indexing algorithms can be applied for determining a set of features, the *selectOneAlgorithm* method is used to select only one algorithm, thus avoiding applying multiple indexing algorithms for determining the same set of features. For instance, this selection could be based on the algorithms execution time.

Finally, when Algorithm 1 stops, it returns a relevant list of indexing algorithms that retrieves the features mentioned in the query. Nevertheless, this algorithm is not complete because the resulted list is related only to the given feature. Hence, when some features remain unmarked, it could be possible that other indexing algorithms identify them, especially algorithms that do not consider the given feature f.

To assure the application of Algorithm 1 for all requested features, we propose Algorithm 2 which includes multiple calls of Algorithm 1 such as to cover as many as possible unmarked features from the query.

When Algorithm 2 stops, one can check the marked features in order to verify if all requested features can be identified by the indexing algorithms in the resulted list:

Algorithm 2. getAllIndexingAlgorithms

Input: A user query Q and a list of indexing algorithms L_A.
Output: A list of the indexing algorithms.
Data: A list of indexing algorithms L which gather all results returned by the call of *indexingAlgorithmsSelection*.
$L \leftarrow \emptyset$;
foreach f_i *unmarked in* Q **do**
 $L \leftarrow L \cup indexingAlgorithmsSelection(Q, L_A, f_i)$;
return L;

- if all query features are marked, it means that the resulted set of indexing algorithms may identify all query features, thus producing after their executions some metadata solutions to the query.
- if some query features remain unmarked, it means that the resulted set of indexing algorithms cannot identify all query features.

3.2 Concrete Application Domains

As we already announced in Introduction, our solution could be applied during the development phase of a multimedia information retrieval system as well as during the concrete system usage.

The Case of an Information System Development. When an information system is designed, one important task consists in establishing an exhaustive set of possible queries that will be submitted by users. This enable developers to assure that the system will meet the concrete user needs. According to this query set, our proposition provides a solution to automatically determine a list of relevant indexing algorithms to be used during the indexation process. This list corresponds to the implicit indexation process, i.e., executing indexing algorithms over each multimedia contents during the acquisition phase (i.e., when the multimedia content is included in the system). The indexation results will constitute the initial multimedia metadata collection.

For this purpose, our proposed technique consists in applying Algorithm 2 for each query and in unifying the obtained lists of indexing algorithms into a single list. For example, suppose n queries and m indexing algorithms. For each query Q_i, Algorithm 2 produces a list L_i of indexing algorithms, with $1 \leq i \leq n$. The unified resulted list L will be $\bigcup_n^{i=1} L_i$, such that L contains k indexing algorithms with $k \leq m$.

It is important to note that, if some query features remain unmarked, the system may advise developers to collect new indexing algorithms that identify these unmarked features. This sort of dialog enables developers to be assisted during the information system development.

The Case of Querying an Information System. When a user query is submitted to the system, it retrieves a set of solutions into the metadata collection.

If no solution is retrieved, this probably means that other indexing algorithms should be executed in order to find some results. According to a user query, our proposition enables to determine the list of indexing algorithms which could provide the supplementary metadata to be used for retrieving some results. This list corresponds to the explicit indexation process, i.e., executing new indexing algorithms over each multimedia contents. During the explicit indexation process, the user is informed that no results are available for the moment and that he could receive some later.

Thanks to our feature based modeling, one may know before the evaluation of a query if it is possible or not to retrieve some results. Indeed, suppose L_f the list of features covered by an implicit indexation process. If all requested features specified by Q are included in L_f, it means that no supplementary indexing algorithms are required for responding to the query Q. Otherwise, one may execute Algorithm 2 to establish the explicit indexation process.

As could be noticed, the solution proposed in this section considers queries, multimedia metadata and indexing algorithm descriptions as lists of features. However, in many applications and metadata standards, multimedia features (the low-level ones and the semantic ones) are most often organized into a structure, such as XML [4], RDF [16], etc. In the following section, we show that our proposition is still effective for such structured descriptions.

4 Application to Structured Metadata

Previously, we have shown that our framework uses a unified modeling of the query and the list of indexing algorithms which are both based on multimedia features. Features may be specified thanks to keywords or based on specific structures (e.g., RDF, XML). Currently, most of the standardized multimedia metadata vocabularies are XML based. Many of them are already translated in RDF in order to be available for Semantic Web technologies, such as [2] for MPEG-7 and [13] for EXIF. Consequently, we propose to illustrate our proposition on RDF-based descriptions (§4.1). Moreover, we show that our framework can be further optimized thanks to the Semantic Web technologies (§4.2).

4.1 Application to RDF-Based Descriptions

Consider an information system which adopts RDF-based data descriptions. In such a system, indexing algorithm outputs are RDF descriptions or they are translated into RDF descriptions.

In Figure 1, we consider an example set of indexing algorithms described in terms of their outputs through RDF schemes. For instance, the indexing algorithm described in Figure 1(c) accomplishes a vehicle recognition inside a video content. Its description contains blank nodes (e.g., :_url, :_x) which are instantiated when the indexing algorithm is applied on a particular video content.

After the execution of indexing algorithms, the obtained metadata are stored into a metadata collection which is queried using SPARQL [20]. For instance, the

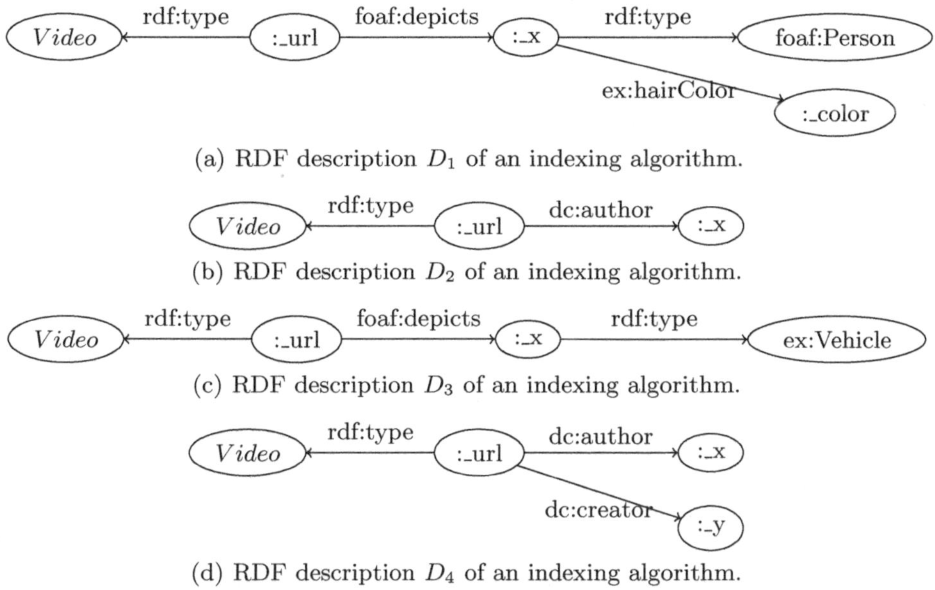

Fig. 1. Examples of RDF-based indexing algorithm descriptions

following SPARQL query Q evaluated against the metadata collection retrieves authors of videos which depict persons:

```
PREFIX rdf: <http://www.w3.org/1999/02/22-rdf-syntax-ns#>
PREFIX foaf: <http://xmlns.com/foaf/0.1/>
PREFIX dc: <http://purl.org/dc/elements/1.1/>
SELECT ?author
FROM <metadataCollection.rdf>
WHERE {
   ?video <rdf:type> <Video> .
   ?video <foaf:depicts> ?person .
   ?person <rdf:type> <foaf:Person> .
   ?video <dc:author> ?author .
}
```

Our proposed framework could be used to determine a relevant set of indexing algorithms which provides information for answering this query. The features from the user query and from the entire list of algorithms should be previously identified. For the former, each triple pattern specified in the SPARQL query (e.g., ?video <rdf:type> <Video>) could be considered as a requested feature. For the latter, each RDF triple specified in the indexing algorithm descriptions (e.g., :_url rdf:type Video) could be considered as an extracted feature.

Figure 2 presents the backtracking search tree of the execution of Algorithm 2 over the query Q and the list of indexing algorithms illustrated in Figure 1. Each

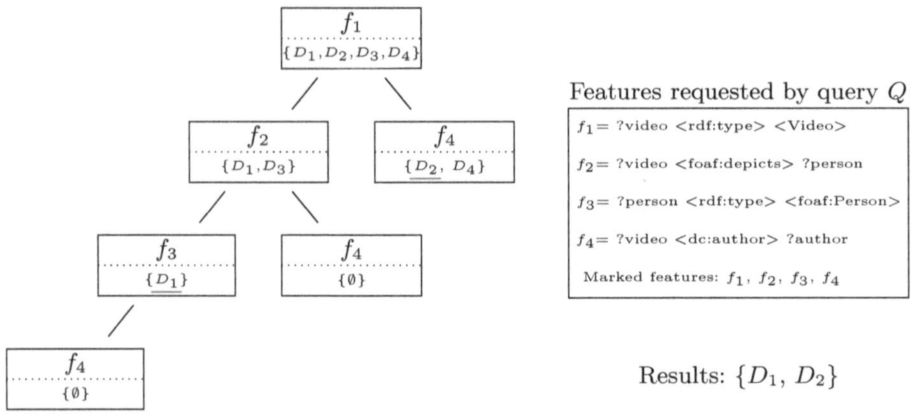

Fig. 2. An execution of Algorithm 2 over the SPARQL query Q and the indexing algorithm descriptions illustrated in Figure 1

node represents a requested feature and a set of indexing algorithms which detect not only this feature but also the features represented by the node's ancestors.

In this figure, D_1 is a good candidate because it extracts three requested features, namely f_1, f_2 and f_3. For f_4, two candidates are found: D_2 and D_4. The *selectOneAlgorithm* method selects only one of them (cf., Section 3). Moreover, all query features are marked which means that the application of the selected indexing algorithms (described by D_1 and D_2) may produce solutions for the given query.

In the following, we propose some optimizations of our framework based on grouping features and using the semantic meaning of features.

4.2 Optimizations and Discussions

In the case of RDF-based descriptions and SPARQL queries, we have shown that features are expressed as triples and triple patterns, respectively. Consequently, for queries involving many triple patterns, Algorithm 2 will test a lot of feature combinations. Indeed, the complexity of Algorithm 2 is determined by the number of features. Moreover, since each triple pattern is viewed as a feature, certain general features (e.g., ?x <foaf:depicts> ?y) could correspond to many indexing algorithms. Consequently, certain tree nodes will include quite big lists of indexing algorithms.

In order to improve the execution of Algorithm 2, we propose to group triple patterns which are related to each other. For example, in Figure 2, the feature f_2 and f_3 could be grouped into one single feature because we are looking for something which depicts a person. This could be achieved by processing the query: since each triple pattern is composed of {subject, predicate, object}, two triple patterns could be grouped if the object of the former is the same with the subject of the latter (as it is the case for f_2 and f_3).

Another improvement consists in using the semantic meaning of features. Actually, in many situations a requested feature does not correspond exactly to the ones in an indexing algorithm description, as shown in the following examples.

Example 1. If a query is searching information about a person, and a list of indexing algorithms identifies only information about man, Algorithm 2 won't produce any result. However, using the fact that a man is a person, one is able to select some indexing algorithms, and consequently produce some results. This reasoning facilities could be achieved by using ontologies which could be also used to determine synonyms and related features.

Example 2. Suppose we add to the query Q, the following statement f_5: `?person <ex:hairColor> "blond"`. Our approach won't retrieve some indexing algorithms which determine this feature (i.e., f_5 will remain unmarked) because no indexing algorithms determine exactly this feature. However, the indexing algorithm associated to the description D_1 extracts information about hair colors. In order to select this algorithm, it is possible to relax the query constraint by replacing "blond" with a variable ?color. Ontologies may also be used to find out that blond is a specialization of hair color.

Adopting ontology could be a solution for optimizing our framework, since it specifies a network of concepts containing general concepts as well as specific ones.

5 Conclusion

Our approach is situated in the context of multimedia information systems where an intensive indexation process is required in order to facilitate the multimedia content retrieval. We propose a generic framework for selecting a relevant set of indexing algorithms according to user queries. Moreover, we have shown that our proposal could be applied on concrete multimedia metadata vocabularies, such as RDF-based descriptions. This application to such language allows the use of Semantic Web technologies which could improve our framework. Furthermore, our solution is general enough for functioning in a context of a local information system as well as inside a distributed information system, or even inside a Web based information system. For the last case, the solution of implementing the indexing algorithms as Web services is to be considered.

Our solution is going to be integrated and tested in the context of the LINDO project[6]. This project is focused on managing the multimedia indexation process inside a distributed environment, where all the indexing algorithms are centralized on a central server and deployed on demand on the remote servers, according to the user queries. Our solution provides an optimization of the indexing algorithms deployment and indexation processes due to the selection of the most suitable algorithms according to the user queries. Furthermore, in the

---

[6] `http://www.lindo-itea.eu`

distributed environment of the LINDO project, queries might be different for each remote server. Consequently, we will observe if some indexing algorithms are frequently demanded by most of remote servers.

Acknowledgement

This work has been supported by the EUREKA Project LINDO (ITEA2 – 06011).

References

1. Agosti, M.: Information Retrieval and HyperText. Kluwer Academic Publishers, Norwell (1996)
2. Arndt, R., Troncy, R., Staab, S., Hardman, L., Vacura, M.: COMM: Designing a well-founded multimedia ontology for the web. In: Aberer, K., Choi, K.-S., Noy, N., Allemang, D., Lee, K.-I., Nixon, L.J.B., Golbeck, J., Mika, P., Maynard, D., Mizoguchi, R., Schreiber, G., Cudré-Mauroux, P. (eds.) ASWC 2007 and ISWC 2007. LNCS, vol. 4825, pp. 30–43. Springer, Heidelberg (2007)
3. Sarwar, B., Karypis, G., Konstan, J.A., Riedl, J.: Incremental SVD-based algorithms for highly scalable recommender systems. In: Proceedings of the Fifth International Conference on Computer and Information Technology (2002)
4. Bray, T., Paoli, J., Sperberg-McQueen, C.M., Maler, E., Yergeau, F.: Extensible markup language (XML) 1.0, 5th edn. Recommendation, W3C (2008)
5. Brin, S., Page, L.: The anatomy of a large-scale hypertextual web search engine. Computer Networks and ISDN Systems 30(1-7), 107–117 (1998)
6. Buckland, M.K., Plaunt, C.: On the construction of selection systems. Library Hi Tech. 12, 15–28 (1994)
7. Chen, S.-C., Ghafoor, A., Kashyap, R.L.: Semantic Models for Multimedia Database Searching and Browsing. Kluwer Academic Publishers, Norwell (2000)
8. Chrisment, C., Sèdes, F.: Media annotation. In: Multimedia Mining: A Highway to Intelligent Multimedia Documents (Multimedia Systems and Applications Series), pp. 197–211. Kluwer Academic Publishers, Dordrecht (2002)
9. Devlin, B.: MXF – the Material eXchange Format. EBU Technical Review, Snell & Wilcox (July 2002)
10. Dönderler, M.E., Şaykol, E., Arslan, U., Ulusoy, Ö., Güdükbay, U.: BilVideo: Design and implementation of a video database management system. Multimedia Tools and Applications 27(1), 79–104 (2005)
11. Foote, J.: An overview of audio information retrieval. Multimedia Systems 7(1), 2–10 (1999)
12. Japan Electronics and Information Technology Industries Association: Exchangeable image file format for digital still cameras: Exif Version 2.2 (April 2002)
13. Kanzaki, M.: EXIF vocabulary workspace – RDF Schema. W3C (2003), `http://www.w3.org/2003/12/exif/`
14. Lambolez, P.Y., Queille, J.P., Chrisment, C.: EXREP: A generic rewriting tool for textual information extraction. Ingéniérie des Systèmes d'Information 3, 471–487 (1995)
15. Lancaster, F.W.: Information Retrieval Systems. Wiley, New York (1979)
16. Manola, F., Miller, E.: RDF primer. Recommendation, W3C (2004)

17. Martínez, J.M.: MPEG-7 Overview v.10. ISO/IEC JTC1/SC29/WG11/N6828 (2004), http://www.chiariglione.org/mpeg/standards/mpeg-7/mpeg-7.htm
18. Micarelli, A., Sciarrone, F., Marinilli, M.: Web document modeling. In: Brusilovsky, P., Kobsa, A., Nejdl, W. (eds.) Adaptive Web 2007. LNCS, vol. 4321, pp. 155–192. Springer, Heidelberg (2007)
19. National Information Standards Organization: The Dublin Core Metadata Element Set. ANSI/NISO Z39.85 (May 2007)
20. Eric Prud'hommeaux and Andy Seaborne. SPARQL Query Language for RDF. Recommendation, W3C (January 2008)
21. Salton, G., Mcgill, M.J.: Introduction to Modern Information Retrieval. McGraw-Hill, Inc., New York (1986)
22. Viola, P., Jones, M.: Robust real-time object detection. International Journal of Computer Vision (2001)
23. Yang, M.-H., Kriegman, D.J., Ahuja, N.: Detecting faces in images: a survey. IEEE Transactions on Pattern Analysis and Machine Intelligence 24(1), 34–58 (2002)

Empirical Analysis of Errors on Human-Generated Learning Objects Metadata

Cristian Cechinel[1], Salvador Sánchez-Alonso[2], and Miguel Ángel Sicilia[2]

[1] Computer Engineering Course, Federal University of Pampa, Caixa Postal 07, 96400-970, Bagé (RS), Brazil
contato@cristiancechinel.pro.br
www.unipampa.edu.br

[2] Computer Science Department, University of Alcalá, Ctra. Barcelona km. 33600, 28871 Alcalá de Henares (Madrid), Spain
{salvador.sanchez,msicilia}@uah.es
www.uah.es

Abstract. Learning object metadata is considered crucial for the right management of learning objects stored in public repositories. Search operations, in particular, rely on the quality of these metadata as an essential precondition for finding results adequate to users requirements and needs. However, learning object metadata are not always reliable, as many factors have a negative influence in metadata quality (human annotators not having the minimum skills, unvoluntary mistakes, lack of information, for instance). This paper analyses human-generated learning object metadata records described according to the IEEE LOM standard, identifies the most significant errors committed and points out which parts of the standard should be improved for the sake of quality.

Keywords: Metadata errors, IEEE LOM, learning objects.

1 Introduction

In the context of Learning Objects (LOs), metadata could be defined as records which describe the most important features of a LO. These descriptions may consider several aspects of the technology, such as authorship, technical and educational information, preferably according to the specifications of some adopted metadata standard or model [7]. The main purpose of LOs metadata (and its standards) is to support indexation and search of LOs in retrieval systems, such as the Learning Objects Repositories (LORs) [2]. Thus, providing good quality metadata is key to succeed on discovering and selecting desirable and relevant material on any retrieval system. As stated by Currier et al. [3], "poor quality metadata can mean that a resource is essentially invisible within the repository and remains unused". Although, some researches in the field have been studying and proposing alternatives to provide automatic extraction and generation of good LO metadata (for instance [8]), certain information has a strong subjective

F. Sartori, M.Á. Sicilia, and N. Manouselis (Eds.): MTSR 2009, CCIS 46, pp. 60–70, 2009.

component and thus requires direct human involvement [3]. Identifying the most common errors on human-generated LOs metadata helps one to understands the main difficulties on the process of describing this information, facilitating at the same time the creation of guidance material for specific topics of LOs metadata standards yet to be properly understood. Moreover, information about the errors can lead one to develop better skills on the professionals and researchers of this field, which will certainly produce richer quality metadata.

This paper analyses human-generated LOs metadata records described according to the IEEE LOM standard [5], identifying and describing the most significant errors observed. The rest of the paper is structured as follows. Section 2 presents background information about other studies involving metadata quality and correctness. Section 3 gives introductory information about the analysed data. Section 4 describes the methodology used on this study and the errors which were identified. Section 5 presents a statistical overview of the observed errors, as well as a discussion on these findings. Section 6 points out some conclusions and further work on this field.

2 Quality of Learning Objects Metadata

Metadata quality has been discussed in several works from different perspectives and points of view. In [4], Farance criticized some aspects of IEEE LOM claiming the standard was not ready for a fully adoption due to inconsistencies, ambiguities and imprecise definitions about its data elements and categories. Farance provided a list of 16 data elements (from a total of 45 existent fields) which he believes to be the most problematic ones, and he suggested "either to improve the definitions, or remove the data elements, or identify which data elements aren't useful for IT interoperability".

On [1], Barton et al. exposed some areas where problems on metadata arise, such as: 1) spelling and abbreviations; 2) author and other contributor fields; 3) title; 4) subject; and 5) date. As these authors claim that the errors are strongly impacted by "who creates metadata and how", they presented preliminary figures on the variations of the classification data when annotators from different groups provide metadata for the same LO. These data were the basis of their final suggestion on the use of a collaborative approach to metadata creation.

Another relevant effort to quantify the amount of errors committed on LOs metadata records was reported in [6]. In his study, Caceres classified the errors as follows: missing (no data were recorded for the studied metadata element), syntactic (the metadata values recorded do not conform to what the IEEE LOM standard specifies in syntactic terms), and semantic (the metadata value of the metadata element do not match the expected information). In his findings, Caceres reported that most fields in IEEE LOM are left blank, which reinforces Farance's point of view on the dubious importance of some fields of this standard [4].

3 Data Description

From 2006 to 2008, students of the discipline *Languages and Techniques for Digital Content Design* of the Computer Science Master Course of the University of Alcalá, were asked to select one LO publicly available on the Internet, to use it for a while, and to provide a complete metadata description of it according to the IEEE LOM standard. In order to do this activity, the students should use a LO Metadata Editor (the LomPad editor[1] was recommended), and fill in all the elements of each one of the nine categories of IEEE LOM (even considering some metadata information was not posible to provide, filling every data elements in was mandatory). In total, 44 xml files containing information about several LOs were collected and analyzed. From these, 1 (2,27%) offered no information at all and was disregarded. The remaining 43 (97,73%) records were used for this first introductory study.

4 Methodology

The study was divided into the following steps: 1) to open all XML files containing the metadata records, taking notes and commenting on anything relevant for the subsequent analysis of data; 2) to create different groups of errors according to the observations; 3) to classify the observed errors into the observed groups; and 4) to extract descriptive statistics about the encountered errors. Steps 1 to 3 were run in parallel, given that the groups of errors had to be created during the study of each record. These groups of errors were revised and refined throughout and until the end of the study of the last record.

4.1 Groups of Errors

Errors were divided into two groups: *General* and *specific*. The *general* group contains errors that can be found in any IEEE LOM category, such as a missing field. The second group contains errors that can be observed only in a given IEEE LOM category or data element. An error about the wrong use of some taxonomy, for instance, will either appear in category 9.Classification, but not any other categories.

Table 1 describes the 8 different situations that were identified for the *general* category of errors. It is important to remark that errors prompted by missing information –those labelled as G-2 and G-3– are not exactly errors, so in this sense they are a subset within this category. Also, observations of patterns of wrong usage of data elements and categories were the input for the creation of several specific categories of errors as we will discuss later in this section.

As it can be noticed, there is a high level of subjectivity on the definition of these groups, and some of them could even be grouped into a bigger one, as it is the case of G-7 and G-8, or of G-1, G-3 and G-4(this could be usefull

[1] http://sourceforge.net/projects/lompad/

Table 1. General errors

Label	*Error*	*Comments*
G-1	No error	There was no error in the evaluated data
G-2	Missing data element	Some information about the data element should be given and it was not
G-3	Missing data element, but impossible to identify which its value should be	There was no information about the data element. However, it could be considered unnecessary or even impossible to provide it. Example: the attribute 4.7 Technical.Duration is useful for sounds, movies or animations, but is not applicable to text-only LOs.
G-4	Wrong use of metadata editor	The XML file has not the expected structure because of a wrong use of the metadata editor which resulted in an unexpected structure to be produced in the resulting XML metadata record. Example: the attribute 1.5 General.Keyword does not have a different field for each different keyword and instead several keywords were provided for in the same field.
G-5	Correct, but incomplete information	The information in the data element is only partial, or could be improved. Example: a LO composed of songs, texts and applets, has the attribute 4.1 Technical.Format set to just "text/html".
G-6	Impossible to track the information	Considering that the filled all the data elements in, some fields contain either invented information or information that was not available later (during the period of time our study was being carried out).
G-7	Wrong information (or wrong classification)	The information provided is not correct. Example: the LO supports active learning and the attribute 5.7 Educational.InteractivityType was set to "expositive".
G-8	Unexpected information	The information provided is not what one would expect for that particular data element or category. Example: the attribute 5.7 Educational.InteractivityType was filled in with a website adress in the form of a URL.

depending the statistical method used to analyse the data). Furthermore, if some field is classified as G-6, for instance, it is not possible to guarantee whether the given information is wrong, or it is just not available at the time of the study. Even considering the existence of overlapping groups, this classification depicts a general idea of the most frequently occurrences in the observed records, allowing one to establish some level of students' knowledge about the IEEE LOM data elements.

Considering that many IEEE LOM data elements and categories are interdependent, i.e, the value of one depends on the value of the other, the so-called *specific* errors were sometimes related to a given IEEE LOM category as a whole, but sometimes related to an isolated data element only. Regarding isolated data elements, *specific* errors were identified for 1.6 General.Coverage, 2.3.3 LifeCycle.Date, 4.2 Technical.Size, 4.5 Technical.InstallationRemarks and 4.7. Technical.Duration (see table 2). Regarding categories as a whole, *specific* errors were identified for categories 3.Meta-Metadata, 7.Relation and 9.Classification (see

Table 2. Specific errors for data elements

Data Element	*Label*	*Error*	*Comments*
1.6 General.Coverage	S1.6-1	Confusion with data element 1.5 General.Keywords or with the LO contents	The information provided speaks about the content or subject of the LO. Example: in a LO which on the fundamentals of fiber optic, the value was set to "fiber optic".
	S1.6-2	Confusion with data element 5.5 Educational.IntendedEndUserRole	The information provided is about the possible users of the LO. Example: "learning designers and teachers".
	S1.6-3	Confusion with data element 5.6 Educational.Context	The value set refers to the environment where the LO could be used. Example: "primary education".
2.3.3 LifeCycle.Date	S2.3.3-1	Confusion between the dates of LO contributions and the dates of the LO storage in the repository from which it was reached	Sometimes the date set for some contribution to the LO refers to a date which appears only in the repository where the LO is stored.
4.2 Technical.Size	S4.2-1	The size does not exist, or it is not possible to measure it	This error overlaps error G-6, but due to its high frequency on this data element constitutes a separate group.
4.5 Technical.Installation Remarks	S4.5-1	Confusion with data element 4.6 Technical.Other Platform Requirements	Instead information about how to install the LO, the annotator set software/hardware requirements. Example: "Flash plug-in"
4.7 Technical. Duration	S4.7-1	Confusion with data element 5.9 Educational.TypicalLearningTime	The annotator provides the same information here and in data element 5.9
	S4.7-2	The duration provided does not exist, or it is not possible to measure it	For instance, the LO is constituted only by texts and some duration in terms of time is provided for it. This error overlaps error G-6, but due to its high frequency on this data element constitutes a separate group.

respectively tables 3, 4 and 5). On these last groups, specific errors were also used to estimate the level of student knowledge about the category. The criteria for establishing this level of knowledge was an estimation of how far (or close) the information provided was from full correctness.

5 Results

This section is divided into two parts. Section 5.1 presents a general overview of the errors observed in all the categories, and section 5.2 introduces a descriptive

Table 3. Specific errors for category 3.Meta-metadata

Label	*Error*	*Comments*	*Level of Knowledge*
S3-1	Information about the LO	Instead of providing information about the LO metadata, the annotator set information about the LO itself.	None
S3-2	Information about an external meta-metadata	Instead of providing information about the present meta-metadata, a reference to an external meta-metadata about this LO was given. Example: the author of the meta-metadata was set to a person who created some metadata for this learning object in an external repository such as Merlot.	None
S3-3	Information about the LOR	The information provided was about the Internet location of the repository through which the LO was reached.	None
S3-4	Wrong information about 3.1 Meta-metadata.Identifier and mistakes in some other data element	Data in field 3.1 are wrong (or missing), and the annotator also commited an error in some other field	Very Little
S3-5	Wrong information about 3.1 Meta-metadata.Identifier	This error is very close to S3-4, however the only field wrong (or missing) is 3.1	Little
S3-6	Wrong information about 3.3 Meta-metadata.Metadata Schema	Good meta-metadata quality, the annotator understood the concepts behind 3.1, 3.2 and 3.4, but commited an error on 3.3	Medium

Table 4. Specific errors for category 7.Relation

Label	*Error*	*Comments*	*Level of Knowledge*
S7-1	Information about the LO	The annotator set the data field with a reference to the LO instead of to some relation between the LO and another LO.	Very little
S7-2	The relation does not exist	The relation mentioned in the metadata does not exist.	Very little
S7-3	Wrong nature of the relationship	The annotator choose the wrong type of relationship. Example: the LO is part of other LO, but the annotator set the value to "hasPart"	Little

Table 5. Specific errors for category 9.Classification

Label	*Error*	*Comments*	*Level of Knowledge*
S9-1	Only the Purpose field was filled in	All fields in the category missing apart from 9.1.Classification.Purpose	None
S9-2	Only the purpose and description were filled in	The annotator selected the LO purpose and provided some explanation on the purpose in the description field. All other fields were left blank	Very little
S9-3	Problems with the *taxonpath* attribute (lack of/invented ID, lack of source entry).	The annotator does not use any taxonomy or uses one of her own	Very little
S9-4	Wrong use of the taxonomy	Some taxonomy is used, however some fields are incorrectly filled in (lack of/invented ID, lack of source entry, etc.).	Little
S9-5	Conflict between the purpose and the taxonpath information	The information set for taxonpath is unexpected at the light of the purpose. Example: purpose was set to "educational level" and the taxonomy used refers to disciplines	Medium

analysis of the errors observed in categories 3.Meta-metadata and 9.Classification. These are considered two of the most problematic categories in IEEE LOM, as a higher percentage of errors are usually put in the elements of these categories. The analysis in section 5.2 relates the errors and the different groups of students which produce them, as well as the level of knowledge of each of these groups.

5.1 General Overview

Figure 1 shows the percentage of errors observed in each IEEE LOM category. This initial analysis disregards the peculiarities of the *specific* errors described before, considering every data element as an unique and independent field (i.e. errors in categories that have interdependent fields were taken into account for each and every data element in the category, and not only for the category as a whole).

Considering (in figure 1) that only the *Missing* and *Some other error* columns reveal a lack of knowledge from the student side, we observe that:

1. the less problematic categories are (in ascending order): 6.Rights, 1.General, 5.Educational, 2.Life Cycle and 8.Annotation; and
2. the most problematic categories are (in descending order): 9.Classification, 7.Relation, 4.Technical and 3.Meta-metadata.

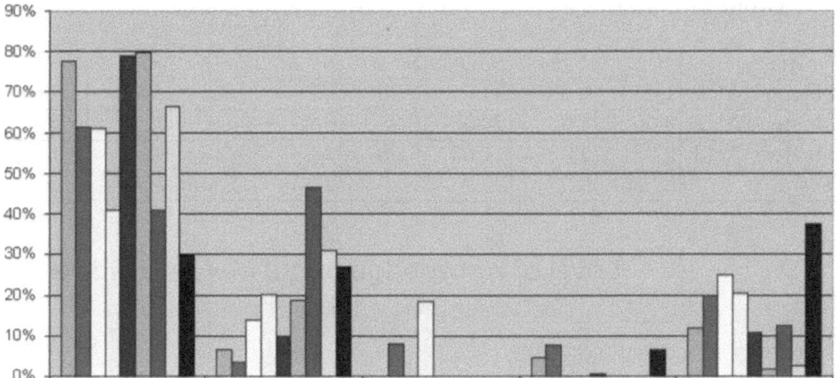

	No Error	Missing	Missing, but not an error	Wrong Use of the Editor	Some Other Error
General	78%	6%	0%	4%	12%
LifeCycle	61%	3%	8%	7%	20%
Meta-metadata	61%	14%	0%	0%	25%
Technical	41%	20%	18%	0%	21%
Educational	79%	10%	0%	0%	11%
Rights	80%	19%	0%	0%	2%
Relation	41%	47%	0%	0%	12%
Annotation	67%	31%	0%	0%	2%
Classification	30%	27%	0%	6%	37%

Fig. 1. General overview of the errors in IEEE LOM categories

Regarding the most problematic categories, categories 9.Classification and 3.Meta-metadata have the highest percentage of errors classified as *Some other error*. This fact can indicate that these categories are the ones with a higher number of occurrences of errors related to *specific* groups of errors.

It is also relevant to see that problems regarding the use of the editor are observed particularly in categories 2.Life Cycle, 9.Classification and 1.General, being 7.Relation the category including the highest percentage of missing fields.

5.2 Errors and Level of Knowledge about Categories

The students participating in the study belonged to 3 different groups of the Master Course on Computer Science, years 2006 (MasterUAH1), 2007 (MasterUAH2) and 2008 (MasterUAH3). The descriptive analysis of this section regards the peculiarities of the *specific* errors observed for the categories 3.Meta-metadata, 7.Relation and 9.Classification. This part of the study considers the categories as a whole, i.e. the errors counted where related to the whole category and not to each data element separately.

Category 3.Meta-metadata: As table 6 shows, only 23.26% of the XML files studied were correctly filled in (G1). The number of correct records decreased significantly from the first group (11.63%) to the last one (2.33%), being the

Table 6. Percentage of errors in the category 3.Meta-metadata (by group)

Group	*G-1*	*G-2*	*S3-1*	*S3-2*	*S3-3*	*S3-4*	*S3-5*	*S3-6*	*Total*
MasterUAH1	11.63	0.00	9.30	0.00	0.00	2.33	2.33	2.33	27.91
MasterUAH2	9.30	4.65	9.30	6.98	0.00	0.00	2.33	0.00	32.56
MasterUAH3	2.33	2.33	11.63	6.98	2.33	9.30	2.33	2.33	39.53
Total	23.26	6.98	30.23	13.95	2.33	11.63	6.98	4.65	100.00

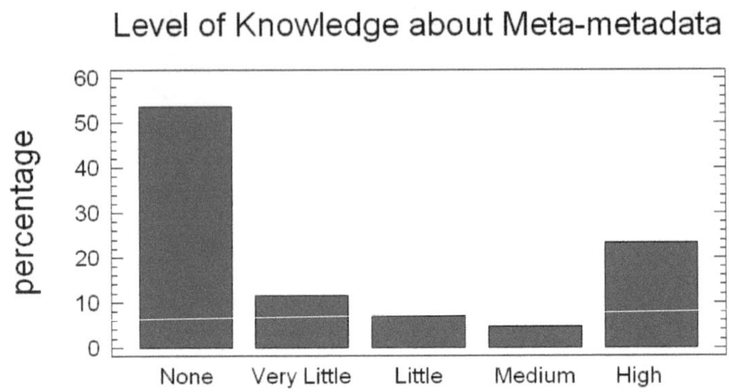

Fig. 2. Level of knowledge about the category 3.Meta-metadata

most significant error S3-1 (information about LO) with 30.23% of occurrences, followed by S3-2 (information about an external meta-metadata) with 13.95%.

Figure 2 shows the general level of knowledge of each group of students for the IEEE LOM category 3.Meta-metadata. The figures show that more than half the students (53.49%) have no knowledge about meta-metadata, whereas only the 23.25% present a high level of understanding about this category.

Category 9.Classification: This is the most problematic category. As one can see in table 7, only 16.28% of the records do not present any error. For those record presenting errors, most of these errors are not general but specific of this category. The most significant error observed is S9-5 (conflict between the purpose and the taxonpath information) with a 25.58% of occurrences, followed by S9-4 (wrong use of the taxonomy) with a 18.6% of occurrences.

Table 7. Percentage of errors in the category 9.Classification (by group)

Group	*G-1*	*G-2*	*G-8*	*S9-1*	*S9-2*	*S9-3*	*S9-4*	*S9-5*	*Total*
MasterUAH1	2.33	4.65	0.00	2.33	0.00	2.33	4.65	11.63	27.91
MasterUAH2	4.65	2.33	2.33	6.98	4.65	0.00	4.65	6.98	32.56
MasterUAH3	9.30	4.65	0.00	0.00	4.65	4.65	9.30	6.98	39.53
Total	16.28	11.63	2.33	9.30	9.30	6.98	18.60	25.58	100.00

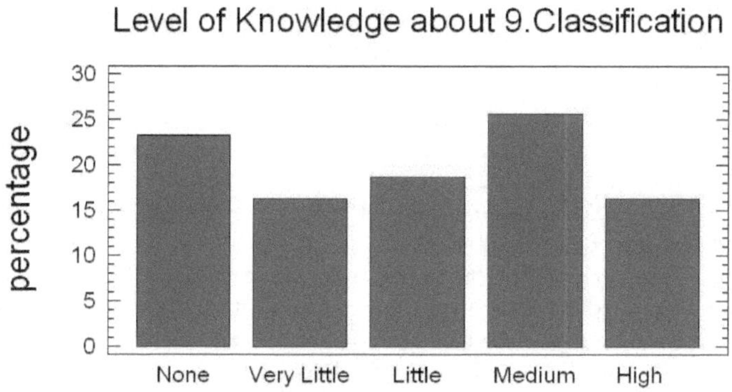

Fig. 3. Level of knowledge about the category 9.Classification

Figure 3 shows the students' level of knowledge about category 9.Classification. As it can be seen, a discouraging 23.26% of the students presented no knowledge, and 16.28% presented very little knowledge.

6 Conclusions and Future Work

This study has shown that several parts of the IEEE LOM standard remain unclear and thus annotators have problems with them. Even assuming that some defects exist in the way that the theoretical basis was presented to students, such a high percentage of errors in groups of mature students suggest that IEEE LOM has still room for improvement. Some of the categories and data elements in which the students faced most difficulties were already mentioned as problematic in a 2003 study by Farance [4]. Remarkably category 3.Meta-metadata, a category that should be relatively easy to fill in, presented a high percentage of errors. Further investigation is still needed to better understand the source of these errors, as well as to know if they are strictly related to the context of the present work, or if they can be extrapolated to some other contexts. The study reinforces the urgency of providing better guidance-material to students, and the improvement of the existent definitions for IEEE LOM data elements.

Acknowledgments. This work was supported by Carolina Foundation through its Mobility Program for Public Brazilian Professors. It also was funded by Comunidad de Madrid and University of Alcala as part of the activities of the MARIA project, reference code: CCG08-UAH/TIC-4178.

References

1. Barton, J., Currier, S., Hey, J.M.N.: Building quality assurance into metadata creation: an analysis based on the learning objects and e-prints communities of practice. In: 2003 Dublin Core Conference: Supporting Communities of Discourse and Practice - Metadata Research and Applications, Seattle, Washington, USA, September 28 - October 2 (2003)

2. Brasher, A., McAndrew, P.: Human-Generated Learning Object Metadata. In: Meersman, R., et al. (eds.) OTM-WS 2004. LNCS, vol. 3292, pp. 723–730. Springer, Heidelberg (2004)
3. Currier, S., Barton, J., O'Beirne, R., Ryan, B.: Quality assurance for digital learning object repositories: issues for the metadata creation process. ALT-J: Research in Learning Technology 12(1), 5–20 (2004)
4. Farance, F.: IEEE LOM Standard Not Yet Ready For "Prime Time". IEEE Learning Technology newsletter 5(1) (2004), `http://lttf.ieee.org/learn_tech/issues/january2003/index.htm`
5. IEEE, IEEE Standard for Learning Object Metadata (2002), `http://ltsc.ieee.org/doc/wg12/`
6. Cáceres Tello, J.: Estudio exploratorio de defectos en registros de metadatos IEEE LOM de objetos de aprendizaje. In: Post-Proceedings of SPDECE 2007 - IV Simposio Pluridisciplinar sobre Diseño, Evaluación y Desarrollo de Contenidos Educativos Reutilizables, Bilbao, Spain, September 19-21 (2007)
7. Sicilia, M., Garcia, E.: On the concepts of usability and reusability of learning objects. The International Review of Research in Open and Distance Learning 4(2) (October 2003)
8. Cardinaels, K., Meire, M., Duval, E.: Automating metadata generation: the simple indexing interface. In: Proceedings of the 14th international Conference on World Wide Web. WWW 2005, Chiba, Japan, May 10 - 14, pp. 548–556. ACM, New York (2005)

Analysis of Educational Metadata Supporting Complex Learning Processes

Jorge Torres[1] and Juan Manuel Dodero[2]

[1] Tecnologico de Monterrey, México
jtorresj@itesm.mx
[2] Universidad de Cádiz, Spain
juanma.dodero@uca.es

Abstract. Educational metadata provide learning objects and designs with required information that is relevant to a learning situation. A learning design specifies how a learning process involves a set of people in specific groups and roles engaging learning activities with appropriate resources and services. These elements are usually described by using structured primitives of an Educational Modeling Language. Metadata records must explicitly provide a representation of the flow of learning activities and how learning resources and services are utilized. We have analyzed a number of common workflow patterns in order to extend current Educational Modeling Languages' primitives used in complex learning flows. The information model of the Learning Process Execution and Composition Language is used as the basis to extend structured metadata required by such learning process descriptions.

1 Introduction

Computer-aided design and execution of learning activities try to describe a learning experience from the point of view of the tasks that participants have to carry out and the learning resources involved. Such learning experiences are generally expressed as a learning design (LD) specification, which describes by means of an Educational Modeling Language (EML) how a learning process involves a set of people in specific groups and roles engaging learning activities with appropriate resources and services [1].

An EML is a semantic information model and a binding to a computer-based description language, used to describe learning resources and processes from a pedagogical perspective, in order to support reuse and interoperability [2]. EMLs represent an important approach to integrate diverse educational aspects, allowing the technology-enhanced design and implementation of learning experiences [2,3]. EMLs make possible integrating educational resources personalized for each student's learning process and promoting the active participation of students. Learning process descriptions are delivered to the students as executable Units of Learning (UoL). The elements implicated in such learning experiences are the learning objectives, results, roles, activities, resources and services. These elements are modeled in a UoL as a structured set of metadata, along with the organization

F. Sartori, M.Á. Sicilia, and N. Manouselis (Eds.): MTSR 2009, CCIS 46, pp. 71–82, 2009.

and coordination mechanisms needed to deploy the learning experience —i.e. activity structures, assignment of roles to the activity structures and rules that determine the material to be delivered—. The temporal sequence in which the various learning activities unfold as a UoL is called the *learning flow* [4].

The information model of current EMLs includes structured primitives to describe the learning flow of a set of activities. For instance, the IMS Learning Design (IMS LD) specification and the Learning Activity Management System (LAMS) provide languages to describe collaborative learning scenarios, in which a number of concurrent users can be engaged to accomplish a number of activities. However, these EMLs do not consider all the range of learning flow primitives required to implement complex workflow structures, with a special view on concurrency control operations. The aim of this work is analyzing which these workflow structures should be, on the basis of common Workflow Control Patterns (WCP) [5], as well as explaining how the Learning Process Execution and Composition Language (LPCEL) can be used to provide extensions to current EMLs such as IMS LD.

The rest of this paper is structured as follows: section 2 analyzes the related work on learning process design and execution from the point of view of a framework characterization of EML-based complex learning process descriptions. Section 3 is dedicated to study how workflow control patterns provide the foundations to implement the required features as an extended set of EML primitives. Section 4 depicts a case study on how the LPCEL information model is used to implement the required extensions. Finally, in section 5 some conclusions and future work are outlined.

2 Design, Execution and Control of Complex Learning Processes

Koper defines a number of requirements for the general class of EMLs, namely: formalization, pedagogical flexibility, explicitly typed learning objects, completeness, reproducibility, personalization, medium neutrality, interoperability and sustainability, compatibility, reusability and life cycle [3]. These features must be implemented as structured metadata in any EML that supports a Complex Learning Process (CLP) learning flow [6,7]. Some authors classify EMLs as *exchange* (i.e., languages with a great expressiveness that include low-level primitives) or *authoring* (i.e. the language primitives are closer to the instructors' needs to describe a learning process) [8]. The EML used to describe a CLP must be expressive enough to be properly designated as an exchange language. Therefore, EMLs must take into account how annotating the process description with the appropriate set of design and control primitives to drive complex learning flows.

2.1 Related Work on Learning Process Design and Execution

Learning design is a fundamental phase before the delivery of a learning process as an executable UoL. However, defining activities and assigning users at design time will result in a too rigid scheme that cannot be adapted to the concrete events coming out from executing the learning sequences. On the contrary, leaving activities and task assignments to the run time will result in a more flexible scheme, but there may be no learning design at all. EMLs often prescribe defining some elements at design time, and preparing others to be managed during the run time. For instance, the LAMS approach involves defining a series of sequenced activities and tools that drive the course flow [9]. The IMS LD specification uses a theater metaphor to specify in advance the required flow of a learning process [4]. On the other hand, LAMS gates and grouping activities or IMS LD level B elements are used to set up design-time solutions prepared to manage runtime issues, such as binding user groups to running activities. However, some educators have found the IMS LD theatre metaphor difficult to understand [10], and found it difficult to design a LD with multiple workflows and optional interactions [11]. It was also difficult to model activities which are not done in a particularly specified sequence.

A study of current EML implementations has been carried out and depicted in Table 1, including the analysis of pedagogical aspects such as content cognitive structure, objectives, prerequisites, assessment, learning processes and learning service support. In this context, LAMS, the IMS LD and LPCEL are the EMLs that define structured metadata to cover most of the analyzed features, especially for the learning process dimension at which this work is aimed:

– The IMS LD specification includes an exchange language to explicitly describe learning experiences as UoL, and provides a conceptual framework for the specification of learning process designs, which have to be afterwards deployed onto a computer-based execution engine.

Table 1. EML coverage of required features for a complex learning process. The qualitative scale, from narrow to wider coverage, is represented as –, ±, +, ++.

Implemented Elements	CDF	TeachML	LMML	PML	PALO	Xedu	LPCEL	IMS-LD	LAMS	E2ML
Metadata	+	±	+	±	+	+	+	+	±	±
Content	+	++	++	+	++	+	+	+	+	±
Content Structure	±	±	+	+	+	–	+	–	–	–
Content Cognitive Structure	±	+	+	+	+	+	+	–	–	–
Objectives	–	–	–	–	±	±	+	+	–	++
Prerequisites	–	–	–	–	±	±	+	+	–	+
Expected Results	–	–	–	–	–	±	+	+	–	+
Assessment	–	–	–	–	–	–	+	±	++	±
Learning Process										
Activity Sequence	–	–	–	–	+	+	+	+	+	+
Activity Parallelization	–	–	–	–	–	–	+	+	+	+
Optional Activity Parallelization	–	–	–	–	–	–	+	–	+	+
Conditions	–	–	–	–	–	–	+	+	+	+
Messages	–	–	–	–	–	–	+	+	+	–
Roles	–	–	–	–	–	–	+	+	+	+
Groups	–	–	–	–	–	–	+	–	+	–
Services										
Monitoring	–	–	–	–	–	–	+	+	+	–
Search	–	–	–	–	–	–	+	+	+	–
Communication	–	–	–	–	–	–	+	±	+	–
External Tools	–	–	–	–	–	–	+	±	–	–
Other Propietary Tools	–	–	–	–	–	–	+	–	+	±

- The LAMS approach has taken a step further by integrating design and execution of learning activities into the same authoring language, design and execution environment.
- LPCEL provides a framework including the appropriate metadata primitives that describe execution-aware learning designs. LPCEL manifests these execution-related issues as emergent to author process-oriented learning designs [12]. In this sense, LPCEL provides both an exchange and an authoring language.

2.2 An Extended Characterization of Complex Learning Processes

To support the features of a CLP, the following characteristics have been defined additionally to Koper's [6]:

1. Pedagogical diversity: EMLs must be able to guide the composition, execution and control of the learning process including various pedagogical mixings and levels of complexity.
2. Learning flow description: EMLs need to be expressive enough to specify complex structures (e.g. activities, dependencies, rules, contents, roles, scenarios and participants).
3. Dynamic and unanticipated composition: in some cases the initial specification of a learning process must be redefined and changed as derived from collaboration and negotiation of learners and instructors.
4. Separation of learning process and services: detailed information has to be included that enables accessing to a set of required support services.
5. Learning service availability and containment: suitable descriptions are needed to achieve the required availability that do not rely only on self-containment of resources.
6. Transactional support: EMLs must be provided with an operational transaction support to execute a learning process that implements long-run activities.

Since the EML metadata required to represent the flow control have to do with workflow structures, we consider that Workflow Control Patterns (WCP) are a good starting point to analyze in detail how the characterization of a CLP should be carried out as learning flow metadata primitives [5].

3 Workflow Control Patterns-Based Analysis of EMLs

Workflows are defined by the Workflow Management Coalition as "the computerized facilitation or automation of a business process" [13]. Workflow Management Systems (WMS) have become a key technology for business process automation, supporting organizational aspects, user interfaces, monitoring, accounting, simulation, distribution and heterogeneity [14]. When designing a WMS, there are two core components to take care of: the workflow engine and the workflow language.

The workflow engine is the basic workflow management control software responsible for process creation and deletion, control of the activity scheduling and interaction with computer-based applications or human resources. It also provides the run time execution environment for a workflow instance and is responsible for part or all of the runtime control environment within an enactment service. The workflow enactment service provides the run-time environment in which process instantiation and activation occurs. The workflow engine controls the execution of a set of instances of a process or sub-process within a defined scope, determined by the range of object types and their attributes specified in a workflow language process definition.

3.1 Workflow Control Patterns

Workflow languages usually define the process flow by an XML notation representing activity dependencies. Several WCP have been identified with the aim of delineating the requirements that arise during business process modeling on a recurring basis and to describe them in an imperative way. So far, 43 WCP have been identified classified as basic control flow, advanced branching and synchronization, multiple instance, state-based, cancelation and force completion, iteration, termination, trigger and disclaimer. A comprehensive description of a workflow process also requires consideration of the data perspective. With this in mind, other 40 patterns in the data perspective have been identified.

Usual learning flows range from basic activity structures such as sequences, parallelism, loops, splits and unions; up to more complex structures such as multiple-choice and multiple-union. The EML evaluation in this paper focuses on 10 WCP including all the basic control flow patterns, some of the advanced branching and synchronization patterns, and one iteration pattern, because of their capability to represent the most common learning scenarios.

Basic control flow patterns capture elementary aspects of process control, namely:

1. *Sequence*: an activity in a workflow process is enabled after the completion of a preceding activity in the same process.
2. *Parallel split*: the divergence of a branch into two or more parallel branches each of which execute concurrently.
3. *Synchronization*: the convergence of two or more branches into a single subsequent branch such that the thread of control is passed to the subsequent branch when all input branches have been enabled.
4. *Exclusive choice*: the divergence of a branch into two or more branches, such that when the incoming branch is enabled, the thread of control is immediately passed to precisely one of the outgoing branches based on the outcome of a logical expression associated with the branch.
5. *Simple merge*: the convergence of two or more branches into a single subsequent branch. Each time an incoming branch is enabled, results in the thread of control being passed to the subsequent branch.

Advanced branching and synchronization patterns characterize more complex branching and merging concepts. Although relatively commonplace in practice, these patterns are often not directly supported or even able to be represented in current systems. The patterns of this category included in the analysis are the following:

1. *Multi-choice*: the divergence of a branch into two or more branches, such that when the incoming branch is enabled, the thread of control is passed to one or more of the outgoing branches based on the outcome of distinct logical expressions associated with each of the branches;
2. *Structured synchronizing merge*: the convergence of two or more branches (which diverged earlier in the process at a uniquely identifiable point) into a single subsequent branch, such that the thread of control is passed to the subsequent branch when each active incoming branch has been enabled;
3. *Multi-merge*: the convergence of two or more branches into a single subsequent branch. Each enablement of an incoming branch results in the thread of control being passed to the subsequent branch;
4. *Structured discriminator*: the convergence of two or more branches into a single subsequent branch following a corresponding divergence earlier in the process model. The thread of control is passed to the subsequent branch when the first incoming branch has been enabled. Subsequent times incoming branches that are enabled do not result in the thread of control being passed on. Finally the discriminator construct is reset when all incoming branches have been enabled

Iteration patterns deal with capturing repetitive behavior in a workflow. *Structured loop* is the only iteration pattern included in this study, and it is described as the ability to execute an activity or sub-process repeatedly. The loop has either a pre-test or post-test condition associated with it that is evaluated at either the beginning or the end of the loop to determine whether it should continue. The looping structure has a single entry and a single exit point.

3.2 Learning Process Analysis Based on Workflow Control Patterns

Considering the selected group of WCP, the learning process design and control constructions to be considered in an EML to manage execution issues are the following:

1. *Composite activities*: When designing a learning process, pattern-based groups of activities that can be sequenced, selected and concurrently executed, are often needed. In IMS LD, activity-structures can be used, but they can be only sequenced or selected. If concurrent activities are needed, they can be defined at the play level. An alternative is defining different role-parts within the same act, which does not make sense when a single learner is executing the activities. A third alternative is managing activities' completion status properties, but this is too burdensome when a large number of activities are considered.

LAMS provides sequenced, parallel, and optional activities that comprise the complex activity type, which makes easier to describe such concurrent composite activities.

2. *Conditional branching and looping*: The usual learning flow of activities must be branched sometimes, according to a special condition that often depends on the learner profile or the results of previously executed activities. Other times, remedial learning activities may require executing an activity or activity group an indefinite number of times, until the remediation condition is achieved by the learners. With IMS LD, to branch an activity an ad hoc property and a set of conditions must be defined to drive the learning flow through the branched activities. When LAMS is used, conditional branching is expressed through branching activities, which drive the learning flow depending on the teacher selection, grouping objects, or the output of previous tool-based activities.
3. *Parallel branching on users' learning flow*: Sometimes is needed to fork the learning flow of one learner, a group of learners, or all learners, to define a branching of parallel but different activities, which are executed by different groups of learners. This is achieved in IMS LD by defining roles or properties plus conditions, but once the execution of an IMS LD run has started, the association of users to roles cannot change. On the contrary, LAMS provides a grouping activity type that permits expressing that association during run time. The combination of a LAMS branching activity with a grouping activity enables forking user groups on a set of activities during run time, without requiring defining additional properties and conditions.
4. *Synchronization points*: Setting some synchronization points can be useful in a learning flow of activities, when parallel running activities must join on a given instant before proceeding with the following ones. The IMS LD specification states that plays always run concurrently, whilst acts always run in sequence. Synchronization points can be only marked on completion of the activities, according to user choices or time limits, or when a property is set. With LAMS, gate activities can be used to synchronize learning flows of activities, holding concrete semantics of concurrent access.
5. *Split running activities*: A special case of learning flow synchronization occurs when two flows of activities (possibly run by different users) must synchronize in certain points along a number of iterations. IMS LD does not provide adequate primitives to describe this pattern. In LAMS gates and flows can be used to describe that complex structure, although it is not a straightforward learning design task.

A summary of the common WCP found as structured metadata in the vocabulary of the three EMLs analyzed is presented in Table 2.

IMS LD defines learning flows by means of a script theater metaphor, represented by play, act and activity constructions, which are used to control the synchronization of a scenario. In every act, the role-part associates a role with an activity (i.e. learning activity, support activity or activity structure). Acts may contain diverse role-parts, which are concurrent and independent, but syn-

Table 2. Analysis of common workflow patterns as found in IMS LD, LAMS and LPCEL educational modeling languages

Workflow Control Pattern	IMS-LD	LAMS	LPCEL
Basic Control Flow			
1. Sequence	+	+	+
2. Parallel Split	±	±	+
3. Synchronization	–	±	+
4. Exclusive Choice	–	–	+
5. Simple Merge	–	±	+
Advanced Branching and Synchronization			
6. Multi-Choice	–	–	+
7. Structured Synchronizing Merge	–	–	+
8. Multi-Merge	–	–	+
9. Structured Discriminator	–	–	+
Iteration			
10. Structured Loop	–	–	+

chronous. The learning activities of any activity structure under a role-part must be executed sequentially. As shown in Table 2, IMS LD only has full support for the sequence pattern. A parallel split can be expressed by defining more than one concurrent plays, but this leads to a loss of sense of the script metaphor, because plays are meant for coarse-grained learning processes and not for fine-grained activities. Also there is no possibility of synchronizing the plays later. IMS LD provides the option to choose a number of activities. This does not mean that exclusive choice, multi-choice, synchronization, simple merge, multi-merge or structured discriminator patterns are supported, because activities are then executed in a rigid sequence manner.

In the LAMS LD data model, full support for the Sequence pattern is provided through the SequenceActivity construction. The parallel split, synchronization and simple merge patterns are thus supported, but restricted to a set of activities previously defined. The support for synchronization and simple merge patterns is provided by gateways such as SynchGateActivity, ScheduleGateActivity and PermissionGateActivity. The Parallel Split pattern is implemented by a ParallelActivity element. The situation for the rest of the patterns is the same as in IMS LD.

In the LPCEL language, the complex-learning-process element is used to represent a CLP, including the definition of diverse component-CLPs. Each one of these can be a complex-component or a basic-component, which allows designing complex and unanticipated learning structures as specializations of the former ones. A basic-component contains a collection of component-activity elements to execute activities such as learning-activity, assessment-activity or context-activity; they can refer to different specifications of resources to be used in the activity. The resource can invoke local or remote resources (e.g. SCORM contents) and services on a remote application (e.g. a project management tool).

To define a learning flow, LPCEL makes reference to a full-fledged collection of complex-component elements such as sequence, parallel, choice, switch, while, do-while, join, split and compensate-CA; and also to a collection of basic-component elements such as action, flow, delay, invoke, gateway, terminate, assign, create and compensate-BA; including their relationships and a number of criteria to indicate the start and termination conditions of the component-CLP. This supports for a vast number of WCP including all the basic ones, but not limited to them.

4 Case Study: Wrapping a Learning Design with LPCEL Metadata Primitives

In this section we show an instance of how a complex learning process can be described by wrapping IMS LD with LPCEL metadata primitives. The following listing is an excerpt of the learning design of an actual project-oriented software engineering course:

```
<imsld:learning-activity identifier="LA-Project-Planning">
  <imsld:title>Project Planning</imsld:title>
</imsld:learning-activity>
<imsld:learning-activity identifier="LA-Domain-Analysis">
  <imsld:title>Problem Domain Analysis</imsld:title>
</imsld:learning-activity>

<!-- ... declaration of all learning activities ... -->

<lpcel:assessment-activity identifier="AA-Eval-Elaboration-Criteria">
  <imsld:title>Evaluate Elaboration Criteria</imsld:title>
  <lpcel:split>
    <lpcel:guard>
      <imsld:conditions>
         <imsld:if>
           <imsld:is>
              <imsld:greater-than>
                 <imsld:property-ref ref="LR-Use-Case-Completed"/>
                 <imsld:property-value>80</imsld:property-value>
              </imsld:greater-than>
           </imsld:is>
         </imsld:if>
      </imsld:conditions>
    </lpcel:guard>
    <lpcel:components>
       <lpcel:component-activity-ref ref="LA-Project-Planning"/>
       <lpcel:component-activity-ref ref="LA-Domain-Analysis"/>
       <lpcel:component-activity-ref ref="LA-Software-Architecture"/>
    </lpcel:components>
  </lpcel:split>
</lpcel:assessment-activity>
...
<imsld:activity-structure identifier="AS-Elaboration-phase"
                          structure-type="complex-activity">
  <imsld:title>Elaboration phase</imsld:title>
  <lpcel:join>
    <lpcel:parallel>
      <lpcel:component-activity-ref ref="A-Project-Planning"/>
      <lpcel:component-activity-ref ref="A-Problem-Domain-Analysis"/>
      <lpcel:component-activity-ref ref="A-Use-Case-Modelling"/>
      <lpcel:component-activity-ref ref="A-Software-Architecture"/>
    </lpcel:parallel>
    <lpcel:assessment-activity-ref ref="AA-Eval-Elaboration-Criteria"/>
  </lpcel:join>
</imsld:activity-structure>
```

The design of this course requires structuring the project in phases, such that "the elaboration phase will be completed after project planning, problem domain analysis, software architecture and use case modeling activities are completed and use cases are 80% described." The combination of primitives of the two EMLs is solved through defining disjoint XML namespaces for LPCEL and IMS LD.

The scarce expressiveness of IMS LD to describe these requirements has been solved by wrapping a simple IMS LD learning activity-structure inside an LPCEL split kind of complex-component synchronized activity, which in turn is nested in an assessment-activity. Parallel running activities of the IMS LD activity-structure are also nested in a LPCEL join composite activity that uses a reference to the assessment activity as synchronization condition.

From this case we found that EMLs such as IMS LD do not include the primitives that build up the concurrency model necessary to express even the more basic workflow control patterns required. Although IMS LD is classified as an exchange EML, we find that it still lacks expressiveness, since their workflow primitives and semantics should be extended in order to describe exchangeable units of learning that include concurrent activities.

LPCEL supports a number of language primitives enabling to describe concurrent activities in the workflow description within a UoL. Advanced forms of concurrency control are provided by LPCEL, built on top of the basic concurrency language primitives. For instance, advanced transactional models are provided that mark the difference between a workflow transaction and a learning transaction [15]. The failure of a workflow transaction is due to the unavailability of resources, incorrect input formats, internal application failures, etc. Learning transaction failures can be produced if the learning objectives of a set of activities is not attained (i.e. a student's grade is below the minimum, the user playing a role is not available, etc.) These situations cannot be easily managed by current EMLs, but LPCEL provides the basis for describing, authoring and exchanging UoLs including such learning transactions.

5 Conclusions

This paper analyzes how current EMLs provide a structured set of metadata prepared to design and control usual learning flow situations. The study is realized over IMS LD and LAMS and is framed on the analysis of usual workflow control patterns. As the main conclusions of the study, we found that IMS LD needs to implement basic concurrency control structures; and both IMS LD and LAMS LD need to extend learning structures with more complex ones, in order to provide full support for CLP workflows. It is important for the new generations of EMLs to take into account the characteristics of the LPCEL framework, which were inspired by WCP. It enables representing the more complex requirements of a learning flow, in which the completion of a learning activity does not guarantee the successful fulfillment of a learning objective. However in a workflow, the execution of a process activity assures a state change of the software system.

This motivates the extension of LPCEL metadata primitives for transactional support, which is the main line of the future works.

Acknowledgements

This work is funded by the Distributed and Adaptive Systems Lab for Learning Technologies Development, DASL4LTD (C-QRO- 17/07) from the Tecnológico de Monterrey, México; by the Avanza projects TSI-020301-2008-19 and TSI-020501-2008-53 from the Spanish Office of Industry and Commerce.

References

1. Olivier, B., Tattersall, C.: The learning design specification. In: Koper, R., Tattersall, C. (eds.) Learning Design: A Handbook on Modelling and Delivering Networked Education and Training, pp. 21–40. Springer, Berlin (2005)
2. Rawlings, A., van Rosmalen, P., Koper, R., Rodríguez-Artacho, M., Lefrere, P.: Survey of educational modelling languages (EML). Technical report, CEN/ISSS WS/LT (September 2002)
3. Koper, R.: Modeling units of study from a pedagogical perspective: the pedagogical metamodel behind eml. Technical report, Open University of the Netherlands (June 2001)
4. Koper, R.: Learning Design, A Handbook on Modelling and Delivering Networked Education and Training. Springer, Berlin (2005)
5. Russell, N., ter Hofstede, A., van der Aalst, W., Mulyar, N.: Workflow control-flow patterns: A revised view. Technical report, BPM Center (2006)
6. Dodero, J.M., Torres, J., Aedo, I., Díaz, P.: Beyond descriptive EML: Taking control of the execution of complex learning processes. In: Simposio Pluridisciplinar sobre Diseño, Evaluación y Descripción de Contenidos Educativos Reutilizables, Barcelona (2005)
7. Torres, J., Dodero, J.M., Aedo, I., Díaz, P.: Designing the execution of learning activities in complex learning processes using LPCEL. In: Proc. of the 6th ICALT, Kerkrade, The Netherlands, pp. 415–419 (2006)
8. Martínez-Ortiz, I., Moreno-Ger, P., Sierra-Rodríguez, J.L., Fernández-Manjón, B.: A flow-oriented visual language for learning designs. In: Li, F., Zhao, J., Shih, T.K., Lau, R., Li, Q., McLeod, D. (eds.) ICWL 2008. LNCS, vol. 5145, pp. 486–496. Springer, Heidelberg (2008)
9. Dalziel, J.R.: Implementing learning design: The learning activity management system (LAMS). In: Crisp, G., Thiele, D., Scholten, I., Barker, S., Baron, J. (eds.) Proc. of the 20th ASCILITE, Adelaide, Australia (2004)
10. Hagen, K., Hibbert, D., Kinshuk, P.: Developing a learning management system based on the IMS Learning Design specification. In: Proc. of the 6th ICALT, Kerkrade, The Netherlands, pp. 420–424 (2006)
11. Gutiérrez-Santos, S., Pardo, A., Delgado-Kloos, C.: Authoring courses with rich adaptive sequencing for IMS learning design. Journal of Universal Computer Science 14(17), 2819–2839 (2008)
12. Dodero, J.M., Torres, J.: Awareness of execution in designing learning activities with LAMS and IMS LD. In: European LAMS Conference, Cádiz, Spain (2008)

13. Hollingsworth, D.: The workflow reference model - issue 1.1. Specification TC00-1003, Workflow Management Coalition, Winchester, Hampshire, UK (January 1995)
14. Bhiri, S., Perrin, O., Godart, C.: Extending workflow patterns with transactional dependencies to define reliable composite web services. In: Advanced International Conference on Telecommunications and International Conference on Internet and Web Applications and Services, p. 145 (2006)
15. Torres, J., Juárez, E., Dodero, J.M., Aedo, I.: Advanced transactional models for a new generation of educational modelling language engines. In: Proc. of the 8th ICALT, Riga, Latvia (2008) (to be published)

A Fine-Grained Metric System for the Completeness of Metadata

Thomas Margaritopoulos, Merkourios Margaritopoulos, Ioannis Mavridis, and Athanasios Manitsaris

Department of Applied Informatics, University of Macedonia,
156 Egnatia Street 54006, Thessaloniki, Greece
{margatom,mermar,mavridis,manits}@uom.gr

Abstract. Metadata quality is an issue that can be approached from different aspects. Among the most essential properties characterizing a quality metadata record is its sufficiency to describe a resource, which is expressed as the completeness of the record. The paper presents a fine-grained metric system for measuring metadata completeness that is capable of following the hierarchy of metadata as it is set by the metadata schema and admeasuring the effect of multiple values of multi-valued fields. Moreover, it introduces the aspect of the representation level of semantically equivalent information that should be taken into account when measuring completeness. The proposed metric system, based on the definition of completeness of a field, treats several deficiencies of the traditional coarse metrics and offers the ability of targeted measures of completeness throughout the metadata hierarchy.

Keywords: Metadata completeness, metadata quality, metrics.

1 Introduction – Related Work

The purpose of metadata is to provide adequate, correct and relevant information to their user so as to obtain a true picture of the content of a resource without having to access it. The more thorough the description of a resource is, the less vague is its picture. Completeness of metadata refers to their sufficiency to fully describe a resource covering all its possible aspects. Given the definition for metadata [8], as structured data about an object that supports functions associated with the designated object, it is clear that sufficiency to fully describe a resource is directly associated with the particular activities or processes that metadata are intended to support. As [22] notes, "...each kind of activity or process can be said to require a number of concrete metadata elements". Thus, completeness of metadata depends on the specific functionality or usage expected by the metadata application.

Numerous metadata standards have been established in an attempt to define sufficient descriptions from different perspectives and satisfy diverse functionalities. Theoretically, a sufficient description exists when all metadata elements of the standard are populated with values. However, in practice, this is not what

F. Sartori, M.Á. Sicilia, and N. Manouselis (Eds.): MTSR 2009, CCIS 46, pp. 83–94, 2009.

happens in the real world. Relevant surveys [6], [9], [19] have shown that indexers tend to fill out only particular metadata elements that could be considered "popular", while they ignore other elements of less popularity. The creation of metadata is a task requiring major labor and financial cost and, most important, the involvement of knowledgeable and experienced people [1], [13]. Since all these requirements are, generally, difficult to be fully met, it is rather common, in the majority of digital repositories, to have incomplete metadata. The issue of incomplete metadata records is rather problematic, especially, in collections resulting from harvesting from metadata databases [5] or applying automatic metadata generation techniques [7], [14], [20]. On the other hand, when searching for information, users of metadata limit their search criteria by using only a very small percentage of metadata elements, as it is highlighted by [18]. This fact, as [23] notes, shows that completeness is, often, in conflict with simplicity. However, this conflict between completeness and simplicity of a metadata standard is attempted to be balanced by application profiles [10]. Application profiles, among other things, select certain metadata fields of one or more standards and account them as mandatory or as optional, based on the needs of the particular users' community. This distinction is directly associated with the definition of completeness in relation to the selection of the metadata fields.

The concept of completeness of metadata, as an object of study, is integrated in the more general concept of metadata quality. The research community considers completeness of metadata as a fundamental quality characteristic. Several dimensions or characteristics of metadata quality have been proposed by researchers [2], [15], [24] in an effort to define quality and provide the necessary means for its assessment and improvement. Among the numerous suggested quality dimensions, completeness is one of the most essential. Indeed, it can be considered as a prerequisite to assess quality, since incomplete records are in any case not of quality due to lack of essential information [22]. Among all the constituent elements of metadata quality, completeness is the easiest to be quantified and measured by automatic means without human intervention. It is assessed based on the presence or absence of values in metadata fields of some metadata standard. These fields are selected by the application that uses the metadata based on their importance for the specific process or activity handling the metadata.

The final objective of the relevant research on metadata quality is to provide credible means for measuring quality so as to perform quality control and diagnose deficiencies and errors for correction and improvement. Several researchers have created metric systems to measure metadata quality by computing indicators of quality and, among them, the completeness indicator of a metadata record [3], [11], [17], [21]. All these systems define completeness at the record level by providing formulas to compute completeness of a record based on the existence or absence of values in the metadata fields of a standard and their relative importance expressed as the weight of each field. Although such metrics seem to be providing meaningful measures of completeness of a metadata record, several issues are overlooked. Multiple values of multi-valued fields or the internal structure of aggregate fields are not taken into account when measuring completeness. Moreover, the presence of multiple representation forms of the same value in a metadata field as a possible factor that might influence completeness is not examined, as well.

In what follows, these issues are put into consideration leading to the definition of a new fine-grained metric system for the completeness of metadata. Since, as it is already stated, the purpose of metadata is the provision of adequate, correct and relevant information for a resource, a sufficient description of the resource refers to the "adequate information" and not the "correct or relevant information". In the rest of the paper, completeness is only examined in terms of the existence of values in metadata elements and is not concerned with the values themselves. In order to avoid any misunderstanding, we need to emphasize that the evaluation of completeness as a metadata quality parameter based on the existence of values in metadata elements (regardless of the actual content of these values) does not, in any way, underestimate or overlook this aspect. According to [15], the evaluation of the content of a value is an issue assigned to the other two dimensions of metadata quality, i.e. correctness and relevance – providing that a value is already there. This condition is evaluated and measured by completeness. In fact, completeness undertakes and measures the quantitative part of the quality of metadata.

2 A Metric System for Completeness

In an effort to measure completeness, what should be made clear is that completeness of a record can only be measured against the specifications of the metadata standard or application profile used by any given application. The standard or the application profile prescribes a certain number of fields as necessary placeholders of information to describe a resource. Completeness can be considered as the degree of coverage of these placeholders and its values may swing in a continuum from its total presence to its absolute absence. Thus, a requirement for a metric system to measure completeness would be its ability to assign values in a continuous range between these two endpoints. Intuitively, a range of values in the closed interval [0, 1] seems to adequately satisfy this requirement, since non-existence of completeness would map to the value of 0 and total completeness would map to the value of 1, offering the convenience of easily translating the values to percentages. Limiting the range value of completeness in this condensed interval should not restrict the ability of the metric system to detect any slight variation in the amount of information contained in a metadata record. Thus, a critical requirement for the system would be to be able to differentiate the measure of completeness so as to reflect such variations.

In the following subsections a fine-grained metric system for measuring metadata completeness is presented, introducing the concept of completeness of a field.

2.1 Introducing Completeness at the Field Level

The measurement of completeness of a metadata record is a matter coarsely covered in the relevant literature. In [21], completeness is defined as the number of fields that contain a non-null value divided by the total number of fields of the record according to the metadata standard, as in the following equation (1).

$$\text{com} = \frac{\sum_{i=1}^{N} P(i)}{N} \tag{1}$$

where P(i) is 1 if the i-th field has a non-null value, 0 otherwise. N is the number of fields. In cases where the fields of a record are not of equal importance, any particular application might assign weights of importance to each field. Therefore, the measure of "weighted completeness" is computed [21] as in equation (2).

$$\text{com} = \frac{\sum_{i=1}^{N} a_i * P(i)}{\sum_{i=1}^{N} a_i} \tag{2}$$

where a_i is the weight of field i. Thus, a field is considered either as complete or as non-complete, being assigned only two discrete values.

However, such an approach seems to be over-simplistic because it raises some obvious concerns. A source of concern is the case of multi-valued fields (fields with cardinality greater than 1). As [21] notes: "In the case of multi-valued fields, the field is considered complete if at least one instance exists". This proposition overlooks the fact that a multi-valued field can be loaded with a variable amount of information. As a consequence, such variations are not reflected to the measure of completeness of its record, yielding degraded results. For example, a Dublin Core [4] record describing a multi-language document would have the same measure of completeness regardless of the number of values the field "DC.Language" had. The field is considered complete if at least one instance is present. Yet, this is not right, since a complete field should be populated with all the languages present in the document and not just one.

Another major source of concern is the case of aggregate fields. For example, metadata standards like IEEE LOM [12] and METS [16] define a hierarchy of data elements, including aggregate data elements and simple data elements (leaf nodes of the hierarchy). Only leaf nodes have individual values defined through their associated value space and datatype. Aggregates do not have individual values. If completeness is measured taking into account only the simple fields that have individual values, then the whole structure of the metadata schema is ignored. There is no meaning in assigning a 0 or 1 as an existence indicator of value when measuring completeness in the 58 simple fields of the LOM metadata schema and ignoring the hierarchy and interdependence of the fields. It is rather pointless to have a date filled in the "2.3.3 Lifecycle.Contribute.Date" field of LOM and, at the same time, have no values in the fields "2.3.1 Lifecycle.Contribute.Role" and "2.3.2 Lifecycle.Contribute.Entity". The importance of the presence of any value in this "date" field should be assessed against the aggregate field "2.3 Lifecycle.Contribute" and not against the whole record.

The above discussion brings out the obvious need to define completeness at the field level. Such a definition would satisfy all of the above requirements and take into account all the pieces of information included in multi-valued or aggregate fields. Completeness of a field will be defined in the general case of hierarchical structure of a metadata record, starting from the top level, which is the root node representing the whole record and going down recursively to the lowest level, which is the leaf nodes representing the simple fields of the record.

2.2 Measuring Completeness at the Field Level

Just like completeness of a metadata record is defined as the weighted average of the existence or non-existence of values in its fields (according to the metadata standard), completeness of a metadata field is defined following the same logic and taking into account the nature of the field (simple, aggregate, multi-valued). Thus, completeness of a field f, represented as COM(f), will be given a disjunctive definition.

$$\mathrm{COM}(f)=\begin{cases}\sum_{k=1}^{m(f)}[a_k(f)*P_k(f)] & \text{if f is a simple field}\\ \sum_{k=1}^{m(f)}[a_k(f)*\mathrm{COM}_k(f)] & \text{if f is an aggregate multi-valued field}\\ \sum_{k=1}^{nsf(f)}[a_k(f)*\mathrm{COM}(sf_k(f))] & \text{if f is an aggregate single-valued field}\end{cases} \quad (3)$$

In equation (3), f represents any node (field) belonging to the hierarchy of the metadata schema. For example, for the LOM metadata schema, f may be a field like "5.2 Educational.Learning resource type", a category like "1 General", or even the whole record, which is the root node of the hierarchy.

The first case of the split definition of equation (3) defines completeness for a simple field, either single-valued, or multi-valued. The minimum number of values field f should have in order to be considered complete is represented by m(f). Considering the values of field f to be stored in m(f) placeholders, P_k(f) is defined as an existence indicator marking the presence (equal to 1) or absence (equal to 0) of value in the k-th placeholder. Then, completeness of a simple field is defined as the weighted average of the existence indicators of its values. In the special case of a single-valued field, obviously, m(f) is equal to 1. In a multi-valued field, m(f) is a number depended on the semantics of the field. Sometimes it is defined by reality itself, for example, the number of languages contained in a multi-language document. In cases where m(f) is not a real life fact, it should be defined by the application in which the metadata are used, for example, the number of third-party annotations for a resource. For this latter case LOM standard defines "the smallest permitted maximum value". This value, in a sense, could serve as the minimum number of values a multi-valued field should have in order to be considered complete.

The second case of the split definition of equation (3), is similar to the first one with the difference that the values of an aggregate field are structured fields themselves, bearing a measure of completeness of their own, other than the two district values of the existence indicator $P_k(f)$. Thus, completeness of a multi-valued aggregate field is computed as a weighted average of the completeness measures of its values (instances), with $COM_k(f)$ be the completeness measure of the k-th value of the field. For example, completeness of the category "7 Relation" of LOM will be computed as the weighted average of the measures of completeness of each "7 Relation" instance. In this particular case, m(f) will be the real number of relations of the learning object being described with other ones in a repository.

In the first two cases of equation (3), $a_k(f)$ represents the weights of importance of the values of a multi-valued field. In order to limit the measures of completeness in the closed interval [0, 1], we force these weights to have a sum of 1. The values of a multi-valued field are generally considered to be equally weighted, that is, no value is regarded as more important than the rest. However, there might be applications where this default assumption does not fit. For example, an application might wish to assign progressively reduced weights to the contributors of a resource being registered in a metadata record, as it may not consider the additional information as much important as the previous. The application considers any additional contributor less important than the previous one. Another special case of weight assignment is the assignment of conditional weight to an aggregate field, based on the value of a specific subfield of its. For example, referring to LOM fields, an application might wish to assign a bigger weight to an instance of "2.3 Lifecycle.Contribute", in case that its "2.3.1 Lifecycle.Contribute.Role" of the contributor is "author", than the weight to a different instance where the role of the contributor is "editor".

In the last case of the split definition of equation (3), the completeness of a single-valued aggregate field, is computed from the measures of completeness of its subfields, referring to the fields of the next lower level down the metadata hierarchy. The symbol nsf(f) represents the number of subfields of field f according to the metadata standard or application profile used and $sf_k(f)$ is the k-th subfield of field f. For example, for the single-valued aggregate field "1 General" of LOM nsf("1 General")=8, which is the number of the next lower level subfields – 1.1 through 1.8 – and sf_2("1 General") is "1.2 General.Title". The variable $a_k(f)$ represents the weight of each subfield, which, in this case, is its relative importance to its parent field. Again, the sum of these weights is considered to be equal to 1.

The definition of equation (3) for the completeness of a field has a clear recursive nature, since it uses previous values from lower field levels to create new ones. For example, completeness of the category "1 General" of LOM is computed as the weighted average of the measures of completeness of its eight subfields (1.1 through 1.8). Seven of these eight subfields are simple fields. In case they are single-valued, such as "1.7 General.Structure", completeness is equal to the existence indicator of the value. In case they are multi-valued, such as "1.4 General.Description", their measure of completeness is computed as the weighted average of the measures of completeness of their instances. The completeness of each instance of the multi-valued simple fields is, again, equal to the existence indicator of its value. Field "1.1 General.Identifier" is an aggregate field, so its completeness will be computed as the weighted average of the measures of completeness of its subfields ("1.1.1 General.Identifier.Catalog" and "1.1.2 General.Identifier.Entry"), which, in turn, are simple fields and their completeness is equal to the existence indicators of their values. Thus, traversing the sub-tree of the category "1 General" of LOM down to its leaves, we compute the completeness of this aggregate field. In general, in order to compute the completeness of any node, we start from this node and traversing the hierarchy of the metadata schema we recursively compute the measures of completeness of the nodes of its sub-tree, applying equation (3). The base of the recursion is always the existence indicator of a single value of a simple field. Completeness of the whole record is computed the same way – starting from the top level (the root node) of the hierarchy. The record is considered as an aggregate field.

2.3 Extending Completeness Metrics at the Representation Level

When traversing the nodes of a metadata hierarchy in order to measure completeness of a record or of a specific field, one could say that the end of this process is the lowest level of the hierarchy or a field of a leaf node with no further structure of subfields. Yet, the measure of completeness cannot just be computed by checking the presence of values in a simple field because such a field may still contain additional information that should be taken into account when measuring completeness. This additional information belongs to the representation level of the values. It is possible that the values of certain datatypes may be represented in different multiple forms which are semantically equivalent. Basically, the idea is that the same value is represented in different forms, e.g. a particular text in different languages, such as in the datatype "langstring" of LOM. The existence of different representations should affect the measure of completeness of the respective field as being valuable additional information. Following the same logic, other forms of representation that could be set by a metadata standard, might be different visual or audio versions of the same piece of textual information, which are simultaneously present in a metadata field.

The consideration of the representation level does not allow us to think of the completeness of a single value of a simple field as having only the two district values of the existence indicator $P_k(f)$. The completeness of such a field should be computed as the weighted average of its existent forms of representation. Thus, if we consider the different forms of representation of a single value of a simple field, completeness of a simple field is formulated as in equation (4).

$$COM_{representations}(f) = \sum_{k=1}^{m(f)} [a_k(f) * \sum_{j=1}^{L} [b_j * R_j(k,f)]] \quad (4)$$

where L is the number of all the possible different representation forms, b_j is the weight of the j-th representation form and R_j(k,f) is the existence indicator of this specific representation form of the k-th single value of field f. The sum of the weights b_j is again set to 1. The minimum number of the possible different representations – so as the represented value to be considered complete – and the weight of each is an issue determined by the application. In the case of the different languages of the same textual information, a minimum number of languages that would make the value of the textual field to be considered complete might be the number of languages of a certain geographical region (e.g. the European Union languages) or the number of languages used in a particular educational institution. As for the weights of the languages, apparently, mother languages of a particular place or languages most frequently used would be assigned a bigger weight.

Equation (4) seems to adequately cover the issue of the completeness of a simple field from all possible aspects. However, an essential problem emerges. If a field is a multi-valued one, then, applying the formula of equation (4), the measure of completeness will be computed taking into account all the instances of the different representation forms, regardless of the total number of values they express. For example, let's assume a situation where the completeness of field "1.5 General.Keyword" of LOM is to be measured. The maximum number of values m(f) is set to 5 and the maximum number of languages L is set to 3, while we assign equal weights for all the different values (a_k=1/5, k=1,. . . ,5) and all the different languages (b_j=1/3, j=1,2,3) to express these values. We will consider two different cases. In case a the field is populated with one keyword expressed in three different languages, so the measure of completeness (according to equation (4)) will be (1/5)*(1/3)*[(1+1+1)+(0+0+0)+(0+0+0)+(0+0+0)+(0+0+0)]=0.2. In case b the field is populated with two keywords each expressed in one language, so completeness of the field would be (1/5)*(1/3)*[(1+0+0)+(1+0+0)+(0+0+0)+(0+0 +0)+(0+0+0)]=0.13. Apparently, in terms of completeness, one keyword outperforms two keywords just because this keyword is expressed in more languages than the two keywords, each of which is expressed in only one language.

Consequently, we need to make a distinction between measurement of completeness that takes into account only the number of values of a field (which, obviously, is the most important) and measurement of completeness taking into account the number of the different representations of these values. The former is the one expressed by the first case of the split definition of equation (3), from now on called COM_{values}(f). The latter is the one expressed by equation (4). The above example involving keywords expressed in different languages is presented in a graphical form in Figure 1.

The requirement, in order to maintain the importance of the different values against their different representations, is that more values must always result in a higher completeness score than fewer values, regardless of the number of representation forms the values are expressed in. If we compute completeness as the weighted average of COM_{values}(f) and $COM_{representations}$(f) and assign a bigger weight to COM_{values}(f), then this requirement is met. Hence, the measure of completeness of a simple field – to substitute the first case of equation (3) – becomes

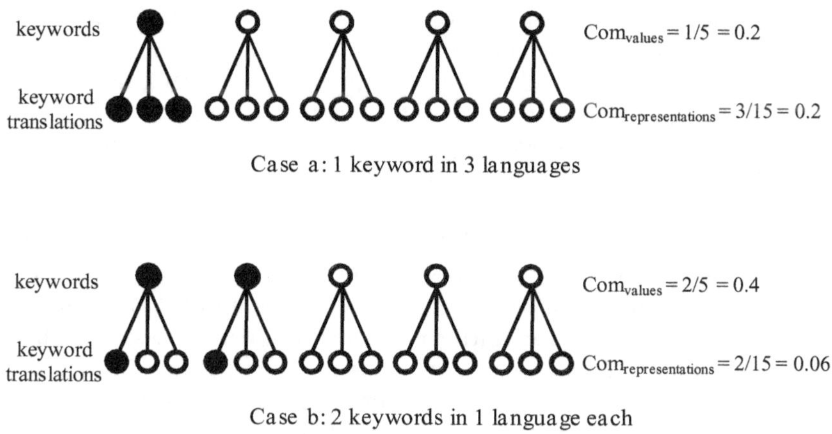

Fig. 1. Completeness based on values or representations

$$COM(f) = c * COM_{values}(f) + (1 - c) * COM_{representations}(f) \quad (5)$$

The value of the weight c expresses the maximum measure of completeness of a complete field (taking into account only its values), regardless of their representations. The value of the weight 1-c expresses the additional amount of completeness, attributed to the representations of the values, so as to reach the maximum value of completeness (the value of 1) for a complete field from all possible aspects. In the example with the keywords expressed in different languages, if we set c=0.8 (a value bigger than 0.5) then for case a we have $COM(f)=0.8*COM_{values}+(1\text{-}0.8)*COM_{representations}=0.8*1/5+0.2*3/15=0.2$, while for case b we have $COM(f)=0.8*COM_{values}+(1\text{-}0.8)*COM_{representations}= 0.8*2/5+0.2*2/15=0.35$. The weight c acts as expected, forcing the second case to have a bigger measure of completeness than the first one, since it employs two keywords compared to one, regardless of the number of different languages these keywords are expressed in.

3 A Discussion of the Proposed Metrics

The way the above metric system for metadata completeness was designed offers the ability of differentiating the measure of completeness for any slight variations in the amount of information loaded into the metadata fields acting as placeholders of values. This is a result of admeasuring the effect of any multiple values a multi-valued field might have to the total measure of completeness. For example, let's assume that a Dublin Core record of a resource contains three different instances of the element DC.Identifier, which provide three references to the resource conforming to three different formal identification systems. In a coarse metric system there is no way to differentiate the measure of completeness for this enriched record, since completeness of the record would be computed taking into account only one instance of this element. No matter how many identifiers of this resource are registered into its metadata record, this desired state of excessive quality of the record is not reflected to the measure of its completeness. Thus, it is impossible to detect and quantify completeness problems due to missing values in multi-valued fields. This deficiency is treated by the proposed metric system.

Another significant advantage of the proposed metric system is its ability to give targeted measures of completeness at specific fields of a record containing some internal structure such as the case with aggregate fields is. This way the metric system can spot and quantify problems of completeness anywhere in a hierarchy of metadata fields. Completeness is measured taking into account the internal structure of the fields. When measuring completeness, the metric system computes existence indicators of values of the fields and weighs these indicators according to the importance of each field in relation to its parent node.

Moreover, the proposed metric system adds a new dimension in the measurement of completeness by taking into account the representation level of a single value of a simple field. This level constitutes information, which, in many cases, might be of significant importance for the metadata application and influences the completeness of the field according to the weight of importance it is assigned by the application (the value of 1-c of equation (5)). Extending the analysis for the completeness of a simple field taking into account the representation level of single values, we could argue that this analysis can be extended to aggregate fields, as well, in cases where whole aggregate fields could be represented in different forms. For example, the whole category "7 Relation" of LOM might be represented in different forms of equivalent content (e.g. in a graphical form showing relations of a learning object instead of the representation in the established textual form). The proposed metric system provides the ability for this additional information, which enhances the quality of the particular metadata field, to be reflected to the measure of its completeness.

Measuring the completeness of metadata using the proposed fine-grained metric system, beyond the above described benefits and advantages it provides, it can offer valuable help in fulfilling the specific requirements set by the context of use. The specific process or activity handling the metadata determines the pragmatics of measuring and defines the exact purpose of measuring. For example, measuring the completeness of metadata can be used as an important tool to evaluate automatic metadata generation methods and techniques. Another potential application of completeness measuring might be in cases where targeted measures of completeness can be used as additional criteria to filter the results produced by a search engine. For example, when searching for learning objects with specific keywords, a teacher preparing a lesson might wish to make sure that the metadata of the returned results will contain a certain amount of educational information. Hence, the teacher might set a threshold value for the completeness measure of the metadata of the returned learning objects at specific elements (educational fields) and filter the results according to this criterion.

4 Conclusion – Future Work

In this paper, a metric system for measuring the completeness of metadata was presented. In an effort to treat the inadequacies of the traditional approach which is based on measuring completeness of a metadata record by generally counting the presence or absence of values in fields, the proposed system defines completeness at the field level in a recursive way following the hierarchy of the metadata schema. Multi-valued and aggregate fields were put into consideration, as well as the representation level of semantically equivalent information in the metadata fields. The result is a set of metrics that takes into account the needs and requirements of the application level (determined by weighting factors specified by the particular process or activity of the metadata application) and can be easily implemented by automated means.

The next step following this work includes an experimental implementation of the metric system and its application on a working database of metadata in a digital repository. This can be done providing the weights of importance for each field, particular values or specific forms of representation were determined in advance. The results for various targeted measures of completeness of particular fields are expected to provide valuable information in order to draw interesting conclusions about completeness, and quality of metadata in general, which would otherwise be impossible to reach.

References

1. Barton, J., Currier, S., Hey, J.M.N.: Building Quality Assurance into Metadata Creation: an Analysis based on the Learning Objects and e-Prints Communities of Practice. In: International Conference on Dublin Core and Metadata Applications: Supporting Communities of Discourse and Practice, pp. 39–48 (2003)
2. Bruce, T.R., Hillmann, D.I.: The continuum of metadata quality: defining, expressing, exploiting. In: Hillmann, D.I., Westbrooks, E. (eds.) Metadata in Practice, pp. 238–256, ALA Editions, Chicago (2004)
3. Bui, Y., Park, J.: An assessment of metadata quality: A case study of the national science digital library metadata repository. In: Moukdad, H. (ed.) CAIS/ACSI 2006 Information Science Revisited: Approaches to Innovation (2006)
4. Dublin Core Metadata Initiative, `http://dublincore.org/`
5. Dushay, N., Hillmann, D.: Analyzing metadata for effective use and re-use. In: International Conference on Dublin Core and Metadata Applications: Supporting Communities of Discourse and Practice, pp. 1–10 (2003)
6. Friesen, N.: International LOM Survey: Report (Draft), `http://dlist.sir.arizona.edu/403/01/LOM%5FSurvey%5FReport2.doc`
7. Greenberg, J., Spurgin, K., Crystal, A.: AMeGA Automatic Metadata Generation Applications) Project, Final Report, University of North Carolina (2005), `http://www.loc.gov/catdir/bibcontrol/lc_amega_final_report.pdf`
8. Greenberg, J.: Metadata and the World Wide Web. In: Dekker, M. (ed.) Encyclopaedia of Library and Information Science, pp. 1876–1888 (2003)
9. Guinchard, C.: Dublin Core use in libraries: a survey. OCLC Systems & Services 18(1), 40–50 (2002)

10. Hillmann, D.I., Phipps, J.: Application profiles: exposing and enforcing metadata quality. In: International Conference on Dublin Core and Metadata Applications: Application Profiles: Theory and Practice, pp. 52–62 (2007)
11. Hughes, B.: Metadata quality evaluation: Experience from the open language archives community. In: Chen, Z., Chen, H., Miao, Q., Fu, Y., Fox, E., Lim, E.-p. (eds.) ICADL 2004. LNCS, vol. 3334, pp. 320–329. Springer, Heidelberg (2004)
12. IEEE. 1484.12.1: Draft Standard for Learning Object Metadata. Learning Technology Standards Committee of the IEEE (2002), `http://ltsc.ieee.org/wg12/files/LOM_1484_12_1_v1_Final_Draft.pdf`
13. Liddy, et al.: Automatic Metadata Generation & Evaluation. In: Proceedings of the 25th annual international ACM SIGIR conference on Research and development in information retrieval, pp. 401–402 (2002)
14. Margaritopoulos, M., Margaritopoulos, T., Kotini, I., Manitsaris, A.: Automatic metadata generation by utilising pre-existing metadata of related resources. Int. J. Metadata, Semantics and Ontologies 3(4), 292–304 (2008)
15. Margaritopoulos, T., Margaritopoulos, M., Mavridis, I., Manitsaris, A.: A Conceptual Framework for Metadata Quality Assessment. In: International Conference on Dublin Core and Metadata Applications: Metadata for Semantic and Social Applications, pp. 104–116 (2008)
16. Metadata Encoding & Transmission Standard (METS), `http://www.loc.gov/standards/mets`
17. Moen, W., Stewart, E., McClure, C.: Assessing metadata quality: Findings and methodological considerations from an evaluation of the US Government information locator service (GILS). In: Proceedings of the Advances in Digital Libraries Conference, pp. 246–255. IEEE Computer Society, Los Alamitos (1998)
18. Najjar, J., Ternier, S., Duval, E.: User behavior in learning object repositories: an empirical analysis. In: Proceedings of the ED-MEDIA 2004 World Conference on Educational Multimedia, Hypermedia and Telecommunications, AACE, pp. 4373–4379 (2004)
19. Najjar, J., Ternier, S., Duval, E.: The Actual Use of Metadata in ARIADNE: an Empirical Analysis. In: Proceedings of ARIADNE Conference, pp. 1–6 (2003)
20. Ochoa, X., Cardinels, K., Meire, M., Duval, E.: Frameworks for the Automatic Indexation of Learning Management Systems Content into Learning Object Repositories. In: Proceedings of the ED-MEDIA 2005 World Conference on Educational Multimedia, Hypermedia and Telecommunications, pp. 1407–1414 (2005)
21. Ochoa, X., Duval, E.: Quality Metrics for Learning Object Metadata. In: Proceedings of the ED-MEDIA 2006 World Conference on Educational Multimedia, Hypermedia and Telecommunications, AACE, pp. 1004–1011 (2006)
22. Sicilia, M.A., García, E., Pagés, C., Martínez, J.J., Gutiérrez, J.: Complete metadata records in learning object repositories: some evidence and requirements. Int. J. of Learning Technology 1(4), 411–424 (2005)
23. Stvilia, B., Gasser, L., Twidale, M., Shreeves, S., Cole, T.: Metadata quality for federated collections. In: Proceedings of the 9th International Conference on Information Quality, pp. 111–125 (2004)
24. Stvilia, B., Gasser, L., Twidale, M., Smith, L.: A Framework for Information Quality Assessment. Journal of the American Society for Information Science and Technology 58(12), 1720–1733 (2007)

Unified Semantic Search of Data and Services

Domenico Beneventano[2], Francesco Guerra[2], Andrea Maurino[1], Matteo Palmonari[1], Gabriella Pasi[1], and Antonio Sala[2]

[1] Università di Milano Bicocca
Viale Sarca 336, 20126, Milano, Italy
{maurino,palmonari,pasi}@disco.unimib.it
[2] Università of Modena and Reggio Emilia,
Via vignolese 905, 41100 Modena, Italy
firstname.lastname@unimore.it

Abstract. The increasing availability of data and eServices on the Web allows users to search for relevant information and to perform operations through eServices. Current technologies do not support users in the execution of such activities as a unique task; thus users have first to find interesting information, and then, as a separate activity, to find and use eServices. In this paper we present a framework able to query an integrated view of heterogeneous data and to search for eServices related to retrieved data. A unified view of data and semantically described eServices is the way in which it is possible to unify data and service perspectives.

1 Introduction

The research on data integration and service discovering has involved from the beginning different (not always overlapping) communities. As a consequence, data and services are described with different models, and different techniques to retrieve data and services have been developed. Nevertheless, from a user perspective, the border between data and services is often not so definite, since data and services provide a complementary vision about the available resources: data provide detailed information about specific needs, while services execute processes involving data and returning as well an informative result.

Users need new techniques to manage data and services in a unified manner: both the richness of information available on the Web and the difficulties the user faces in gathering such information (as a service result, as a query on a form, as a query on a data source containing information extracted from web sites) make a tool for querying data and services at the same time, with the same language, really necessary.

Integration of data and services can be tackled from different perspectives. In the most common perspective, access to data is guaranteed through Service Oriented Architectures (SOA), and Web services are exploited to provide information integration platforms [8,16]. In a second perspective, the goal is to provide a global view on the data sources managed in a peer and on eServices

F. Sartori, M.Á. Sicilia, and N. Manouselis (Eds.): MTSR 2009, CCIS 46, pp. 95–107, 2009.

available in the peer or even on the Web, in order to support the access to the two complementary kind of resources at a same time. This paper addresses this last perspective, which is very different from the first one, and, to the best of our knowledge, completely new in the literature.

To make the research statement more precise, we take on the following assumptions. We assume to have a mediator-based data integration system which provides a global virtual view of data sources and is able to query data based on SQL queries expressed in the global virtual view terminology (a Dot notation will be used for the queries throughout the paper). In the context of a semantic peers environment, we refer to this global virtual view as the Semantic Peer Data Ontology (SPDO). We also assume to have a set of semantically annotated service descriptions; these descriptions may refer to services developed within the peer, but they can be also descriptions retrieved on the Web about services provided by other peers and addressing many different domains. Ontologies used in the service descriptions can be developed outside the peer and are not known in advance, in the integration process.

In the following paper we propose a semantic approach to perform data and service integration according to the following principle: given a SQL query expressed in the terminology of the SPDO, retrieve all the services that can be considered "related" to the query on the data sources.

The approach developed to address the above research question is based on the integration of a mediator-based data integration system, namely the MOMIS system (Mediator envirOnment for Multiple Information Sources)[1] [3,2], and of a service retrieval engine based on Information Retrieval (IR) techniques performing semantic indexing of service descriptions and keyword-based semantic search. The integration of the two is achieved by: (i) building the SPDO (a functionality already provided by MOMIS), (ii) building a Global Service Ontology (GSO) consisting of the ontologies used in the service semantic descriptions, (iii) defining a set of mappings between the SPDO and the GSO, (iii) exploiting, at query time, query rewriting techniques based on these mappings to build a keyword-based query for service retrieval expressed in the GSO terminology starting from a SQL query on the data sources.

The outline of the paper is the following: Section 2 introduces a motivating scenario that will be adopted as running example. Section 3 describes the approach developed for publishing unified contents of data and eServices, Section 4 provides the description of the technique for retrieving data and related services. In section 5 some related works are discussed, and finally, some conclusions and future work are sketched in Section 6.

2 Motivating Scenario and Running Example

Assume that a virtual touristic district is composed of a set of touristic companies (including travel agency, hotels, local public administrations, touristic promotion

[1] Further publications about MOMIS are available at http://www.dbgroup.unimo.it/Momis

agencies) creating a semantic peer in which they want to share and expose an integrated view of touristic data and services. The semantic peer wants to supply the tourist promoters and travelers with all the information about a location by means of only one tool managing data and services provided by different web sources.

Let us introduce as an example three information systems providing information about Italian locations that may be integrated to create a larger information source available for touristic purposes:

- `BookAtMe` provides information about more than 30.000 hotels located in more than 8.000 destinations. For each hotel, information is provided about facilities, prices, policies, multimedia contents ... Some services are also available for checking the availability of a room and booking it.
- `Touring` provides information about Italian hotels, restaurants and cities. By means of three different forms, it is possible to find the available hotels, restaurants (described with some specific features) and main monuments for each city.
- `TicketOne` provides information about Italian cultural events. For each event, a description including place, price and details is provided. The information system offers services to check the ticket availability of a particular event and, also, to buy tickets.

A global view of the data sources provided by each of these information system is created by means of a data integration system (see Figure 1). It is composed of four classes: *accommodation*, *restaurant*, *event* which are related to each other by means of the *city* class. Moreover, another set of WS descriptions has been made available starting from public repositories available on the Web; such services are related to different domains such as Economy, Communication, Education, Food, Medical, Travel and Weapon.

Now let us consider that a user asks the query $Q1$ below:

```
select Accommodation.Name, Accommodation.City, Accommodation.Country
from   Accommodation
where  Accommodation.City='Modena'
```

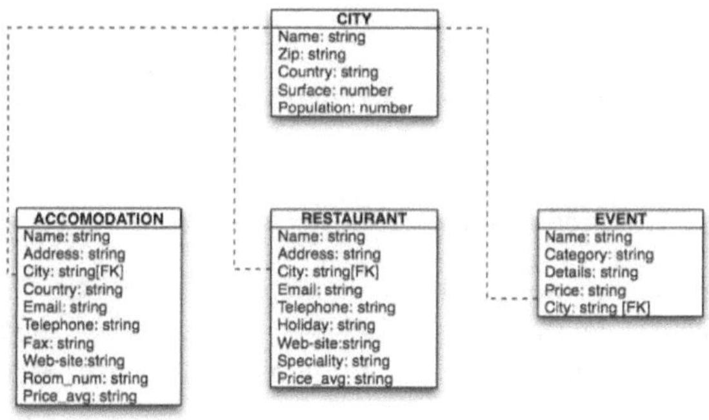

Fig. 1. The tourist integrated schema

The problem we address in this paper is to retrieve, among the many services available possibly related to the many domains mentioned above, the ones that are possibly related to Q1, according to the semantics of the terms involved in the query.

3 Building the Global Data and Service View at Set-Up Time

In our approach, data sources and services are grouped into semantic peers. Each semantic peer generates a Peer Virtual View (PVV), i.e. a unified representation of the data and the services held by the sources belonging to the peer. A PVV is made up of the following components, shown in Figure 2:

- a **Semantic Peer Data Ontology (SPDO)** of the data, i.e. a common representation of all the data sources belonging to the peer; the SPDO is built by means of MOMIS as described in Section 5.
- a **Global Light Service Ontology (GLSO)** that provides, by means of a set of concepts and attributes, a global view of all the concepts and attributes viewed for the descriptions of the Web services available in the peer;
- a set of **mappings** which connects GLSO elements to SPDO elements.

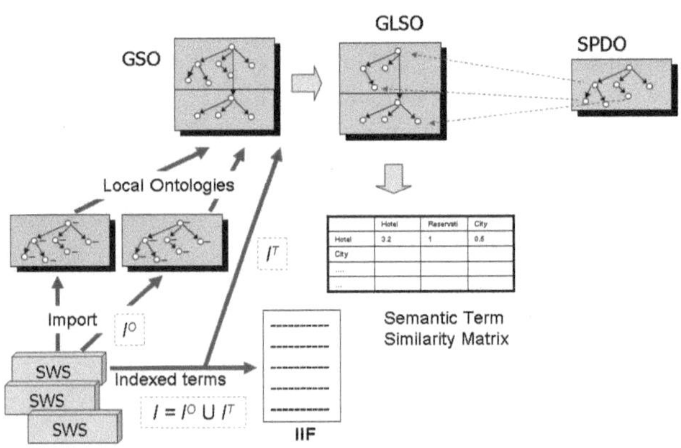

Fig. 2. A sketch of the overall process for building the GLSO

3.1 Building an Integrated Representation of eServices: The GLSO

The global light service ontology is built by means of a process consisting of three main steps: (i) service indexing, (ii) Global Service Ontology (GSO) construction, (iii) Global Light Service Ontology (GLSO) construction and Semantic Similarity Matrix (SSM) definition.

Service Indexing. In order to inquiry services, an Information Retrieval approach is applied to the semantic descriptions of Web services has been developed. We consider OWL-S as the semantic Web Service description language where OWL-S semantic descriptions are OWL ontologies and refer to OWL domain service ontologies (SOs). The approach can be easily generalized to other lighter semantic annotation languages for Web services such as SAWSDL [12]. Using OWL-S as reference semantic description languages, we can consider available to the peer, among other ones, the large set of service descriptions given in the OWL-S Service Retrieval Test Collection 2.2. (OWLS-TC)[2]; referring to this set of services we stress out that many services may be collected from the Web or from available repositories (OWLS-TC provides more than 1000 for the OWL-S 1.1 language), that many ontologies are referred to in the service descriptions (OWLS-TC provides 43 ontologies), and that these ontologies may concern different domains (OWLS-TC provides services from seven different domains).

The IR approach requires a formal representation of the services descriptions stored in the repository and it is based on full text indexing which extracts terms from six specific sections of the service description: *service name, service description, input, output, pre-condition and post-condition*[3]. As an example, the service "City_Hotel_Service" from the OLWS-TC collection imports two domain ontologies (Travel and Portal) and is described by: the name "CityHotelInfoService", the text descriptions "This service returns information of a hotel of a given city", and the ontology concepts *City* (from the Portal ontology) and *Hotel* (from the Travel ontology) as input and output respectively. While the service name and description consist of short text sections, input and output refer to domain SOs, namely, a portal and a travel ontology (pre-condition and post-condition are represented analogously but missed in the example).

According to this process a set of index terms I that will be part of the dictionary is extracted. We call I^O, where $I^O \subseteq I$, the set of index terms consisting of ontology resources (e.g. *City* and *Hotel* in the example); we call I^T, where $I^T \subseteq I$, the set index terms extracted from textual descriptions (e.g. "information", "hotel", "city"). Index terms are not repeated, i.e. $I^O \cap I^T = \emptyset$, and where the identifier of a term $i \in I^O$ is equivalent to the string identifying a term $j \in I^T$, the term $j \in I^T$ is discarded (in the example the terms "hotel" and "city" extracted from the service descriptions are discarded, and $I = \{``information", Hotel, City\}$).

The indexing structure is based on a "structured document" approach, where the structure of the document consists of the six aforementioned sections. The inverted file structure consists of (i) a *dictionary* file based on I, and (ii) a *posting file*, with a list of references (for each term in the dictionary) to the services' sections where the considered term occurs. The posting file is organized to store,

---

2 http://projects.semwebcentral.org/projects/owls-tc/

3 More precisely, ontological references for inputs and outputs are defined by *process:Input* and *process:Output*; moreover, URIs are not completely specified in the paper for sake of clarity.

for each index term $i \in I$, a list of blocks containing (i) the service identifier, and (ii) the identifier of the service's section in which the term appears.

Each block has a variable length, as a term appears in at least one section of a service in the posting list, but it may appear in all sections. In the usual IR approaches to text indexing an index term weight is computed, which quantifies the informative role of the term in the considered document. We propose an approach to sections weighting, in order to enhance the informative role of term occurrences in the distinct sections of the service. These importance weights will be used in the query evaluation phase in order to rank the services retrieved as a response to a user's query as explained in Section 4.1

GSO Construction. The GSO is built by

- loosely merging each service ontology O such that $i \in O$ for some $i \in I^O$ (e.g. the portal and travel ontologies in the reference example) - the process take into account ontology imports recursively.
- associating a concept C_i with each $i \in I^T$, introducing a class $Terms$ subclass of *Thing* in the GSO and stating that for every $i \in I^T$, C_i is subclass of $Terms$ (e.g. the term "reservation" occurring in the *service name* or *service description* sections).

With "loosely merging" we mean that SOs are merged asserting that their top concepts are all subclasses of *Thing* without attempting to integrate similar concepts across the different integrated ontologies. Therefore, if the source SOs are consistent, the GSO can be assumed to be consistent because no axioms establishing relationships among the concepts of the source SOs are introduced (e.g. axioms referring to the same intended concept would actually refer to two distinct concepts with two different URI). Loose merging is clearly not the optimal choice with respect to ontology integration, but since the existent techniques do not allow to integrate ontologies in a completely automatic way [13], this is the only technique that guarantees consistency, without requiring a further user intervention. Moreover, since the XIRE retrieval component is based on approximate IR techniques and semantic similarity rather than on input/output matching algorithms, approximate solutions to the ontology integration problem can be considered acceptable; instead, the whole GSO building process need to be fully automatized.

GLSO Construction and Semantic Similarity Matrix. The GSO may result extremely large in size, which makes the semi-automatic process of mapping the GSO to the SPDO more expensive; moreover, only a subset of the terms of the ontologies used and recursively imported by the SWS descriptions are actually relevant to the SWS descriptions. To solve this problem a technique to reduce the ontology size is exploited and a GLSO (Global Light Service Ontology) is obtained. The technique adopted is based on the logic-based and tool-supported ontology module extraction approach described in [10], which is the only domain-independent and (completely) automatic approach to extract subontologies from OWL-DL source ontologies [7];

According to this approach, we extract from the GSO, the subontology that preserve the meanings of the terms explicitly used in the service descriptions, namely, the set of the index terms I. Formally, the GLSO is interpreted as a syntactic locality-based **S**-module of the GSO, where **S**=**I**, that is, the extraction interface is given by the set $I \in GSO$. Let us observe that after this process $I \subseteq GLSO$ still holds.

Another output created along with the GLSO is the Semantic Similarity Matrix (SSM), which is exploited later on for query expansion at query time. In particular, the dictionary used by the IR engine is built from the GLSO, and the SSM is defined by analyzing the GLSO structure. Let Sig^O be the signature of an ontology O, i.e. the set of all concepts, property and instance names occurring in O; $SSM = \{Sig^{GLSO} \times Sig^{GLSO}\}$ is a matrix whose values represent the semantic similarity between two terms in the GLSO; the function $sim = \{Sig^{GLSO} \times Sig^{GLSO}\} \rightarrow [0,1]$, where $sim(x,x) = 1$, is a similarity function based on the structure of the ontology; the function is defined according to the formulas provided in [4] and takes into account subclass paths, domain and range restrictions on properties, membership of instances, and so on. Observe that loose coupling of SOs can lead to *false dissimilarities* in the SSM: two terms that have a same meaning can be considered dissimilar in the SSN because terms coming from different ontologies are connected only by long paths passing through the root concept. However, in the following we show how this problem is solved when the GLSO is mapped to the SPDO and in the query rewriting process.

3.2 Mapping of Data and Service Ontologies

Mappings between the elements of the SPDO and the GLSOs are generated by exploiting and properly modifying the MOMIS clustering algorithm. The clustering algorithm takes as input the SPDO and the GLSO with their associated metadata and generates a set of clusters of classes belonging to the SPDO and the GLSO.

Mappings are automatically generated exploiting the clustering result. The following cases are possible:

- **A cluster contains only SPDO classes:** It is not exploited for the mapping generation; this cluster is caused by the selection of a clustering threshold less selective than the one chosen in the SPDO creation process.
- **A cluster contains only GLSO classes:** It is not exploited for the mapping generation; it means that there are descriptions of Web Services which are strongly related.
- **A cluster contains classes belonging to the SPDO and the GLSOs:** This cluster produces for each SPDO class a mapping to each GLSO class. Mappings between the attributes of the classes are generated on the basis of the relationships held in the MOMIS Common Thesaurus.

The user may set some parameters in order to choose between *rough mappings* (large clusters are created with classes which are also not strictly related) and *precise mappings* (the cluster generation produces many groups of few classes that strongly match).

As an example, consider the SPDO described in Figure 1, and a piece of the *GLSO* concerning the class *Hotel* and the attributes this class is domain of; using a dotted notation in the form of *"concept.property"* this piece of ontology is represented as follows:

```
Hotel
    Hotel.Denomination
    Hotel.Location
    Hotel.Country
```

The following mappings are generated with the application of our technology:

```
Accommodation  -->  Hotel

Accommodation.Name     -->  Hotel.Denomination
Accommodation.City     -->  Hotel.Location
Accommodation.Country  -->  Hotel.Country
```

Mappings between the class *City* in the SPDO and the concept *City* in the GLSO and the related attributes are defined in an analogous way.

Observe that clusters can be exploited to correct false dissimilarities in the SSM. However, clusters provide also another solution to the false dissimilarities problem: when two terms t_1 and t_2 of the GLSO are clustered together, they are mapped to the same term s of the SPDO; when a query formulated in the SPDO terminology contains the term s, both t_1 and t_2 will be extracted as keywords and used to retrieve services.

4 Data and eService Retrieval

Let us introduce a query expressed in the SQL language:

```
select <select_attribute_list>
from   <from_class_list>
where  <condition>
```

The answer to this query is a data set from the data sources together with a set of services which are potentially useful, since they are related to the concepts appearing in the query and then to the retrieved data.

The query processing is thus divided into two steps, that are simultaneously executed:

- a data set from the data sources is obtained with a query processing on an integrated view
- a set of services related to the query is obtained by exploiting the mapping between SPDO and GLSOs and the concept of relevant service mapping

Data results are obtained by exploiting the MOMIS Query Manager (see [1] for a complete description) which rewrites the global query as an equivalent set of queries expressed on the local schemata (local queries); this query translation is carried out by considering the mapping between the SPDO and the local schemata. Since MOMIS follows a GAV approach, the query translation is thus performed by means of query unfolding. Results from the local sources are then merged exploiting the reconciliation techniques. As the query processing on an integrated view is already well described in literature, in the following we focus our attention on the queries for services.

Services are retrieved by the XIRE (eXtended Information Retrieval Engine) component, which is a service search engine based on the vector space model [5], and implemented with the open source libraries Lucene [9]; in particular, the approach takes as input a vector where each term has a relevance weight associated. In order to provide XIRE with this input, the set of query processing phases represented in Figure 3 need to be defined. The process starts extracting the terms of the SPDO appearing in the SQL query; then, the query manager checks which elements of the GLSO are mapped on these terms of the SPDO; these keywords are expanded w.r.t. the SSM, and a weighted terms query vector is provided to XIRE which will retrieve the related services as described in details in Section 4.1.

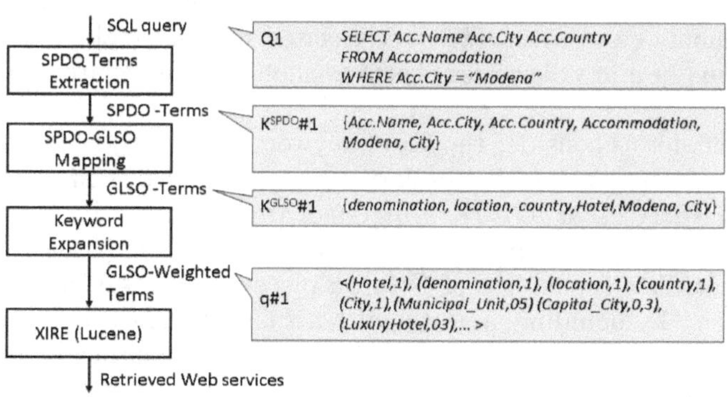

Fig. 3. The query processing steps and their application to the reference example

4.1 eService Retrieval

Keywords Extraction. Given a SQL query expressed in the SPDO terminology, the set of keywords K^{SPDO} extracted consists of: all the classes given in the "FROM" clause, all the attributes and the values used in the "SELECT" and "WHERE" clauses and all their ranges defined by ontology classes. As an example, the set of keywords extracted from the query Q1 introduced in Section 2, consists of the set $K^{SPDO}\#1$ represented in Figure 3.

Keywords Rewriting and Expansion. The set of keywords K^{SPDO} extracted from the query inserted by users are rewritten in a set of keywords K^GLSO exploiting the mappings between the SPDO and the GLSO. Let us define a data to service ontology mapping function $\mu = Sig^{SPDO} \rightarrow \mathcal{P}\left(Sig^{GLSO}\right)$. The function, given a term $s \in SPDO$ returns a set of terms $T \subseteq Sig^{GLSO}$ iff every $t \in T$ is in the same cluster of s. Given a set of keyword $K^{SPDO} = \{k_0, ..., k_m\}$, each keyword k_i with $0 \leq i \leq m$ is replaced by the set of keywords returned by $\mu\left(k_0\right)$. Assuming the mappings described in Section 3.2, and assuming $\mu\left(Modena\right) = Modena$, and $\mu\left(City\right) = City$, the set of keywords obtained in the reference example is the set $K^{GLSO}\#1$[4] represented in Figure 3.

Semantic similarity between GLSO terms defined in the SSM is exploited to expand the K^{GLSO} set into a weighted terms vector $q =< (k_1, w_1), ..., (k_n, w_n) >$, where for $1 \leq i \leq n$, $k_i \in Sig^{GLSO}$, and w_i are weights that represent the relevance of every terms w.r.t. the specified keywords. The vector q is obtained associating each keyword with a weight equal to 1, and adding a set of terms that are similar to the given keywords up to a given threshold weighted according to their similarity w.r.t. the given keywords.

More formally, let $simsets_s(t) \subseteq Sig^{GLSO}$ be the set of terms of the GLSO such that their similarity w.r.t. t is greater than a given threshold s. Given a set of keywords $K^{GLSO} = \{k_0, ..., k_m\}$, the vector q is obtained as follows: all the keywords $k_i \in K^{GLSO}$, with $1 \leq i \leq m$, are inserted in q and are associated with a weight $w_i = 1$; for every $k_i \in K^{GLSO}$ the set $simset_s\left(k_i\right)$ is inserted in q, and each elements $e \in simset_s\left(k_i\right)$ is associated with a weight $w_e = sim\left(k_i, e\right)$; duplicate terms in q are discarded, keeping the terms associated with a greater weight.

For example, let us consider the set of keywords $K^{GLSO}\#1$ given in the reference example. Assume to set the similarity threshold $s = 0,3$; $simset_{0,3}\left(City\right) \subseteq \{Municipal\_Unit, Capital\_City\}$, $sim\left(City, Municipal\_Unit\right) = 0,5$ ($City$ is subclass of $Municipal\_Unit$ in the Travel ontology) and $sim\left(City, Capital\_City\right) = 0,3$ ($Capital\_City$ is subclass of $City$); a piece of the resulting weighted term query vector $q\#1$, including $Municipal\_Unit$,$Capital\_City$ and $LuxuryHotel$ (added in an analogous way based on the ontology including $Hotel$), is represented in Figure 3.

eServices Retrieval. Query evaluation is based on the vector space model[5]; by this model both documents (that is Web Service descriptions) and queries (extracted queries) are represented as a vector in a n-dimensional space (where n is the total number of index terms extracted from the document collections). Each vector represents a document, and it will have weights different from zero for those keywords which are indexes for that description. The value of such weight is computed according to the weights of the six sections of the service description in which the keyword appears. We assume that the implicit constraint specified in a user query, when selecting a query term (a single keyword) is that it must appear in *at least one* section of a service description in order to

[4] Remind that all the GLSO terms are URIs, although parts of the URI specification are omitted for sake of clarity.

retrieve that service. Based on the above assumptions the weight which at query evaluation time is associated with a keyword and a service description is equal to the maximum of the weights of the service sections in which the keyword appears.

Relevance weights introduced in the previous section are used to modify the weights in the list resulting from keyword evaluation process described above. In particular, we consider a relevant weight, as a "fine-tuning" value of the keyword weights in the modified posting list. The modification is performed by a product between the keyword weight and the relevant weight. The process returns an ordered list of services (results for Q1 include the service described in the reference example).

5 Related Work

To the best of our knowledge, this is the only approach that aims at integrating data and services to provide a unified view about available resources, and therefore providing query answering functionalities across multiple data sources, together with service discovery. Some very preliminary ideas about our research have been published in [14]. In this paper, we extend and complete those ideas, in particular, introducing a different and more effective way for querying services exploiting IR techniques; moreover, the extraction of a subontology from the GSO, and the concept similarity matrix are novelties of this paper.

As for the data integration approach exploited in the paper and based on the system MOMIS, and the comparison to other data integration approaches, we refer to previous works where the approach has been presented in details [1,2,3]. Here we just, recall the main functionalities of MOMIS w.r.t. data integration. The MOMIS (Mediator envirOnment for Multiple Information Sources) is a framework to perform information extraction and integration from both structured and semistructured data sources. Information integration is performed in a semi-automatic way, by exploiting the knowledge in a Common Thesaurus (defined by the framework) and descriptions of source schemas with a combination of clustering techniques and Description Logics. This integration process gives rise to a virtual integrated view of the underlying sources (the Global Schema) for which mapping rules and integrity constraints are specified to handle heterogeneity.

For what concerns other service discovery and retrieval engines, this paper proposes a solution aimed at being integrated with MOMIS. From this perspective the aim is not to outperform current Web service discovery engines; however it is worth noting that nor other systems based on IR techniques such as Woogle [6], nor other discovery based on OWL-S service descriptions such as OWLS-MX [11] could be trivially integrated with the MOMIS system to provide data and service integration. Woogle does not allow to search for semantic Web services, and, in particular for OWL-S descriptions; it is based on WSDL descriptions that have not references to external ontology terms to be mapped with the SPDO ontology. OWLS-MX has been recognized as one of the most prominent discovery engines for OWL-S service descriptions; it combines logic-based techniques to address the semantic input/output annotations and IR techniques based on syntactic similarity measures to consider the textual part of the descriptions. The main difference consists in the fact that OWLS-MX is a service matchmaker, whose goal is not only to retrieve a set of services related to a set of terms, but a set of services matching w.r.t. a set of given input/output specifications (taking into account logic-based matching relationships such as "exact", "plug-in", and other specific relationships). Syntactic similarity measures are exploited to improve the recall of the matchmaking algorithm. From this point of view the OWLS-MX approach is different form the one presented here: in our approach a semantic term similarity measure based on the structure of the GLSO is exploited; moreover, we provide a strategy to merge all the ontologies available in a unique global ontology. The work discussed in [15] is based on the same principle of OWLS-MX, which outperform w.r.t. efficiency. However, a more detailed comparison of the approach to keyword-based service retrieval described here and the approach introduced in [11] is ongoing.

6 Conclusion and Future Work

In this paper we introduced a technique for publishing and retrieving a unified view of data and services. Such unified view may be exploited for improving the user knowledge of a set of sources and for retrieving a list of web services related to a data set. The approach is semi-automatic, and works jointly with the tools which are typically provided for searching for data and services separately.

Future work will be addressed on evaluating the effectiveness of the approach in the real cases provided within the NeP4B project, and against the OWLS-TC benchmark. Moreover, some work will be devoted to the refinement of the techniques for generating the unified view and for translating a query for data retrieval into services retrieval.

Acknowledgments

The work presented in this paper has been partially supported by the European IST project n. 27347 SEEMP - Single European Employment Market-Place and the Italian FIRB project RBNE05XYPW NeP4B - Networked Peers for Business.

References

1. Beneventano, D., Bergamaschi, S.: Semantic Web Services: Theory, Tools and Applications. In: Semantic Search Engines based on Data Integration Systems. Idea Group Publishing, USA (2007)
2. Beneventano, D., Bergamaschi, S., Guerra, F., Vincini, M.: Synthesizing an integrated ontology. IEEE Internet Computing 7(5), 42–51 (2003)
3. Bergamaschi, S., Castano, S., Vincini, M., Beneventano, D.: Semantic integration of heterogeneous information sources. Data Knowl. Eng. 36(3), 215–249 (2001)

4. Bernstein, A., Kaufmann, E., Buerki, C., Klein, M.: How similar is it? towards personalized similarity measures in ontologies. In: Ferstl, O.K., Sinz, E.J., Eckert, S., Isselhorst, T. (eds.) Wirtschaftsinformatik, pp. 1347–1366. Physica-Verlag, Heidelberg (2005)
5. Christopher, P.R., Manning, D., Schütze, H.: Introduction to Information Retrieval. Cambridge University Press, Cambridge (2008)
6. Dong, X., Halevy, A., Madhavan, J., Nemes, E., Zhang, J.: Similarity search for web services. In: VLDB 2004: Proceedings of the Thirtieth international conference on Very large data bases, pp. 372–383. VLDB Endowment (2004)
7. Doran, P., Palmisano, I., Tamma, V.: Somet: Algorithm and tool for sparql based ontology module extraction. In: Sattler, U., Tamilin, A. (eds.) WORM 2008. CEUR Workshop Proceedings, vol. 348, CEUR-WS.org (2008)
8. Hansen, M., Madnick, S.E., Siegel, M.: Data integration using web services. In: Bressan, S., Chaudhri, A.B., Li Lee, M., Yu, J.X., Lacroix, Z. (eds.) CAiSE 2002 and VLDB 2002. LNCS, vol. 2590, pp. 165–182. Springer, Heidelberg (2003)
9. Hatcher, E., Gospodnetic, O.: Lucene in Action. Manning Publications (December 2004)
10. Jiménez-Ruiz, E., Grau, B.C., Sattler, U., Schneider, T., Llavori, R.B.: Safe and economic re-use of ontologies: A logic-based methodology and tool support. In: Bechhofer, S., Hauswirth, M., Hoffmann, J., Koubarakis, M. (eds.) ESWC 2008. LNCS, vol. 5021, pp. 185–199. Springer, Heidelberg (2008)
11. Klusch, M., Fries, B., Sycara, K.: Automated semantic web service discovery with owls-mx. In: AAMAS 2006: Proceedings of the fifth international joint conference on Autonomous agents and multiagent systems, pp. 915–922. ACM, New York (2006)
12. Kopecký, J., Vitvar, T., Bournez, C., Farrell, J.: Sawsdl: Semantic annotations for wsdl and xml schema. IEEE Internet Computing 11(6), 60–67 (2007)
13. Noy, N.F.: Semantic integration: A survey of ontology-based approaches. SIGMOD Record 33(4), 65–70 (2004)
14. Palmonari, M., Guerra, F., Turati, A., Maurino, A., Beneventano, D., Valle, E.D., Sala, A., Cerizza, D.: Toward a unified view of data and services. In: Proceedings of the 1st International International Workshop on Semantic Data and Service Integration, Vienna, Austria (2007)
15. Skoutas, D., Sacharidis, D., Kantere, V., Sellis, T.: Efficient semantic web service discovery in centralized and p2p environments. In: Sheth, A.P., Staab, S., Dean, M., Paolucci, M., Maynard, D., Finin, T., Thirunarayan, K. (eds.) ISWC 2008. LNCS, vol. 5318, pp. 583–598. Springer, Heidelberg (2008)
16. Zhu, F., Turner, M., Kotsiopoulos, I., Bennett, K., Russell, M., Budgen, D., Brereton, P., Keane, J., Layzell, P., Rigby, M., Xu, J.: Dynamic data integration using web services. In: ICWS 2004: Proceedings of the IEEE International Conference on Web Services, Washington, DC, USA, p. 262. IEEE Computer Society, Los Alamitos (2004)

Preliminary Explorations on the Statistical Profiles of Highly-Rated Learning Objects

Elena García-Barriocanal and Miguel Ángel Sicilia

Information Engineering Research Unit
Computer Science Dept., University of Alcalá
Ctra. Barcelona km. 33.6 – 28871 Alcalá de Henares (Madrid), Spain
{elena.garciab,msicilia}@uah.es

Abstract. As learning object repositories grow and accumulate resources and metadata, the concern for quality has increased, leading to several approaches for quality assessment. The availability of on-line evaluations in some repositories has opened the opportunity to examine the characteristics of learning objects that are evaluated positively, in search of features that can be used as a priori predictors of quality. This paper reports a preliminary exploration of some learning object attributes that can be automatically analyzed and might serve as quality metrics, using a sample from the MERLOT repository. The bookmarking of learning objects in personal collections was found to be a potential predictor of quality. Among the initial metrics considered, the number of images has been found to be also a predictor in most of the disciplines and the only candidate for the Art discipline. More attributes have to be studied across disciplines to come up with automated analysis tools that have a degree of reliability.

Keywords: Learning objects, metadata, repositories, quality, metrics, evaluation, LORI, MERLOT.

1 Introduction

The widespread adoption of e-learning has lead to a proliferation of learning resources available on the Web. This has been in part fostered in the last years by the emergence of the paradigm of "open educational resources" (OER), which is closely related to popular initiatives as Open Courseware (OCW) and reflects a general concern for the open sharing of learning resources for the benefit of the public (Guntram, 2007). Resource repositories have appeared in the last years serving the role of specialized or generalist portals providing enhanced descriptions and evaluations along with metadata-based search mechanisms, mediating between the creators or providers of the contents and the users that seek resources appropriate for some concrete needs (McGreal, 2008). This has been combined also with the notion of "learning objects" as self-standing, modularized resources described by standardized metadata, which has led to many finer granularity resources that have the potential to be more reusable (Nash, 2005).

F. Sartori, M.Á. Sicilia, and N. Manouselis (Eds.): MTSR 2009, CCIS 46, pp. 108–117, 2009.

The abundance of resources covering the same educational needs has in turn raised the concern for quality, and some techniques and instruments have already been devised and applied. Notably, the *Learning Object Review Instrument* (LORI) provides a statistically validated technique for assessing quality (Vargo et al., 2003). However it is not the sole approach to measure quality that has been explored. Another example is the *Connexions* repository[1] that approaches quality in terms of a kind of explicit endorsement of individuals or organizations (Kelty, Burru and Baraniuk, 2008). In other direction, the MERLOT repository[2] approaches quality with several mechanisms, from informal ratings to a process of expert review that results in extensive assessments with detailed textual reviews (Cafolla, 2006). These are examples of *evaluative metadata* (Vuorikari, Manouselis and Duval, 2008) which may eventually be standardized through shared data models for ease of collection. The approaches to assess quality are thus diverse, requiring different levels of effort from experts or repository users, but they bear some similarity in the aspects considered. Concretely, the work on Bayesian networks for learning object rating by Han et al. (2003) supports the hypothesis that some quality attribute ratings considered in MERLOT and LORI are correlated, which would open the opportunity to reuse evaluations across repositories.

In any case, the availability of large databases of learning object evaluations has opened new possibilities to seek for metrics that could complement existing techniques and instruments requirement extensive human effort with others using fully automated analysis tools, useful for getting an inexpensive early indicator of the quality of learning objects. Such approach has been studied previously to automatically analyze the usability of Web sites (Ivory and Hearst, 2001). Learning object quality can be considered a more complex construct than usability as the latter is included in existing instruments as LORI as one out of the several attributes considered. Further, quality attributes of learning objects as reusability are deemed as context-dependent, in the sense that the attribute can only be measured for a particular educational context description (Sicilia and García-Barriocanal, 2003).

Even though automated analysis cannot replace inspection techniques, it has the potential to provide an inexpensive mechanism to explore the quality of resources a priori, complementing the other techniques and instruments mentioned. This paper reports preliminary results on a first study aimed at finding patterns in some of the attributes that can be automatically extracted in highly ranked MERLOT resources. Preliminary results point to bookmarks in personal collections as a good predictor, and to a less extent, the number of images. Other metrics and their combinations require further study to develop models with a degree of reliability that makes them useful for automated analysis.

The rest of this paper is structured as follows. Section 2 explores potential learning object measures that could be derived automatically from the digital contents conforming the objects. Then, Sections 3 and 4 report preliminary results on a study concerning the statistical profiles of learning objects using a sample of resources found in the MERLOT repository. Finally, conclusions and outlook are provided in Section 5.

---

[1] http://cnx.org

[2] http://www.merlot.org

2 Quantitative, Measurable Aspects of Learning Objects

Some repositories have their own internal format for learning resources (as CNXML in *Connexions*), or are based on the use of some specification as IMS LD (Lundgren-Cayrol et al., 2006). However, most of them store only metadata associated to a principal URI (these are often called "*referatories*" instead of repositories) or store contents in any Web format with no special requirements. This latter case is nowadays the most common and reuse of learning objects today should assume that we are dealing with conventional Web contents (Wiley et al., 2004). Taking this as a point of departure, measurable aspects or indicators can be extracted from metadata records and also from the contents themselves. Here we are mainly concerned with the latter, following the methodology and insights Ivory and Hearst (2002) applied to the automated analysis of the usability of Web sites. The basic procedure is that of contrasting measures and data extracted automatically from the content of the resources with the evaluations of the learning objects that are stored in the repository pointing to them. For example, in the case of Ivory and Hearst study, they found the number of links and link clusters to be two of the significant predictors for high rating by experts. These measures can also be directly applied to learning objects as they are in most cases Web content as the one that was subject to Ivory and Hearst study. However, it cannot be taken from granted that the same correlations found by them are still applicable to ratings of learning objects, as usability and learning object quality are two very different, multi-faceted elements (even though it may be hypothesized that the former affects the latter to some extent). So a first step in finding statistical profiles for highly rated learning objects is exploring evidence on potential predictors of learning object quality, taking as a point of departure some of the ones that were identified for usability.

It should be noted that the learning object attributes examined here refer to characteristics that are intrinsic to the objects. Other approaches use instead usage data or other kind of relationships (Ochoa and Duval, 2008a), but here we are concerned with data that is available from the objects themselves, as automated analysis will typically be used for objects that have still not been shared to the public and are under preliminary assessment In those cases, metadata may in some cases be incomplete (Sicilia et al., 2005) or even contain inaccurate descriptions. Therefore, we use a particular kind of referential data (bookmarks in personal collections) only as a way to contrast with the metrics extracted from the contents of the resources.

3 Preliminary Empirical Study

This section reports on the method and exploratory results for the initial study carried out by using data downloaded from the MERLOT repository with the help of a Web crawler developed *ad hoc* for that purpose.

The Multimedia Educational Resource for Learning and Online Teaching (MERLOT) is an international project that aims at facilitating the evaluation and use of online learning objects. MERLOT is structured in collections: Arts, Business, Education, Humanities, Mathematics and Statistics, Science and Technology, and Social Sciences, each category having several sub-categories. Most sub-categories have an editorial peer review board consisting of a group of higher education faculty members that are experts in that concrete field. The editorial boards will assess new materials that are submitted to MERLOT, contact the author about the prospect to review their material, and then assign two reviewers for each item. Afterwards, one of those peer reviewers will create a single composite review. The reviewing process addresses the aspects: *Quality of Content*, *Ease of Use for Teachers and Learners*, and *Potential Effectiveness as a Teaching Tool.* The peer review is then sent to the author with some comments and then posted with the author's approval. In addition to formal peer review, MERLOT users are allowed to post comments about the resources, optionally providing a rating in the scale from 1 to 5, with 5 the best rating. Ochoa and Duval (2008b) found evidence that MERLOT is an average size learning object repositories (referatory), with a large base of contributors, so that it provides a relevant base of materials for empirical analysis.

3.1 Method

The method followed for the exploratory study reported here was the development of a preliminary statistical profile. The goal of profile development is to derive statistical models for classifying learning objects into some classes (e.g. good, average, and poor) based on their quantitative measures. The thresholds for the categories are the terciles, so that objects with ratings below the first threshold are the "poor" category[3] and objects below the second threshold determine the "average" category. Table 1 provides the discipline breakdown of reviewed and rated objects in the database.
The analysis was done by discipline to avoid potential differences in raters related to their background and topical community. Data from a total of 20601 objects was gathered (May 2009), but some of them were classified into more than one category. However the overlap is small, so we decided to retain the objects in more than one discipline category. Most resources have zero or one peer reviews, and only a few have two or three. The distribution is very different with comments, as some resources have up to 101, but the average is less than 3.

Table 1 provides the thresholds for average peer review and average comment ratings that divide data in three approximate subsets that we will call "good", "average"

Table 1. Sample used and breakdown in categories per discipline

Discipline	Size	Peer-reviewed	Commented	Thresholds-peer reviews	Thresholds-comments
Arts	867	161	119	4/4.5	4/5
Business	3115	668	417	4/4.5	4/4.5
Education	4337	645	989	4/4.75	4/5
Humanities	3262	822	636	4.25/5	4/5
Mathematics and Stat.	1688	500	350	4.25/5	4/4.5
Science and Tech.	7793	1257	2250	4/4.75	4/4.25
Social Sciences	1741	305	472	4/4.75	4/4.5
Totals	22803	2605	3181	-	-

[3] It should be noted that labels as "poor" here are simply tags for the comparative analysis, and they do not entail that the resources falling in that category are actually of bad quality.

and "poor", considering only objects with at least one review or comment. These subsets are the basis for exploring statistical profiles. An important conclusion of this study is that the distribution of ratings both for reviews and comments tend to have increasing relative frequency histograms, so that most ratings tend to be positive (i.e. above the intermediate rating 'three'). In consequence, the practical focus of this analysis is on distinguishing the attributes of "highly rated" learning objects, and not establishing categories of "good" and "bad" resources.

3.2 Results

As the distributions of the metrics computed could not be assumed to be normal, Mann-Whitney Wilcoxon (M-W) test and a Kolmogorov-Smirnoff (K-S) test were used to compare the medians and distributions of the samples determined by the upper threshold level, i.e. "good" resources are compared against the rest of them. Table 2 provides the results of the analysis per each of the metrics: number of links (considering only the first HTML page), size in bytes of the main page, images in the first page. Total image size was discarded as it was found to be strongly correlated with image number. The number of times the object was bookmarked in a personal collection of a MERLOT member was also included to contrast the effectiveness of the other measures with a typical metric of social prominence of the item.

Table 2. Sample used and breakdown in categories per discipline

Discipline	Links avg(std)	C/R?	Size avg(std)	C/R?	Images avg(std)	C/R?	Collections avg(std)	C/R?
Arts	37(67)	N/N	20693(75621)	N/N	13(30)	N^*/Y	5(7)	N^*/ N^*
Business	47(91)	N^*/ N^*	61809(698921)	N/ N^*	14(24)	N^*/ N^*	4(8)	N^*/Y
Education	42(118)	Y/ N^*	28390(129007)	N/ N^*	16(29)	Y/ N^*	7(14)	N^*/Y
Humanities	40(85)	N^*/ N^*	21744(119469,0)	N/N	14(27)	N^*/ N^*	6(9)	Y/Y
Math.& Stat.	32(80)	Y/Y	20971(68858,3)	Y/Y	12(35)	Y/ N^*	6(22)	Y/Y
Science & Tech.	31(74)	Y/ N^*	74394(1230610)	Y/ N^*	12(25)	Y/Y	5(12)	Y/Y
Social Sci.	45(94)	N^*/ N^*	40825(321305,0)	N^*/Y	15(27)	N^*/Y	6(15)	Y/Y

Table 2 columns labeled "C/R?" report if the attribute immediately left in the Table was found to be significant discriminator of the "good" category as defined above considering comments (C) and peer reviews (R) respectively.

In Table 2, "N" represents no significant difference for the median of the two samples, and "N^*" stands for no significant difference for the medians but significant difference in distribution. Finally "Y" stands for both differences at the same time.

4 Discussion

The first important conclusion of the results presented is that the number of times a resource appears in personal collections appears to be the more consistent and reliable metric overall. Figure 1 depicts the histograms of personal collection bookmarks for resources poorly rated by reviewers at the bottom and resources rated above 3 by reviewers for the *Science & Technology* category. Poorly rated resources average around two personal collection bookmarks while those with better ratings average around nine personal collections. Further, all the resources with very high number of personal collection bookmarks are in the "non-poor" category. All the resources included in more than 10 personal collections are rated 3 or higher by reviewers. Consequently, personal collections have the potential to become a good predictor for quality in the case of not reviewed resources.

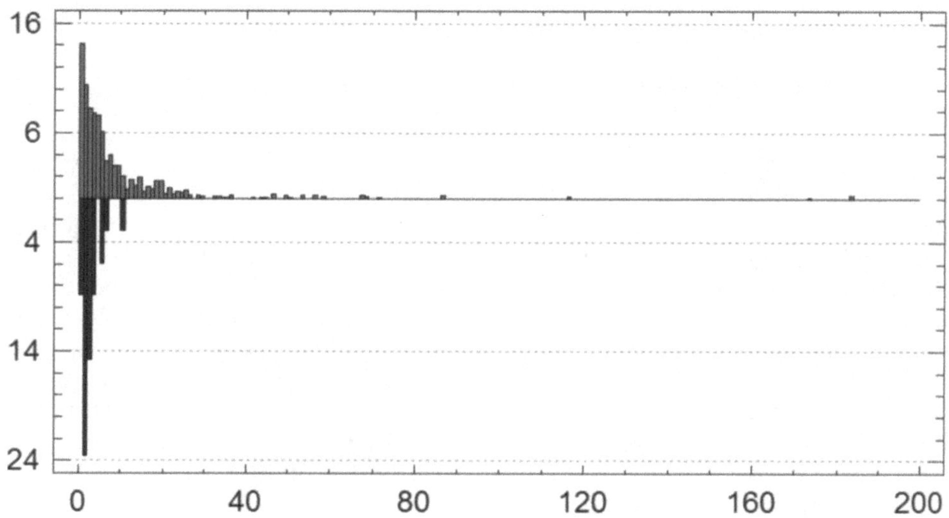

Fig. 1. Comparison of personal collection bookmark distribution between objects rated 3 or higher (upper part) and the rest (bottom part)

Size, number of links and number of images have different profiles for the different disciplines. In the *Math and Statistics*, objects highly rated by reviewers have significantly larger sizes, but the opposite is true for and the *Social Science* category. Size is controversial as metric, as consistent differences have not been found in comments and/or ratings for several areas. Number of links is higher for highly rated resources in the *Math and Statistics*, but again this is not consistent with other categories. Both in the case of links and size, the current analysis is not enough to draw conclusions from these metrics. Images have either differences in distribution or both differences in median and distribution across all the disciplines, and the only inconsistency is in the social sciences discipline. A difference that deserves further study is that for *Arts*, the more significant metric is number of images.

Table 3 provides a summary of the tendency of the metrics in the subset of "good" pages, for the disciplines in which they have been found to have at least differences in distribution in both comments and peer reviews. Small differences have been considered as equal (=). Ratings and peer reviews show coherent differences with respect to directionality for each discipline except for number of links.

Table 3. Tendency of the metrics for the "good" pages in the categories with significant differences in distribution for comments and peer reviews (C/R)

	Links	Size	Images	PC
Arts	-	-	↑/↑	↑/↑
Business	↓/↓	↓/↓	-	↑/↑
Education	↑/↓	↓/=	↑/=	↑/↑
Humanities	↓/↑	-	↑/↑	=/↑
Math.& Stat.	↑/↑	↑/↑	↑/↑	↑/↑
Science & Tech.	↑/=	↓/↓	↑/↑	↑/↑
Social Sci.	↓/=	↓/↓	=/↓	↑/↑

From the above, it seems that the number of images can be considered a quality indicator for most disciplines, and also it becomes apparent that there are differences between disciplines, for example, Math and Statistics resources of high quality are in general larger and with more links, while the opposite occurs in the Business Discipline.

An alternative analysis of the influence of parameters can be done with the help of decision trees. For example, Figure 2 shows a decision tree for the Art discipline. The random decision tree was generated by discretization of the comment ratings in to the three categories showed in Table 1, and preprocessing data with a distance-based outlier detection. As can be appreciated in the Figure, the main criterion the algorithm chooses for the decision is the number of bookmarks in personal collections (pc). For objects that are not included in collections (left branch of the tree), the number of images determines the "poor" (*below average*) category, and to some extent the "average" and "good", with seven images as a rough frontier. For objects included in personal collections, it is that number of bookmarks that determine the category, but the distinctions are not clearly delimited, so that a sharp decision cannot be made, possibly since there are other attributes that need to be added to the model.

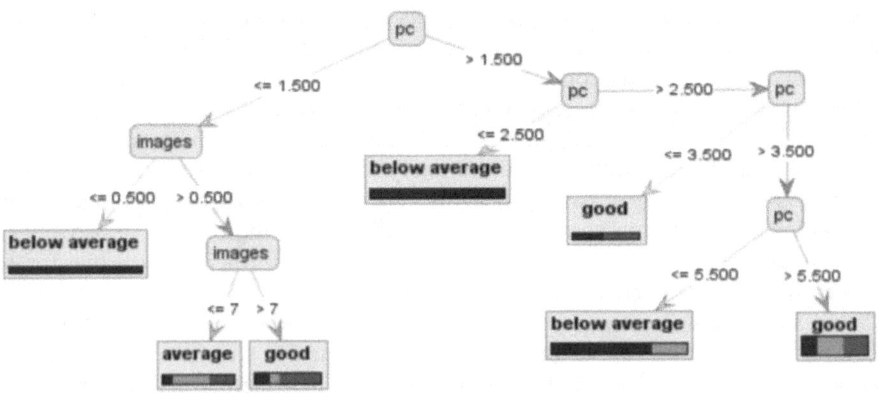

Fig. 2. An example model based on a decision tree for the Art discipline

Decision trees or rules learned from data are the final classification mechanisms implementing the automated analysis, however, the results of the preliminary exploration reported here reveal that further analysis is needed, possibly including additional metrics.

The analysis described in this paper has several important limitations that need to be considered. First, the bias towards high rankings may be a manifestation of the fact that "only good resources are good the effort of reviewing", in which case, the real "poor" category may be lying in non-rated objects. However, the absence of ratings can also be attributed to other issues (e.g. objects that have been contributed recently are still not rated), which makes that analysis difficult. Another limitation is the uneven distribution of ratings among objects. While this is not an issue for peer reviews (as it is uncommon that an object has more than a couple of them), the distribution shows a "long tail" of objects with few reviews and some of them having much more. Figure 3 shows the distribution of number of comments related to the rank (being the lower rank the object with more comments) for one of the categories.

Another limitation is the heterogeneity of the format of learning objects, as some are simpler, low granularity elements, while others are actually link collections. This limitation might be overcome dividing the collections of objects using the granularity metadata element, if this metadata element would be available and accurate for a significant number of objects.

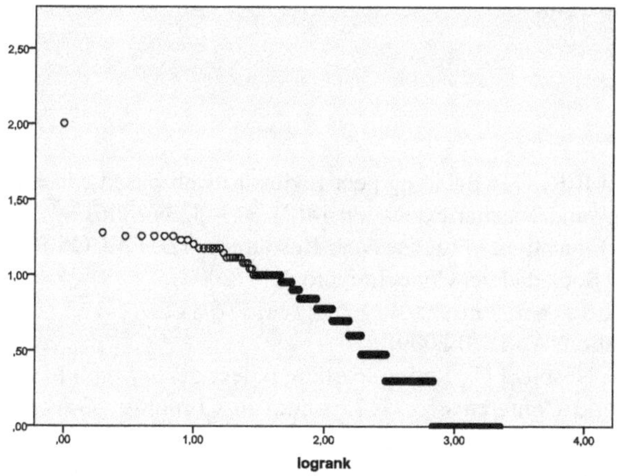

Fig. 3. Scatterplot of rank of learning objects and number of comments in log scale for the *Science and Tech.* category

5 Conclusions and Outlook

The growth of learning object repositories has opened new opportunities to gain insight on what learning object quality is from an empirical perspective. Concretely, repositories as MERLOT provide reviews and ratings that can be used to seek for statistical profiles characterized learning objects that are considered of good quality.

This paper has reported a preliminary study on some basic metrics that could be used to develop such an statistical profile, based on the analysis of reviewed and rated learning objects in MERLOT, starting from measures that were used in previous studies on the automated analysis of the usability of Web sites. The inclusion of the learning objects in the bookmark collections of the users has been found to be an important quality predictor. Number of images is a second candidate, but it is less clearly relevant, and its interpretation is not consistent in the case of the discipline of social sciences. The other metrics have not been found to have different interpretations depending on the discipline, so that further analysis is required.

Future work will expand the present study to cover more metrics that can be automatically derived from the contents of the learning object or from its metadata, and it will use data from other repositories to contrast the findings. If significant profiles are ultimately found, the next step would be that of constructing an analysis tool that uses these profiles to get a priori assessments of the quality of learning objects, which may be useful as an inexpensive evaluation prior to more time and effort-consuming evaluation techniques.

Acknowledgements

The results presented in this project have been partially funded by the University of Alcalá and the CAM (Comunidad de Madrid), as part of project MARIA, code CCG08-UAH/TIC-4178.

References

Cafolla, R.: Project MERLOT: Bringing peer review to web-based educational resources. Journal of Technology and Teacher Education 14(2), 313–323 (2006)

Guntram, G.: Open Educational Practices and Resources: The OLCOS Roadmap 2012. Revista de Universidad y Sociedad del Conocimiento 4(1) (2007), `http://www.uoc.edu/rusc/4/1/dt/eng/geser.pdf` [Date of consultation: April/10/2009]

Han, K., Kumar, V., Nesbit, J.C.: Rating learning object quality with bayesian belief networks. In: E-Learn: World Conference on E-Learning in Corporate, Government, Healthcare, & Higher Education, Phoenix, AZ (2003)

Ivory, M.Y., Hearst, M.A.: The state of the art in automating usability evaluation of user interfaces. ACM Computing Surveys 33, 470–516 (2001)

Ivory, M.Y., Hearst, M.A.: Statistical profiles of highly-rated web sites. In: Proceedings of the SIGCHI Conference on Human Factors in Computing Systems: Changing Our World, Changing Ourselves CHI 2002, Minneapolis, Minnesota, USA, April 20 - 25, pp. 367–374. ACM, New York (2002)

Kelty, C.M., Burrus, C.S., Baraniuk, R.G.: Peer Review Anew: Three Principles and a Case Study in Postpublication Quality Assurance. Proceedings of the IEEE 96(6), 1000–1011 (2008)

Lundgren-Cayrol, K., Marino, O., Paquette, G., Léonard, M., de la Teja, I.: Implementation and Deployment Process of IMS Learning Design: Findings from the Canadian IDLD Research Project. In: Proc. of the IEEE International Conference on Advanced Learning Technologies 2006 (ICALT 2006), pp. 581–585 (2006)

McGreal, R.: A Typology of learning object repositories. In: Adelsberger, H.H., Kinshuk, Pawlowski, J.M., Sampson, D.G. (eds.) International Handbooks on Information Systems. Springer, Heidelberg (2008)

Nash, S.S.: Learning objects, learning object repositories, and learning theory: Preliminary best practices for online courses. Interdisciplinary Journal of Knowledge and Learning Objects 1, 217–228 (2005), http://ijklo.org/Volume1/v1p217-228Nash.pdf

Ochoa, X., Duval, E.: Relevance Ranking Metrics for Learning Objects. IEEE Trans. Learn. Technol. 1(1), 34–48 (2008)

Ochoa, X., Duval, E.: Quantitative Analysis of Learning Object Repositories. In: Proceedings of World Conference on Educational Multimedia, Hypermedia and Telecommunications 2008, pp. 6031–6048. AACE, Chesapeake (2008)

Sicilia, M.A., García-Barriocanal, E.: On the Concepts of Usability and Reusability of Learning Objects. International Review of Research in Open and Distance Learning 4(2) (2003)

Sicilia, M.A., García-Barriocanal, E., Pagés, C., Martínez, J.J., Gutiérrez, J.M.: Complete metadata records in learning object repositories: some evidence and requirements. International Journal of Learning Technology 1(4), 411–424 (2005)

Vargo, J., Nesbit, J.C., Belfer, K., Archambault, A.: Learning object evaluation: Computer mediated collaboration and inter-rater reliability. International Journal of Computers and Applications 25(3), 198–205 (2003)

Vuorikari, R., Manouselis, N., Duval, E.: Using Metadata for Storing, Sharing and Reusing Evaluations for Social Recommendations: the Case of Learning Resources. In: Social Information Retrieval Systems: Emerging Technologies and Applications for Searching the Web Effectively, pp. 87–107. Idea Group Inc., New York (2008)

Wiley, D., Wayers, S., Dawson, D., Lambert, B., Barclay, M., Wade, D.: Overcoming the limitations of learning objects. Journal of Educational Multimedia and Hypermedia 13(4), 507–521 (2004)

A Semantic Web Based System for Context Metadata Management

Svetlin Stefanov[1] and Vincent Huang[2]

[1] DSV - Department of Computer and Systems Sciences,
Forum 100, SE-164 40 Kista, Sweden
smst@kth.se
http://dsv.su.se/en
[2] Service Layer Technologies, Ericsson research,
Färögatan 6, SE-164 40 Kista, Sweden
vincent.a.huang@ericsson.com
http://www.ericsson.com

Abstract. With the increasing usage of embedded systems and sensors in our surroundings, a new type of information systems – context aware systems – are gaining importance. These user-centric systems acquire context information which describes the state of the user and the user environment, and offer adaptable and personalized services based on the user context information. The central part of a context aware system is the context model used for describing user context information. The context information originates from a multitude of heterogeneous sources, such as personal calendars, sensors attached to the users or to the user's environment and Web based sources, such as social networking sites. The information from these sources is typically on different abstraction levels and is organized according to different data models. This work proposes a Semantic Web based context metadata management system. The first part of the work develops an ontology model for user context. The user context model integrates information from multiple and heterogeneous sources which are modeled largely by reusing existing well accepted ontologies. The second part of the work proposes a method to infer and reason about additional user context information based on the available context information using rules and ontologies. We instantiate and evaluate the proposed system by performing a social networking case study called `meetFriends`. In this application, information is collected from various sources such as sensors attached to the users and public web sources - YellowPages. The moods of the users are inferred from a set of rules. A meeting between two users can be set up based on the moods, locations and preferences of the users. The results indicate that Semantic Web technologies are well suited for integrating various data sources, processing of user context information, and enabling adaptable and personalized services.

Keywords: Semantic Web, Context Aware Systems, Knowledge Management, Ontology Engineering, Ontology Reuse, Inference engines, Reasoning, Metadata.

F. Sartori, M.Á. Sicilia, and N. Manouselis (Eds.): MTSR 2009, CCIS 46, pp. 118–129, 2009.

1 Introduction

With the increased usage of mobile devices, such as cell phones and PDA's, pervasive computing is gaining popularity. The term *pervasive* (or ubiquitous/ubicomp), refers to the seamless integration of devices into people's lives [1]. A major part of pervasive computing deals with the development of context aware systems. These systems acquire user and environment context information from various sensors (e.g. location or temperature sensors) and offer adaptable and personalized services based on the acquired context information. The architecture of a such system has been proposed at Ericsson [2]. The system defines enablers for sensor and actuator based services in the consumer space by introducing a broker and data processing entity. The broker is referred to as the M2M (Machine-to-Machine) support provider, between service providers and individual or public sensor network providers. The data processing entity collects information from different information sources and provides high level context information to applications.

A central part of each context aware system is the context model. The context model is used for structuring context information. The context information originates from physical (e.g. location sensor in room), virtual (e.g. digital phone book), and/or logical (combination of physical/virtual) data sources [3].

In [4] various types of context models are discussed. Ontology based models are identified as the optimal context modeling approach for context aware systems and this is the approach we follow in this work. While various ontology representation languages exist, aiming for system interoperability and for compliance with well-accepted standards, we use W3C's recommendation for representation languages, namely W3C Semantic Web Activity [5].

Our work extends the current system to provide support for virtual and logical data sources. Semantic web technologies are used to obtain extended, reusable and machine-understandable context information and knowledge. The context information and knowledge will enable the provision of more sophisticated, adaptable and personalized applications and services, bridging physical, virtual, and logical data sources.

The rest of the paper is organized as follows. In section 2, we review other context aware systems using Semantic web technologies. Section 3 is devoted to the contribution of this work – A Semantic web based system for context metadata management. In section 4, we evaluate the proposed architecture with a case study. Finally, we end with discussions and suggestions for future work in section 5.

2 Related Work

An extensive account of different types of context aware systems is presented in [6]. The systems differ in among others the data model, used for representing the context model and in the way they process the context information. Some distinctive characteristics of Semantic Web-based context aware systems are the ontology based context model, and the usage of logical inference (reasoning) for context processing.

One of the first context aware systems is introduced in [7]. It has a broker agent architecture. The context information of all parties constituting a context aware system, is centrally managed by a context broker agent. The broker agent uses OWL ontology [8] for representing the context information. For inferencing, the system uses the object-oriented rule based reasoning language Flora-2, providing ontology (OWL Lite) and rule (RDF and OWL) based reasoning.

In [9], the first service oriented context system is introduced. The context model ontology has four general classes: CompEntity, Activity, Person, and Location. CompEntity class has subclasses: Device (e.g. TV, CellPhone), Application, Agent, etc. Activity class has two subclasses: ScheduledActivity (e.g. Party) and DeducedActivity (e.g. Dinner). Location class is further divided into IndoorSpace (e.g. Room), or OutdoorSpace (e.g. Garden). Some general properties are `use` (between Activity and CompEntity), `own` (between Person and CompEntity), `locatedIn` (e.g. between Person and Location), and `engagedIn` (between Person and Activity). The system uses ontology and rule based reasoning, and makes inferences about users' locations and current activities.

Another system, implementing a hybrid peer-to-peer architecture, is introduced in [10]. It consists of three peer types: ContextSource (having ContextWrapper and ContexReasoner), ContextProvider (connecting ContextSource and context aware applications) and UserManager (storing privacy information about users, devices, and applications). The context model is distributed between different ContextSources, but the authors do not provide any detailed description of the system. The system uses OWL DL reasoner for making inferences about context information.

In [11] another broker based system is proposed. There are three main parts: Repository – for storing the context model and related rules, Context manager – having Context broker able to reason with run-time and historical context information, and Agent – using the context information, and having its local context manager, which communicates with the Context manager about the Agent's context information needs. The context model has three main classes: User profile (divided into Domain dependent and Domain independent profiles), Environment (subdivided into Location, Ambient condition, Time, and Actions), and Channel (Device, Network, Application protocol, and Network interface). The authors do not present the details about the reasoning/inference functionality.

All systems use ontology based context model, often expressed in OWL DL. The context models are sometimes application specific (e.g. IndoorSpace / OutdoorSpace); that leads to a limited reuse of context information. The context models are always developed from scratch, even though they have common concepts, such as `Person`, `Location`, or `Time`; these common concepts on the other hand can be reused. Another observation is that few systems make use of both ontology and rule based reasoning. Yet another observation is that the systems offering rule based reasoning do not make use of standard rule representation languages (such as SWRL).

3 Semantic Web Based System for Context Metadata Management

This section contains the proposed semantic web based context management system. We first present the system architecture by describing the functionalities that support context storage, processing, and retrieval. Then we describe our ontology based context model which integrates physical, virtual, and logical data sources and reuses existing ontologies.

3.1 System Architecture

The section describes the system architecture, enabling the management of ontology based on user context information and knowledge. We propose the following architecture (Figure 1). The modules make use of Semantic web technologies.

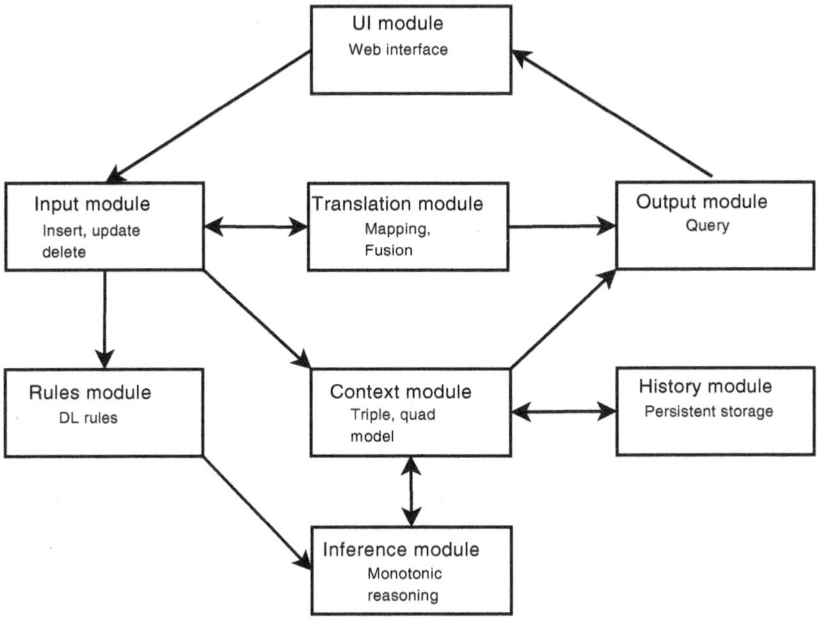

Fig. 1. System architecture

Context module. The module creates triples and so called quad representation (discussed below) of the user context model (described in section 3.3). The triple context model is based on standard RDF triples. The quad context model is based on named graphs. Named graphs [12] extends the traditional RDF graphs, by adding a fourth element to the RDF triples – the name of the RDF graph. We use the quad model for expressing propositions and beliefs about user context information. For example a user's location information at a certain point of time could originate from both GPS and RFID sensors. By modeling the location information from these sensors with two named graphs we are able to select more accurate data (e.g. RFID). The quad model also enables the implementation of user privacy and context information security, because each RDF triple can be signed with a key issued by a certificate authority. The key can be modeled with another named graph.

History module. The module communicates with Context module and provides relational database storage and retrieval services for both triple and quad models. The temporal nature of history data enables various types of processing, such as statistical processing and trends analysis. Another usage of the history data is in the cases of missing current user context information.

Rules module. The module manages the storage of rules and supports two DL rule formats: SWRL and Jena rules. These could be both user and application specific. SWRL is supported by popular DL reasoners, such as Pellet and Racer[13], but one disadvantage is the inability to represent non-monotonic inference. To overcome the problem, we additionally choose to support Jena rules, allowing for invalidating assertions.

Inference module. The module handles communication between the Context and Rules modules and provides reasoning services on top of the triple based context model. It support monotonic reasoning with Pellet, Jena and other DIG[14] compatible reasoners, such as RacerPro. Our system supports both ontology and rule based inference. The reasoning service can be initiated in two modes: explicitly or dynamically, upon query requests.

Input module. The module performs insertion, update and deletion of user context information. It works in the triple based representation (standard RDF) of context information. An update is realized with a sequence of deletion and insertion operations. The module works in push mode, which means that it provides an interface for its methods to applications/services to manipulate the context information.

Output module. The module queries the context model and provides context information. It supports the SPARQL query language and two types of query execution: Pellet and Jena based. Pellet based query execution allows for reasoning upon querying and Jena based query execution supports the explicit reasoning. The module works in pull mode, which means that its functionality is triggered by applications/services that need to retrieve context information.

Translation module. The model contains functionality for changing the format of context related data and information. A change of representation occurs when we want to store context information in RDF/OWL and it is available in other format, such as HTML.

User interface module. The module provides web based interface to the context model. It enables context management operations such as insertions, updates, and deletions, or context retrieval.

3.2 Module Interactions

The main interactions among the modules are shown in Figure 2.

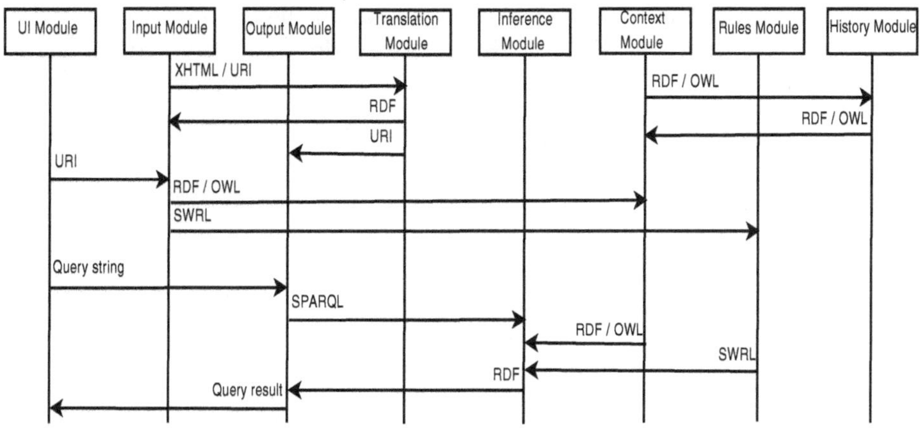

Fig. 2. Interactions Diagram

The user context and reasoning rules are passed to the Context Module and Rules Module respectively through the Input Module. Inference Module infers new information using the existing user context and rules from Context Module and Rules Module. The inferred information then will be passed to the Output Module. The interactions are listed as the following:

- UI module passes URI to Input module.
- UI module passes query to Output module and gets query results.
- Input module passes XHTML/URI to Translation module and gets RDF.
- Translation module passes SPARQL query prefixes (URI) to Output module.
- Input module passes SWRL rules to Rules module.
- Input module passes RDF/OWL to Context module.
- Inference module gets SWRL rules from Rules module.
- Inference module get RDF/OWL from Context module.
- Context module get RDF/OWL from History module and passes RDF/OWL to History module.
- Inference module passes RDF query results to Output module.

Compared to the works outlined in section 2, the proposed architecture provides several benefits. Translation module enables integration and reasoning with other data format than RDF data (e.g. HTML data). Context module provides dual representation of context information, making it possible, among others, to express beliefs about statements (RDF triples) and data privacy and security. Rules module supports multiple rule representation languages (such as Jena and SWRL rules), making it possible to use different reasoning engines. The latter are supported by Inference module, so that one can select a reasoning engine that is most appropriate for the task at hand.

3.3 Context Model

Our context model has layered structure, similar to the one introduced in [15], where there are four types of ontologies: top-level, domain, task, and application

Fig. 3. User context model

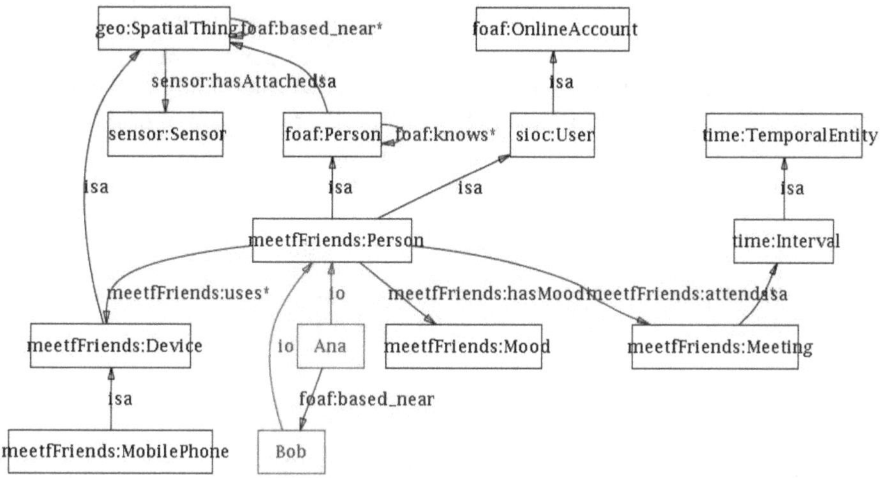

Fig. 4. Application ontology

ontology. The context model has several top layer ontologies, such as `wordnet` and `tzont`. Some domain ontologies are `foaf`, `sioc`, and `sensor`. Finally, there is one application ontology – `meetFriends`. Figure 3 shows a part of the *IS-A* relationship hierarchy of the context model. The upper part, e.g. `Sensor` and `geo` ontologies, model the physical aspects of user's context. The lower part, e.g. `foaf` and `sioc` model user's virtual aspects, such as usernames and web sites. We developed the context model by following the below described iterative steps.

`Step 1.` Having the general goal – the model needs to represent the context of the user, and having a simple use case – to derive user's mood (a logical data source) from available context information, originating from physical and virtual sources, we identified four relevant domains: time, space, social networking and sensors.

`Step 2.` For representation language we choose OWL DL, since we intend to reason with the information stored in the context model. We found relevant ontologies, representing time, space, and social networking ontologies. For time and space we use W3C's time, timezone and geo_pos ontologies, since they are available online, they are well documented and other systems are already using them. Besides, `timezone` is integrated in `time`, and `geo_pos` is integrated in `timezone`. For social networking domain we identified two popular ontologies: `foaf` and `sioc`. The latter is already integrated in the former. Most of these ontologies were represented in OWL Full. We converted these ontologies to OWL DL, by using an OWL validator [16].

`Step 3.` Because we were unable to find appropriate sensor and mood ontologies, we developed them from scratch. We then applied four integration operators, introduced in [17] and exemplified below. That resulted in the final, modular and integrated ontology. Figure 4 shows the application ontology for one of our social networking applications – `meetFriends`, described in details in section 4. We applied the specialization operator on `meetFriends:Person`, so it inherits the properties of `foaf:Person` and `sioc:OnlineAccount`. We took most of the concepts from the identified ontologies as-it-is and modified/extended some of them. For example, in our Sensor ontology we extend geo_pos ontology's `SpatialThing` with one extra relation – `hasAttached`. In such a way we make it possible for the subclasses of `SpatialThing`, such as `meetFriends:Device` and `meetFriends:Person`, to have a `meetFriends:Sensor` attached.

`Step 4.` We deployed the ontology, and evaluated its applicability with a social networking case study (section 4).

4 Case Study

We implemented the proposed system, by integrating several open source Java APIs. The majority of modules are implemented on top of the Jena API. Additionally, in the Inference module, we use both the Jena and Pellet API to implement the reasoning service. In Context module and History module we use Jena and Named Graphs API to implement triple and quad support. Visualization module is implemented with Apache Tomcat [18] API. Finally for designing the context model, we used Protégé editor.

Having implemented the system, in order to evaluate its design, we developed a test application – `meetFriends`. The application suggests meetings for people who are in a positive mood and live near by one another. The application uses data from both physical and virtual data sources, and utilizes the functions of the modules in our proposed architecture (section 3.1).

The physical sources are various types of simulated sensors, such as temperature and heartbeat sensors. These are either attached to the user (heart beat sensor), or are present in the user's environment (temperature sensor at the user's location). We use the data originating from these sensors to infer the user's mood (a logical data source). There are three different moods: positive, negative, and neutral. Each of these is determined by a model, defined with a rule. The inference of user's mood from sensor readings is a challenging problem, and is a subject of active research [19]. We do not attempt to address any of the described challenges, such as sensing and recognizing emotion or affect modeling, but rather to test Inference, Context, and Rules modules with a realistic number of preconditions. Additionally, in order to test Input, Translation, and Output modules, we use live data originated from a virtual data source, the public database Eniro [20]. The available live data are users' location coordinates, addresses and phone numbers.

After populating the context module with the above described data, we successfully inferred meeting possibility between users that are near by and are in positive mood. We thus come to the following conclusions. The context model integrates seamlessly and well physical, virtual, and logical data sources. With an uniform representation of these data sources, we were able to perform real-time ontology and rule based reasoning, allowing for the creation of context-aware and personalized services. The proposed architecture functions well for the task at hand. One distinctive feature of our context model is its three layered structure, making it possible to add and remove new domain and application ontologies. Another distinctive feature is the reuse of existing and well accepted ontologies. That leads to two benefits, namely, decreased ontology development time and interoperability with systems that make use of these ontologies. That becomes increasingly important with the wide spread of social networking sites, such as Facebook and Flickr, and the availability of tools that export user profile and social network data to FOAF and SIOC annotated data. Yet another distinctive feature of our model is the integration of live sensor data with online data in real time. That was possible due to our Translation and Input modules, where the former converts HTML data from Eniro to RDF/FOAF and the latter inserts the resulting data into our context model.

While carrying out the case study we faced several challenges. One challenge is performing of calculations from a rule's body (e.g. calculating the distance between two locations). That is not possible, due to limitations of the existing OWL DL reasoners, supporting SWRL. We expect this limitation to be addressed with the development of the reasoning technology. Another challenge is the inability to invalidate inferred statements. For example once the user status is inferred to be `busy`, it is impossible to execute a rule that sets the status to `free`. One solution to the problem is to avoid storing interred information. If that is not possible, another solution is the usage of non-monotonic reasoners in combination with the context model. Yet another challenge is identifying a suitable update model. That is, deciding on when to update the context model. It is a complex problem, because different data sources have different refresh rates, thus there might be different cases. It might be that we do not necessarily need the latest information, and therefore we need not to update. In other cases, it might be that the context information we need is not valid (being old), or not existing (e.g. user's current location). For example it is not trivial to determine whether we were unable to infer certain information (e.g. user's mood) because the rule's body atoms are invalid or because there is no current information. One solution to the problem is the development of a validity model of the context information, and a update functionality (e.g. as part of input module), which can populate the context model in push, pull, or a combination of these modes.

5 Conclusions and Future Work

Our work was motivated by the question: How to integrate data originating from physical, virtual and logical data sources, by using Semantic Web technologies to obtain extended, reusable and machine-understandable context information and knowledge? We answered this question in Section 3.3, where we propose an ontology-based context model, integrating physical, virtual and logical data sources. Having designed and developed the context model, we used it in our proposed context management architecture (section 3.1), to create extended (due to the interference functionality), reusable (due to the usage of standard representation language), and machine-understandable (due to the high level of formality of the used representation language) information and knowledge. We evaluated the proposed solution in section 4, where we concluded that the context model integrates seamlessly and well physical and virtual data sources. The model provides an uniform view on the data sources, based on which we were able to perform real-time ontology and rule based reasoning, and create context-aware and personalized services. The proposed architecture fulfills our requirements and functions well. Furthermore, we believe that the proposed architecture can be used not only in context aware systems, but also in any other system that uses personalized and machine-understandable user information (e.g. IR system offering personalized search).

We leave for future work throughout performance and scalability evaluation of the proposed architecture. This is a non-trivial problem, due to among others the different types of queries (e.g. queries targeting readily available information, such as cell phone number, and queries targeting information that needs to be inferred, such as user mood), different types of ontologies (e.g. "deep" ontologies, where the subclass hierarchy is long, or shallow ones, where there are many sibling classes on certain abstraction level), different sizes of Tbox and Abox in these ontologies, and different possibilities of combinations with inference engines – Pellet, RacerPro or Jena reasoners among others.

References

1. Weiser, M.: The computer for the 21st century. SIGMOBILE Mob. Comput. Commun. Rev. 3(3), 3–11 (1999)
2. Huang, V., Johansson, M.: Usage of semantic web technologies in a future m2m communication system. In: Proceedings of the 1st European Semantic Technology Conference, Vienna, Austria (May 2007)
3. Indulska, J., Sutton, P.: Location management in pervasive systems. In: ACSW Frontiers '03: Proceedings of the Australasian information security workshop conference on ACSW frontiers 2003, Darlinghurst, Australia, pp. 143–151. Australian Computer Society, Inc. (2003)
4. Strang, T., Popien, C.L.: A context modeling survey (September 2004)
5. W3C semantic web activity (2008)
6. Baldauf, M., Dustdar, S.: A survey on context-aware systems. Tech. rep., Technical University of Vienna (2004)
7. Chen, H., Finin, T., Joshi, A.: Semantic web in the context broker architecture. In: PERCOM 2004: Proceedings of the Second IEEE International Conference on Pervasive Computing and Communications (PerCom 2004), Washington, DC, USA, p. 277. IEEE Computer Society, Los Alamitos (2004)
8. Chen, H., Finin, T., Joshi, A.: An ontology for context-aware pervasive computing environments. Special Issue on Ontologies for Distributed Systems, Knowledge Engineering Review 18, 197–207 (2003)
9. Gu, T., Pung, H.K., Zhang, D.Q.: A service-oriented middleware for building context-aware services. J. Netw. Comput. Appl. 28(1), 1–18 (2005)

10. van Kranenburg, H., Bargh, M.S., Iacob, S., Peddemors, A.: A context management framework for supporting context-aware distributed applications. IEEE Communications Magazine 44(8), 67–74 (2006)
11. Cappiello, C., Comuzzi, M., Mussi, E., Pernici, B.: Context management for adaptive information systems. In: Proceedings of the First International Workshop on Context for Web Services (CWS 2005), pp. 69–84. Elsevier B.V., Amsterdam (2006)
12. Carroll, J.J., Bizer, C., Hayes, P., Stickler, P.: Named graphs, provenance and trust. In: WWW 2005: Proceedings of the 14th international conference on World Wide Web, pp. 613–622. ACM Press, New York (2005)
13. RacerPro is an owl reasoner and inference server for the semantic web (2008)
14. DIG 2.0: The DIG description logic interface (2008)
15. Guarino, N.: Formal ontology and information systems. In: Proceedings of FOIS 1998, pp. 3–15. IOS Press, Amsterdam (1998)
16. OWL ontology validator (2008)
17. Pinto, H.S., Martins, J.P.: A methodology for ontology integration. In: K-CAP 2001: Proceedings of the 1st international conference on Knowledge capture, pp. 131–138. ACM, New York (2001)
18. Apache tomcat (2008)
19. Picard, R.W.: Affective computing: challenges. Int. J. Hum.-Comput. Stud. 59(1-2), 55–64 (2003)
20. Eniro.se - sök företag, kartor, personer, nummer (2008)

An XML Pipeline Based System Architecture for Managing Bibliographic Metadata

Johannes Textor[1] and Benjamin Feldner[2]

[1] Institut für Theoretische Informatik
Univeristät zu Lübeck, Ratzeburger Allee 160, Germany
textor@tcs.uni-luebeck.de
[2] Institut für Multimediale und Interaktive Systeme
Univeristät zu Lübeck, Ratzeburger Allee 160, Germany

Abstract. In our knowledge-based society, bibliographic metadata is everywhere. Although several metadata standards for bibliographic information have been developed and established by the professional librarian community, home-grown ad-hoc solutions are still widespread in small to medium-sized institutions. This paper presents a framework for storing, indexing, and browsing bibliographic metadata that is designed to lower the barrier for metadata standard adoption by facilitating legacy data import and integration into existing infrastructure. These goals are achieved using XML pipelines as a central design paradigm. As a practical use case, we discuss the implementation of the described architecture at a research institute in our university, where it is now in productive use for managing publication lists and the local library.

1 Introduction

Librarians have for many centuries had a unique role in our access to knowledge. At the core of any larger library's infrastructure for storage, indexing, and retrieval of bibliographic items, there is a *bibliographic metadata scheme* that defines structure and semantics of bibliographic resource descriptions. Over the last decades, bibliographic metadata schemes have evolved to adapt to the increasing possibilities and challenges that come with the progress of information technology. Projects like the WorldCat (`www.worldcat.org`), an online catalogue that at the time of writing contains data from more than ten thousand libraries worldwide, impressively demonstrate the key role of metadata standards in the successful transition of librarianship to the age of the Internet.

However, with the increasing accessibility of knowledge, librarians are no longer the only ones who have to cope with bibliographic data. For example, most of us who work at research institutions deal regularly with quite large and complex bibliographies such as publication lists or local paper repositories – and often, we seem unable to build upon the work of librarians, and instead create our bibliography management systems from scratch. Let us illustrate this with an example: At the time of writing, there are 12 institutes in the computer science section of our university, all of which publish publication lists on their websites.

F. Sartori, M.Á. Sicilia, and N. Manouselis (Eds.): MTSR 2009, CCIS 46, pp. 130–140, 2009.

Four institutes use *Typo3* modules to manage these lists, and one uses a *Drupal* module; another one uses *bibtex2html*, and the remaining six rely on different home-grown solutions. Only one of the evaluated institutions used a somewhat standardized metadata format (BibTeX). Certainly, this is not caused by a lack of choice: RIS, EndNote, and Dublin Core are just some example formats that most researchers have heard of. And most of us would agree that a larger prevalence of metadata standards and related software systems would yield unquestionable benefits, such as improved searching capabilities and consistency, easier data exchange among institutions, and increased data lifetime.

Clearly, small and medium-sized institutions can in most cases not simply use the software systems designed for libraries, because these would be far too complex and powerful. On the other hand, the heterogeneity in both format and quality of legacy data that these institutions face is often even larger than in the case of libraries: If an institution's bibliographic database, consisting of thousands of entries, has been managed for several years by the secretary using a Microsoft Excel document, an off-the-shelf software product won't be able to import this data "out of the box" – which raises the barrier of adoption: In sight of having to invest several hours for manual data input, one might well decide to just continue using the Excel file instead of switching to a metadata-based solution. In other words, two important factors that inhibit the spread of bibliographic metadata standards in small to medium institutions are: (1) a lack of user-friendly and easily customizable data im– and export facilities in current software, resulting in (2) a prohibitively high cost of adoption.

In this paper, we describe the design and implementation of a bibliographic metadata management system with the following main design goal: to make import of legacy data and integration into existing environments as easy as possible. We demonstrate how this goal is achieved using an XML-centric approach, based on the Dublin Core and OpenURL standards:

- We integrate an architectural pattern developed by the librarianship community [1], which addresses the common problem of metadata interoperability, into a consistent framework for metadata storage, indexing, searching, and presentation.
- Both for creating new data importers and new data presentation styles (views), only knowledge of XSLT is required. Using the XML pipeline pattern, existing importers and views can be reused, fine-tuned, and plugged together, resulting in a high level of modularity.

To set the scene, we first give a brief review of related work in the following section. Next, we define the core XML format that our solution uses to describe bibliographic resources (Sec. 3), and then give an overview of the system's architecture (Sec. 4). Our prototype implementation, which is now in productive use in our institution, is discussed in Sec. 5. The paper concludes with a preliminary evaluation of our work and perspectives for future research (Sec. 6).

2 Background

Two main questions have to be addressed when approaching the problems highlighted in the previous section: How do we import metadata from heterogeneous sources into a common format, and how do we provide powerful searching, indexing, and formatting capabilities that make these data practically useful? Consequently, there are two main areas of related work: Interoperability of bibliographic metadata, and software systems that manage bibliographic data.

Bibliographic data management software can be grouped in two categories. *Integrated library systems*, such as *Koha* (`koha.org`), provide not only means for indexing and searching bibliographic resource descriptions, but also support the administrative tasks of a large library: Acquisitions management, customer relationship management, billing, and so on. The typical smaller institution needs only a small fraction of these features. On the other hand, *reference management systems* are used by individual researchers or research groups to manage databases of scholarly literature. In this category, we find software such as the commercial *EndNote* and the Open Source *RefDB* (`refdb.sourceforge.net`). RefDB in particular has many properties of our desired solution. However, it relies on external software such as *bibutils* and custom Perl scripts to import bibliographic data into its own format, RIS. Likewise, UNIX programming skills are needed to adapt RefDB's output to, say, a corporate identity design specification.

The subject of metadata interoperability has been broadly addressed by the librarian community. One main problem is defining a *crosswalk* [2], that is, deriving a set of mappings between semantically equivalent elements of the source and target metadata schemes. For sufficiently complex schemes, this can be a highly nontrivial task, which requires attention of a metadata expert. However, this intellectual aspect of the problem will not be addressed in this paper, as we expect to be dealing with metadata of comparably low quality and granularity. We are more interested in the software engineering side of this problem: How can a crosswalk, once constructed, be implemented efficiently and with minimal programming effort?

In a recent paper, Godby et al. [1] introduced a design pattern for crosswalk implementation, which is used at the Online Computer Library Center (OCLC), an online resource that integrates catalogues from different businesses and institutions. The key idea of their solution is to separate syntactical from semantical aspects of the conversion task, so that the metadata expert can focus on the latter. For example, suppose we want to convert a record from the plain-text MARC format to BibTeX. A *MARC reader* component first converts the MARC record to an XML file consisting of *field* and *value* elements. Then, an XML transformation is applied that renames and moves the input elements, and produces an output XML file with the same syntax, but the semantics of BibTeX. This file is fed to a *BibTeX writer*, which takes care of the special encoding rules that TeX requires.

Our work builds on the pattern of Godby et al., and integrates it into a complete system architecture for metadata management. We think that this holistic approach is crucial to achieve wider metadata adoption: Consistent *presentation* of metadata, e.g. on intranet and internet websites, is no less important than metadata standard interoperability in the small to medium-sized institutional setting. This aspect is usually not addressed in librarianship literature (see Tennant [3] for an exception), because it is more a technical than an intellectual challenge. However, we will show that from a software engineering perspective, both problems can be addressed in a consistent manner by means of the XML pipeline architecture.

3 XML Metadata Format

In this section, we define the XML-based metadata format that is used to describe bibliographic resources in our system. Basically, we use the Dublin Core (DC) Metadata Initiative's metadata element set [4] with both some extensions and some simplifications. Following the classification of bibliographic metadata schemes given by Chan and Zeng [5,6], our format is a *switching-across-schema* to which we convert all input data, and from which we generate all presentation formats like HTML and PDF. To represent DC in XML syntax, we use the standard RDF/XML encoding. Because the XML documents are used for communication between our system's internal components, we simplify the syntax by omitting all namespace declarations. This considerably eases writing custom stylesheets to transform the data and saves processing time. For data export to strict RDF/XML, an output filter that adds the correct namespaces can be easily implemented (see below).

Every DC element can contain the attributes *type* to encode DC element refinements (e.g. the subtype *created* of the *date* element [7]), *scheme* to define content encoding schemes (e.g. *HTML* for the *description* element or *ISBN-13* for *identifier*), and *lang* to indicate the language code of elements such as *title* and *rights*. Each element can be omitted or used several times. As recommended by the DCMI, a controlled vocabulary is used for the Element *type* (Fig. 1).

Since our testcase is a system that manages academic literature, we need to augment the Dublin Core to allow for an accurate description of complex citation information, like a reference to an article from a conference proceedings. There was a Citation Working Group at the DCMI to address this problem, which in 2006 by published a definitive guideline for encoding bibliographic citation information that recommends to embed third-party formats within a refined DC *description* element [8]. Following this recommendation, we use the OpenURL framework standardized by ANSI/NISO [9] to encode citation information with a higher granularity. OpenURL metadata can be encoded using either URI key-value pairs or plain XML. We use XML encoding in conformance with the XML schemas from the OpenURL registry [10].

This combination of two metadata standards might appear somewhat odd, even though we are following the DCMI's recommendation. However, this kind of hybrid metadata schema usage is very common in practice. In the terminology of Chan and Zeng [5,6], such a hybrid schema is called an *application profile*,

```
<RDF>
  <description about="235">
    <creator type="AUTHOR">Johannes Textor</creator>
    <creator type="AUTHOR">Juergen Westermann</creator>
    <date type="PUBLISHED" scheme="W3C_DTF">2007</date>
    <description type="BIBLIOGRAPHIC_CITATION"
        scheme="OPEN_URL">
      <proceeding>
        <volume>4628</volume>
        <series>Lecture Notes in Computer Science</series>
        <btitle>6th International Conference on Artificial
            Immune Systems (ICARIS 2007)</btitle>
        <pages>228-239</pages>
      </proceeding>
    </description>
    <identifier
        scheme="URI">http://www.springerlink.com/content/
        [snip]</identifier>
    <publisher>Springer</publisher>
    <title lang="en_EN">Modeling Migration,
        Compartmentalization, and Exit of Naive T Cells in
        Lymph Nodes Without Chemotaxis</title>
    <type>IN_PROCEEDINGS</type>
  </description>
</RDF>
```

Fig. 1. Description of an article in a conference proceedings in our simplified XML syntax. Authors, date of publication, URL, publisher, title, and type are directly expressed in terms of Dublin Core elements, while the information about the proceedings volume and the page numbers are encoded using OpenURL.

because it is created to meet the needs of one specific application – in our case, the need for a higher level of granularity than provided by unqualified DC.

It is interesting to note that the current version of the DC element set defines *bibliographicCitation* as a refinement to the term *identifier*, while the Citation Working Group recommends using the term *description* for such data. We choose to follow the recommendation of the Citation Working Group because the citation does not necessarily contain enough information to uniquely identify the resource (we do not repeat information that is encoded in other elements – such as an ISBN or a title – within the citation information).

4 System Architecture

Conceptually, our system is based on the XML pipeline paradigm from *Cocoon* (Fig. 2), the Apache Software Foundation's Java-based XML presentation framework [11]. Cocoon applications are built from a toolbox of reusable pipeline components that can be plugged together in a Lego-like approach: Generators, transformers, and serializers (Fig. 2). Data is sent through the pipeline in the form of SAX events that describe the structure and content of an XML document. This makes it somewhat cumbersome to write custom generators and serializers, since the SAX API is more optimized for efficiency than for ease of use – however, a more programmer-friendly API like DOM would cause performance problems when processing very large amounts of data. On the other hand, transformers are very easy to write as XSL stylesheets, and Cocoon's philosophy is to provide a large set of generators and serializers out of the box, such that the average user only needs to write transformations. Accordingly, our system provides custom Cocoon components that take care of storing, indexing, searching, and retrieving XML resource descriptions in the format described above. The system's indexing and searching capabilities are provided by Lucene, a Java framework for building search engines [12]. Our components manage the Lucene index and keep it synchronized with the backend XML store (Fig. 3).

The system core manages just two different kinds of objects: *Resources*, which are explicitly identified by their database IDs, and *lists*, which are implicitly identified by a Lucene query, such as `creator:feldner AND date:[1990 TO 2005]`.

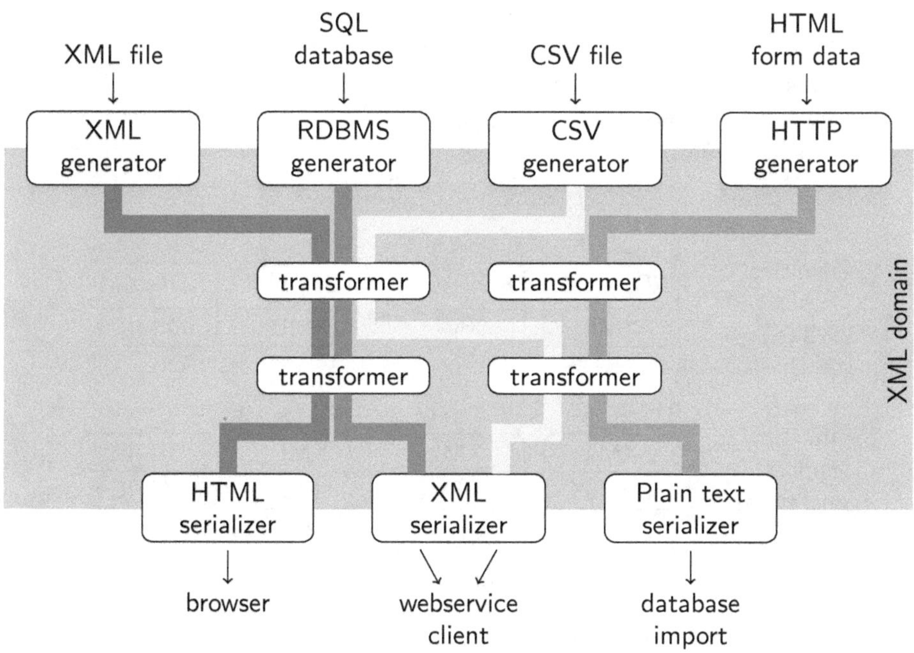

Fig. 2. Illustration of the Apache Cocoon framework: Data from different external sources is fed into an XML pipeline by a *generator* and subsequently modified, extended, and aggregated by *transformers*, which may be written in XSLT or Java. Finally, a *serializer* converts the stream to an output format such as HTML or plain text and delivers the result to the client.

To connect the system's XML-based core to the outside world, we need input and output adapters, which we call *importers* and *views*. Following Cocoon's philosophy, views are created by writing a custom XSL stylesheet that converts the simplified RDF/XML to the desired output format, and combining it with the corresponding serializer (plain ASCII text, XML, PDF, and so on). Since existing pipelines can be used as input sources for new pipelines, one typically does not create a new view from scratch, but rather fine-tune an existing one by adding another stylesheet. Currently implemented views are RDF/XML with namespaces, HTML, and BibTeX.

As stated in the beginning of this paper, importing bibliographic data from external or legacy sources is a crucial function, even if it might be used only once. Following the pattern of Godby et al. [1], we split the import in two

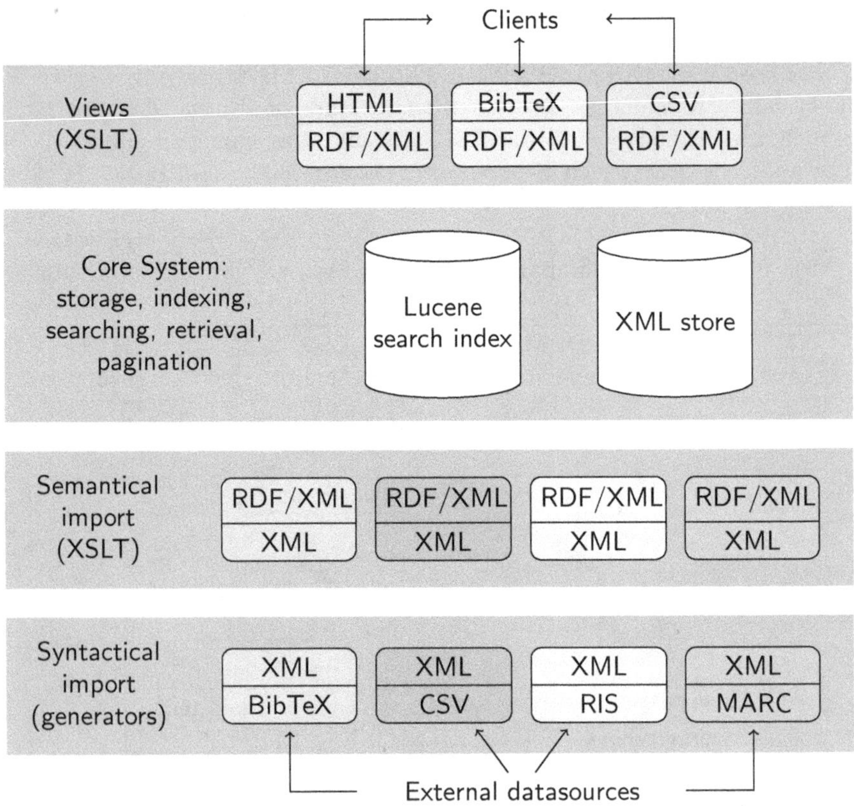

Fig. 3. Overview of the proposed system architecture. The core system uses RDF/XML as its only interface language, and is specialized on efficient storage, searching, and retrieval of documents in this format. Clients will usually not talk to directly to the core system, but use a convenient view instead. Import of existing datasources is split in two stages: First, the data's internal structure is converted to XML (for instance, a CSV file results in a document containing *row* and *column* elements), which is then further processed by XSL stylesheets to produce semantically correct Dublin Core output.

```
<xsl:stylesheet version="2.0"
  exclude-result-prefixes="xs"
  xmlns:xsl="http://www.w3.org/1999/XSL/Transform"
  xmlns:xs="http://www.w3.org/2001/XMLSchema">
  <xsl:template match="csv">
    <RDF><xsl:apply-templates /></RDF>
  </xsl:template>
  <xsl:template match="line">
    <xsl:variable name="title" select="field[position()=3]"/>
    <xsl:if test="$title!=''">
      <description>
        <xsl:variable name="isbn" select="field[position()=1]"/>
        <xsl:variable name="authors" select="field[position()=2]"/>
        <xsl:variable name="publisher" select="field[position()=5]"/>
        <xsl:variable name="year" select="field[position()=6]"/>
        <xsl:variable name="signature" select="field[position()=8]"/>
        <xsl:if test="$isbn!=''">
          <identifier scheme="ISBN_10"><xsl:value-of
              select="$isbn"/></identifier>
        </xsl:if>
        <xsl:for-each select="tokenize($authors,'/')">
          <xsl:if test=".!=''">
            <creator type="AUTHOR"><xsl:value-of select="."/></creator>
          </xsl:if>
        </xsl:for-each>
        <title><xsl:value-of select="$title"/></title>
        <xsl:if test="$publisher!=''">
          <publisher><xsl:value-of select="$publisher"/></publisher>
        </xsl:if>
        <xsl:if test="$year␣castable␣as␣xs:integer?">
          <date type="PUBLISHED" scheme="W3C_DTF"><xsl:value-of
              select="$year"/></date>
        </xsl:if>
        <xsl:if test="$signature!=''">
          <identifier><xsl:value-of select="$signature"/></identifier>
        </xsl:if>
        <type>BOOK</type>
      </description>
    </xsl:if>
  </xsl:template>
</xsl:stylesheet>
```

Fig. 4. XSL stylesheet used for the semantic import of a custom CSV file, which was previously converted to an XML stream consisting of *line* and *column* elements by a Cocoon generator

parts: A syntactical step, where the input data is converted to XML according to its internal syntactic structure; and a semantical step, where the elements of the resulting XML stream are rearranged, post-processed and converted to corresponding Dublin Core elements. While the first step is functionally trivial, it is usually hard to implement, and so the goal is to provide a large toolbox of Cocoon generators that can syntactically preprocess common input formats

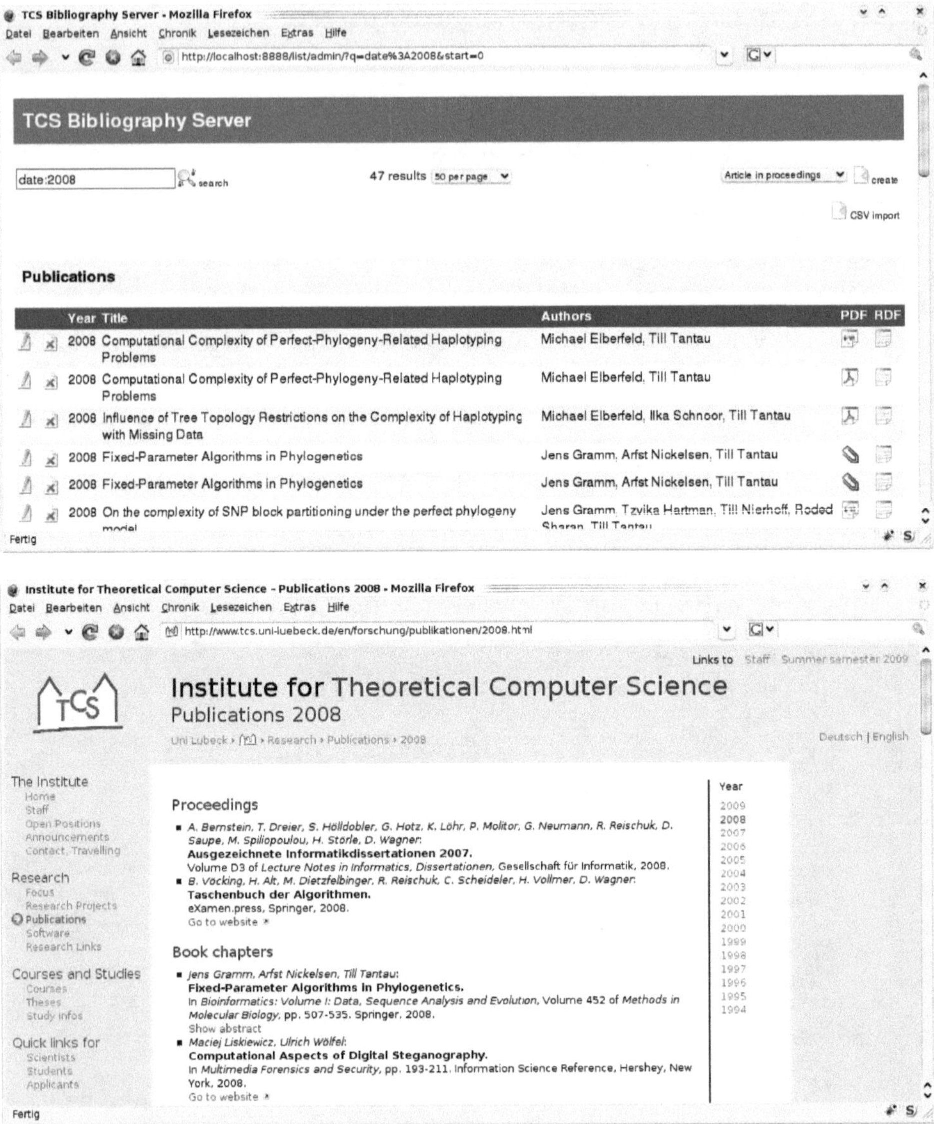

Fig. 5. Screenshots of the implemented system. Two different views of the same resource list, defined by the query `date:2008`, are shown. Top: An administrative interface for managing the resource database. Bottom: A reference list as published on our institute's website.

such as CSV (Excel), BibTeX, RIS and MARC. However, once the data has been fed into an XML stream, the implementation of the actual crosswalk is again merely a matter of writing an XSL stylesheet, or adapting an existing one. Our prototype currently contains generators for CSV, RIS, and BibTeX.

5 Implementing the System in Practice

To illustrate how the described concepts are applied in the real world, we describe the steps that were necessary to implement the system at our faculty's institute for theoretical computer science, where it is now used to manage the publication lists and the institute's internal library. For the publication lists, a view was implemented for HTML form-based editing of the database (Fig. 5). The view consists of some XSL stylesheets for the forms and another one that generates a searchable list of existing publications. Along with some static CSS and image files, these stylesheets are stored in a subdirectory of the application's webapp folder. Some researchers had BibTeX files of their publications available, which were imported directly without using an HTML form interface. To embed the bibliographic information into the institute's website, a view was written that generates bibliographies in HTML format, which was basically a small extension to the standard HTML fragment view to allow for both english and german output. Bibliographic information is inserted from the system in several places at our institute's website (Fig. 5, currently at `http://www.tcs.uni-luebeck.de`). Similarly, a custom view for browsing the local library was created. The library metadata was imported from a legacy CSV file, which was piped through the XSL stylesheet shown in Fig. 4 upon conversion to an XML stream consisting of *line* and *column* elements.

6 Conclusions

Many small institutions nowadays deal with large quantities of bibliographic metadata. Adoption of metadata standards and related software, however, progresses at a rather slow pace, even though the benefits of a wider adoption would be unquestionable. In this paper, we presented an architecture for an XML-based metadata management system that lowers the cost of adopting metadata standards by facilitating data import and integration into an existing environment. We implemented the system, using exclusively Open Source libraries and components, and put it in productive use.

As stated earlier, our work shares several aspects with the metadata interoperability framework designed by Godby et al. [1], in particular the split between syntactical and semantical steps of metadata import. Godby et al., however, argue that XSLT is not appropriate as a transformation language because syntactical and semantical steps have to be merged in a single stylesheet. Instead, they propose a custom XML dialect for expressing semantical equivalence relations. Our work shows that modularity can also be achieved with XSLT by using the pipeline architecture. Since we target small institutions and not library professionals, we consider it beneficial to stick with the widespread XSLT, instead of introducing a new language. This choice also gives us access to many existing resources written in XSLT, such as the collection of stylesheets for conversion to and from the MARC 21 format that is maintained by the Library of Congress [13].

From our own experiences with running the system in practice, we think that the architecture looks promising so far, but certainly there remains much work to be done. First of all, more running implementations and a larger user base are needed for a systematic evaluation of our architecture's efficiency and user-friendliness. A prerequisite for widespread use of the described system would be to hide the fairly complex Cocoon framework, which is known to have a rather steep learning curve, from the end user. Currently, at least a basic knowledge of Cocoon's internals such as the sitemap and configuration of generators and serializers is necessary to write new views and importers. On the other hand, the system is currently tied to our specific RDF/XML implementation of Dublin Core and OpenURL. This could be relaxed to allow storage of arbitrary XML-based metadata formats.

References

1. Godby, C.J., Smith, D., Childress, E.: Toward element-level interoperability in bibliographic metadata. Code4Lib Journal 2 (2008)
2. Baca, M.: Introduction to metadata: pathway to digital information. Getty Research Institute, Los Angeles (2008)
3. Tennant, R.: A bibliographic metadata infrastructure for the 21st century. Library Hi Tech. 22(2), 175–181 (2004)
4. Dublin Core Metadata Initiative Usage Bord: DCMI metadata terms (2008), `http://dublincore.org/documents/dcmi-terms/`
5. Chan, L., Zeng, M.: Metadata interoperability and standardization - a study of methodology. part I: Achieving interoperability at the schema level. D-Lib Magazine 12(6) (2006)
6. Chan, L., Zeng, M.: Metadata interoperability and standardization - a study of methodology. part II: Achieving interoperability at the record and repository levels. D-Lib Magazine 12(6) (2006)
7. Johnston, P.: Element refinement in Dublin Core metadata (2005), `http://dublincore.org/documents/dc-elem-refine/`
8. Dublin Core Metadata Citation Working Group: Guidelines for encoding bibliographic citation information in dublin core metadata (2005), `http://dublincore.org/documents/dc-citation-guidelines/`
9. ANSI/NISO: The OpenURL framework for context-sensitive services. Standard Nr. Z39.88-2004 (2004)
10. ANSI/NISO: The OpenURL registry (2004), `http://www.openurl.info/registry/`
11. Langham, M., Ziegeler, C.: Cocoon: Building XML Applications. Sams (2002)
12. Hatcher, E., Gospodnetic, O., McCandless, M.: Lucene in Action. Manning (2009)
13. The Library of Congress: MARC 21 XML schema, tools & utilities (2009), `http://www.loc.gov/standards/marcxml/xslt/`

DataStaR: Bridging XML and OWL in Science Metadata Management

Brian Lowe

Albert R. Mann Library,
Cornell University,
Ithaca, NY USA 14853
bjl23@cornell.edu

Abstract. DataStaR is a science data "staging repository" developed by Albert R. Mann Library at Cornell University that produces semantic metadata while enabling the publication of data sets and accompanying metadata to discipline-specific data centers or to Cornell's institutional repository. DataStaR, which employs OWL and RDF in its metadata store, serves as a Web-based platform for production and management of metadata and aims to reduce redundant manual input by reusing named ontology individuals. A key requirement of DataStaR is the ability to produce metadata records conforming to existing XML schemas that have been adopted by scientific communities. To facilitate this, DataStaR integrates ontologies that directly reflect XML schemas, generates HTML editing forms, and "lowers" ontology axioms into XML documents compliant with existing schemas. This paper describes our approach and implementation, and discusses the challenges involved.

1 Introduction

DataStaR, short for Data Staging Repository, is a three-year project begun in late 2007 to develop library services to support the local curation of research data.[1] The components of the DataStaR project are described in [1]. The project's chief focus is on so-called "small science" data; that is, smaller sets of data not requiring terabyte-scale storage infrastructure or specialized computational services but which are likely to be of enduring scholarly value. In some cases, an institutional repository may be the only available long-term store of such data; in other instances, one or more domain repositories managed by particular research communities may become long-term custodians.

The DataStaR "staging repository" is designed to serve as an extensible Web-based platform where researchers can upload files containing observational or experimental data to a Fedora[2] repository and create metadata describing these data sets. The staging repository allows access to be limited to research group members until the metadata and/or data are ready to be shared publicly.

[1] http://datastar.mannlib.cornell.edu/
[2] http://fedora.info/

F. Sartori, M.Á. Sicilia, and N. Manouselis (Eds.): MTSR 2009, CCIS 46, pp. 141–150, 2009.

DataStaR is not a permanent data archive: data sets and accompanying metadata are expected to be transferred to one or more long-term repositories for preservation and enduring access. While data files are eventually deaccessioned from DataStaR after successful submission to destination repositories, related metadata are retained indefinitely for discovery through a Web portal, SPARQL[3] query, or Linked Data[4] request. This paper focuses on the approach to managing metadata in DataStaR using Semantic Web technologies, and describes some preliminary results from an alpha prototype implementation.

2 Motivation and General Approach

Cornell's Albert R. Mann Library has employed Semantic Web technologies for a number of years to manage data for various Web portals related to research and scholarly activities, including the VIVO project described in [2]. The flexibility of the Resource Description Framework (RDF) data model and the ability to extend and specialize ontology classes and properties without modifying relational database schemas has been an asset in the management of diverse and open-ended data about Cornell research activities. While these projects have developed domain-specific Web Ontology Language (OWL) ontologies to drive Web portals, DataStaR is focused less on ontology modeling than on operating within the conceptual models already developed by scientific communities, and requires the ability to generate standards-compliant metadata for a variety of scientific domains. These standards, such as Ecological Metadata Language (EML)[5] the Federal Geographic Data Committee's (FGDC) Content Standard for Content Standard for Digital Geospatial Metadata (CSDGM)[6], and Data Documentation Initiative metadata[7] are typically based on records exchanged as XML documents conforming to published schemas.

The DataStaR project operates under the assumption that science metadata communities will make increased use of semantics and ontologies as standards and infrastructure develop. Semantic technologies have been adopted by the biomedical community for a variety of purposes, and billions of RDF triples have been published on the Semantic Web in the Bio2RDF[8] data sets. Ontologies have been developed to formalize the semantics of terminology in scientific disciplines. Two examples are the SWEET project,[9] which publishes ontologies for earth and environmental terminology, and the Marine Metadata Interoperability Project[10] which offers ontologies and metadata tools for marine science. Joshua Madin et al. have described ontology research efforts in ecology [3] and

[3] SPARQL Protocol and RDF Query Language: www.w3.org/TR/rdf-sparql-query/
[4] http://linkeddata.org/
[5] http://knb.ecoinformatics.org/software/eml/
[6] http://www.fgdc.gov/metadata/geospatial-metadata-standards
[7] http://www.ddialliance.org/
[8] http://bio2rdf.org/
[9] http://sweet.jpl.nasa.gov/
[10] http://marinemetadata.org/

have developed a general ontology describing the semantics of scientific observation and measurement [4]. These activities suggest interest and ongoing work in using formal semantics for describing scientific data.

Semantic Web technologies can, however, suffer from a chicken-and-egg problem: useful applications depend on the existence of useful semantic data, but it is difficult to justify the investment in creating these data because of the lack of applications and suitable tools. DataStaR addresses this gap by providing a platform where existing XML metadata schemas can be automatically converted into OWL representations, with metadata "records" edited not as discrete documents but as assertional axioms linked into larger knowledge bases where appropriate and beneficial.

The goal in DataStaR is to store and manage metadata entirely in a form that *supports* explicit semantics without *requiring* that rich semantics be used. At a simple level, DataStaR may be used to create reusable named individuals where the original metadata schema or available tools offer only free-text fields, with the goal of reducing redundant entry or cutting and pasting of metadata. (If, for example, a person, research location, or method is needed in multiple metadata records, individuals can be created once and then referred to by URI.) At an advanced level, complex mappings between auto-generated metadata ontologies and other ontologies may be added using OWL axioms or SWRL rules to effect complete semantic interoperability of conceptual models. In the middle is room for a wide range of semantically-enabled activities involving automation of simpler transformations, such as generation of basic metadata for institutional repositories.

DataStaR includes an ontology for common metadata applicable to all data sets regardless of scientific discipline, including properties such as "title," "abstract," "owner," and "contact." This core ontology includes concepts from the Semantic Web for Research Communities (SWRC)[11] ontology and Friend of a Friend (FOAF)[12], as well as OWL-DL-compatible extensions of Dublin Core properties[13] to enable interoperability with RDF metadata using these established ontologies. Rule mappings between DataStaR's core ontology and ontologies for domain metadata allows us to present users with a consistent form for entering basic metadata about each dataset. Users are then prompted to enter additional metadata once they have signaled intent to deposit their data in a domain repository that uses a particular metadata schema. Such schemas are "lifted" into OWL ontologies for integration with DataStaR.

3 XML Schemas and OWL: Background and Related Work

There is a significant body of research that has been performed on the expression of XML schemas as RDFS or OWL ontologies and the conversion of XML docu-

[11] http://ontoware.org/projects/swrc/
[12] http://www.foaf-project.org/
[13] http://dublincore.org/documents/dcmi-terms/

ments to RDF graphs. Bohring and Auer [5] present XML to RDF conversion as part of a workflow for integrating relational data using ontologies. Though their approach does not depend on existence of a schema, they implement XSD to OWL conversion in XSLT, and generate stylesheets for converting XML documents into RDF instance data. Akhtar et al [6] describe the challenges inherent in XSLT-based conversion and introduce a query language called XSPARQL to be used in bidirectional mappings between XML and RDF. Automated approaches have been developed for lossless "round-tripping" between RDF and XML, such as TopBraid Composer's "Semantic XML."[14]

In DataStaR, we want a direct conversion from an XML schema into an ontology that reflects the syntax of an original schema; additional semantic richness may then be added or not as desired. Though there are variations among techniques for direct automated conversion, Ferdinand et al [7] describe the principles, the fundamental core of which involves XML Schema complex types being converted into corresponding OWL classes, while elements and attributes become properties. Schema constraints such as `minOccurs` or `maxOccurs` are translated into cardinality axioms. This process is somewhat artificial: OWL is based on description logic rather than relational database theory, and includes no mechanisms for integrity constraints. Ongoing research offers possibilities for extending Semantic Web technologies to support schema languages. Boris Motik et al. [8] have proposed "extended DL knowledge bases" wherein certain axioms may be used to impose constraints rather than inferring new information about individuals. Jos de Bruijn et al. [9] have developed "OWL Flight," with semantics grounded in logic programming to support constraint-based modeling.

For our DataStaR prototype we have elected to use Gloze[10], an open-source Java library developed by Steve Battle, which can perform automatic lossless round-tripping between XML and RDF as well as generate OWL ontologies from XSD schemas. Gloze is attractive for our purposes because it does not require that an XSLT mapping or special set of queries be retained, nor does conversion depend on serializing RDF as RDF/XML. Gloze needs only be supplied with a small set of configuration options, and it will lower RDF to XML by consulting the original XSD schema.

4 First Application: Ecological Metadata

Mann Library has already partnered with researchers to describe and share ecological observation data [11] [12]. Ecological Metadata Language (EML) was selected as the most appropriate metadata standard, and the Knowledge Network for Biocomplexity[15] as the principal domain repository destination for datasets and metadata. EML is supported by a suite of tools, including the desktop editor Morpho[16] and the Metacat[13] XML database system developed by the

[14] http://composing-the-semantic-web.blogspot.com/2007/11/xmap-mapping-arbitrary-xml-documents-to.html
[15] http://knb.ecoinformatics.org/
[16] http://knb.ecoinformatics.org/software/morpho/

National Center for Ecological Analysis and Synthesis (NCEAS). (It should be noted that EML supports reusable metadata chunks in certain cases via its own key-reference system.) We have set up local installations of Metacat; our first task for DataStaR has been to create EML metadata as RDF graphs, lower them to XML documents, and deposit these records into Metacat. Because the KNB uses Metacat for its metadata database, this implies that we will be able to transfer from DataStaR directly to the domain repository.

5 System Implementation

The DataStaR metadata management infrastructure is an extension of the Vitro software[17] developed at Mann Library, which combines a Web-based ontology and instance editor with a public display interface. Vitro runs in the Tomcat servlet container and uses the Jena[18] library to create and store OWL ontologies and instance data as RDF graphs. Jena offers a number of built-in rule-based reasoning engines, but by default Vitro uses the complete OWL-DL reasoner Pellet,[19] which also supports reasoning with DL-safe[20] SWRL[21] rules. For small ontologies, the entire set of inferred statements may be extracted from Pellet; where this is too expensive, Vitro can be configured to copy only certain types of inferences into a graph that is queried to serve Web requests.

Gloze, which also uses Jena, connects to Vitro in a straightforward fashion: given an XML schema and a reference to an RDF resource representing a root node (which is automatically generated by DataStaR), Gloze walks the graph to generate a compliant XML DOM. For example, Gloze is aware that `<title>` elements must occur before `<creator>` elements in an EML document. Although the RDF graph offers no intrinsic ordering of the corresponding `eml:title` and `eml:creator` properties, Gloze adds XML element children in the required order.

The process of integrating an XML schema with DataStaR involves using Gloze to generate a corresponding OWL ontology, patching the resulting ontology to ensure compatibility with OWL DL, and extending it as desired. Then, rules are manually supplied to map appropriate parts of this ontology onto elements of DataStaR's core ontology. DataStaR attempts to generate display pages and editing forms automatically for schema-specific metadata, but these may be heavily customized where necessary.

5.1 Generation of Editing and Display Interfaces

As noted earlier, OWL axioms such as cardinality restrictions or domain and range classes are very different from XML schema constraints such as `minOccurs` and `maxOccurs`. We can, however, use these axioms as hints when constructing

[17] http://vitro.mannlib.cornell.edu/
[18] http://jena.sourceforge.net/
[19] http://clarkparsia.com/pellet
[20] http://clarkparsia.com/weblog/2007/08/12/understanding-swrl-part-1/
[21] http://www.w3.org/Submission/SWRL/

editing forms. Although `rdfs:range` or `owl:allValuesFrom` axioms actually inform reasoners that values of certain properties should be inferred to be members of specified ontology classes, we take advantage of them when populating `<select>` lists on forms by including as options those individuals that have been asserted to be members of these classes. Likewise, `owl:someValuesFrom` or `owl:minCardinality` restrictions inform heuristics for indicating "required" properties where a user must enter a value if one cannot be inferred. These techniques allow us automatically to generate editing interfaces needed to describe instances of Gloze-generated ontology classes, which in turn helps ensure sufficient data to create a valid output XML metadata record. This technique alone, however, cannot *ensure* validity, as some XML Schema constructs, such as `xsd:choice` do not manifest themselves in the generated ontology. As DataStaR develops further we will address the most common of such situations with additional automatic tools. Vitro also supports extensive manual customization of individual editing forms where necessary or desirable to improve appearance, enhance usability, ensure that certain properties are marked as required, or add additional validation to submitted entries.

Automatic display generation is another important issue. When the value of a property is an individual with no obvious label, possibly representing an n-ary relationship or a semantically vacuous node, DataStaR walks the graph until stopping at literal values to display. Where the results of this are unsatisfactory, manual tweaks can be applied to generate more concise labels. For example, given a subgraph describing the creator of the following data set,

```
:myDataset        eml:creator          :individual304 .
:individual304    rdf:type             eml:ResponsibleParty .
:individual304    eml:individualName   :individual322 .
:individual322    rdf:type             eml:Person .
:individual322    eml:givenName        "John" .
:individual322    eml:surName          "Barleycorn" .
```

we can use SWRL builtins to insert a simple label during reasoning:

$$\begin{aligned}&\mathit{datastar{:}label}(?responsibleparty, ?fullname) \longleftarrow\\ &\mathit{individualName}(?responsibleparty, ?person)\\ &\wedge\ \mathit{eml{:}surName}(?person, ?surname)\\ &\wedge\ \mathit{eml{:}givenName}(?person, ?givenname)\\ &\wedge\ \mathit{swrlb{:}stringConcat}(?fullname, ?surname, ",\ ", ?givenname)\end{aligned}$$

In this example using the EML ontology a simple label is added to display the name of a person responsible for creating a data set. Adding rules to make consistent use of the label property has the additional benefit of providing cues to the automated form generator: those individuals that have a label are likely to be reused in different contexts, and are placed on `<select>` lists. Individuals lacking labels are assumed, by default, not to make sense in isolation: their properties are displayed in the context of a broader editing form.

5.2 Hidden XML Semantics

EML's notion of a ResponsibleParty is also a good example of interesting semantics that cannot automatically be lifted from an XML Schema. A portion of an EML XML record describing the contact person for a data set might look like this:

```
<contact>
    <individualName>
        <salutation>Mr.</salutation>
        <givenName>John</givenName>
        <surName>Barleycorn</surName>
    </individualName>
    <organization>Barleyco Inc.</organization>
    <positionName>Chief Scientist</positionName>
</contact>
```

The complex content of the contact element is of the type ResponsibleParty, which allows different combinations of the individualName, organizationName, and positionName elements to indicate that an individual, organization, or *position* has responsibility for an aspect of a data set. (For example, if organizationName and positionName are supplied but not individualName, this implies that whoever holds the specified position at the designated organization is the appropriate contact person.) In the Gloze-generated OWL ontology, the ResponsibleParty complex type becomes as class of the same name; instances of this class serve as the objects of the contact property. Thus, the ResponsibleParty class is not equivalent to another ontology's Person, Organization, or even Agent. The use of ResponsibleParty with only an organization and position name actually implies a rule for inferring which individuals can serve as contact persons. Rules of this type may be added to DataStaR to generate useful inferences with metadata. While DataStaR can generate editing forms for creating instances of classes like ResponsibleParty directly, for the simpler uses of ResponsibleParty to represent named individuals or organizations, it is attractive to add rules mapping to DataStaR's core ontology and edit simple object property statements relating data sets to instances of Person or Organization.

6 Ongoing Work

6.1 Multiple Formalisms and DL Compatibility

DataStaR requires enhancing the Vitro software to support multiple graphs in order to expose the correct language constucts to reasoners. To map to Dublin Core or other RDF Schema ontologies, for example, we generate an OWL-DL-compatible "shadow" version that maintains disjointness of object and datatype properties. This form is exposed to Pellet. The subsumption axioms relating the DL shadow ontology to the original RDFS ontology are exposed only to a rule reasoner that generates the remaining inferences. Similar techniques are necessary to massage ontologies generated by Gloze as OWL-Full. Gloze, for example, will create properties that are both object properties and data properties when an XML schema allows both simple and complex content for an element. We expect that DataStaR will be able to generate appropriate DL-compatible subproperties automatically in most cases. A more difficult problem occurs when Gloze creates complex OWL DataRanges involving unions or intersections. Here we can take advantage of the axioms for creating editing forms but must simply omit the DataRange from the axioms fed to Pellet.

6.2 Ordering of Axioms

Semantic knowledge bases, whether considered as RDF graphs or collections of description logic axioms, lack an intrinsic notion of ordering of statements. In XML, order is often significant, and a valid ordering may not be the same as the intended ordering, especially in cases where an element may be repeated. Gloze addresses this by offering the option to use RDF where the statements are reified and sequenced with `rdf:Seq` constructs. While support for generating or maintaining such sequences for Gloze is not yet implemented in the DataStaR prototype, this will be a critical issue as we move to a production phase.

7 Evaluation

DataStaR makes it possible to manage science metadata in a Semantic Web environment using RDF and OWL while facilitating the submission of metadata to repositories that require XML documents conforming to published schemas. DataStaR minimizes the manual effort required to set up a functional Web-based editing environment for researchers. Creation of additional axioms and rules is necessary to exploit the platform's capacity for interoperability with other ontologies. Core infrastructure development will continue through summer 2009; in the fall, DataStaR will be tested by a group of researchers at Cornell. At this time we will gather valuable feedback about the system's usability from the scientist's perspective. We will also be testing the platform with a number of different existing metadata schemas.

Use of DataStaR makes sense when production of semantic metadata is an explicit goal. Projects not desiring Semantic Web interoperability might prefer to use XML technologies such as XSLT, or systems such as Fez[22], which generates editing forms directly from XML schemas. DataStaR is designed to be a general-purpose platform for editing metadata for any number of scientific disciplines. It will not, at least in the near future, accommodate highly specialized editing tools tailored to specific domains.

[22] http://sourceforge.net/projects/fez

7.1 A Particular Advantage of XML Tools – Handling Text Markup

A special challenge in DataStaR is supporting the markup of structure and style for larger sections of text that are needed for various types of metadata for human use, such as abstracts, detailed method descriptions, and rights statements. Vitro uses the TinyMCE[23] Javascript rich text editor to provide a WYSIWYG interface for users to create paragraphs and lists and apply styles: the XHTML produced by TinyMCE is stored as a single literal value in an RDF statement. XML schemas may use any number of ways of structuring text, and these must be followed in the RDF representation. EML, for example, uses the DocBook standard for formatted text. Because the vast majority of rich text sections in the EML documents we have created outside of DataStaR use only single paragraphs without special markup, we can set up editing forms to use standard HTML textareas, and generate text structure individuals each having only a single related paragraph individual. It would not be desirable for our purposes to attempt to maintain complex text passages in RDF as graphs of related paragraphs, headings, and the like. Here, established XML technologies such as XSLT have a clear advantage: the mature DataStaR platform will most likely require that stylesheets be supplied to convert between XHTML and any special text markup used in a particular metadata schema.

8 Conclusion

We have presented a practical application that integrates Web-based RDF graph editing, XML/OWL conversion, and semantic mappings to bridge the gap between today's schema-based metadata standards and the growing Semantic Web of linked data while permitting compatibility with established data repositories. DataStaR builds effectively on existing tools while adding features for automated generation and manual customization of editing interfaces for end users. As we put DataStaR into practice in collaboration with additional research groups, we will be testing the metadata management model with additional schemas, evaluating where resources should be directed in order to enrich the semantics of automatically-generated ontologies and building additional tools to streamline the generation of inter-ontology mappings. While the complexity of the DataStaR infrastructure may not make it appropriate for specialized use cases where production of traditional metadata records is the exclusive goal, we believe the techniques employed in DataStaR will allow us to interoperate successfully with current XML infrastructure as well as the developing Semantic Web.

9 Funding Acknowledgement and Disclaimer

This material is based upon work supported by the National Science Foundation under Grant No. III-0712989. Any opinions, findings, and conclusions or recommendations expressed in this material are those of the author(s) and do not necessarily reflect the views of the National Science Foundation.

[23] http://tinymce.moxiecode.com/

References

1. Steinhart, G.: DataStaR: an institutional approach to research data curation. IASSIST Quarterly (in press)
2. Devare, M., Corson-Rikert, J., Caruso, B., Lowe, B., Chiang, K., McCue, J.: VIVO: connecting people, creating a virtual life sciences community. D-Lib Magazine 13(7/8) (2007)
3. Madin, J., Bowers, S., Schildhauer, M., Jones, M.: Advancing ecological research with ontologies. Trends in Eco. and Evol. 23(3), 159–168 (2008)
4. Madin, J., Bowers, S., Schildhauer, M., Krivov, S., Pennington, D., Villa, F.: An ontology for describing and synthesizing ecological observation data. Ecological Informatics (2), 279–296 (2007)
5. Bohring, H.: Mapping XML to OWL ontologies. In: Leipziger Informatik-Tage, LNI, vol. 72, pp. 147–156. GI (2005)
6. Akhtar, W., Kopecký, J., Krennwallner, T., Polleres, A.: XSPARQL: traveling between the XML and RDF worlds and avoiding the XSLT pilgrimage. In: The Semantic Web: Research and Applications, pp. 432–447 (2008)
7. Ferdinand, M., Zirpins, C., Trastour, D.: Lifting XML schema to OWL. In: Koch, N., Fraternali, P., Wirsing, M. (eds.) ICWE 2004. LNCS, vol. 3140, pp. 354–358. Springer, Heidelberg (2004)
8. Motik, B.: Bridging the gap between OWL and relational databases. In: Proceedings of the 16 th international conference on World Wide Web, pp. 807–816. ACM Press, New York (2007)
9. de Bruijn, J., Lara, R., Polleres, A., Fensel, D.: OWL DL vs. OWL Flight: conceptual modeling and reasoning for the semantic web. In: WWW '05: Proceedings of the 14th international conference on World Wide Web, pp. 623–632. ACM Press, New York (2005)
10. Battle, S.: Gloze: XML to RDF and back again. In: Proceedings of the First Jena User Conference, Bristol, UK (2006)
11. Woodbury, P.B., Howarth, R.W., Steinhart, G.: Understanding nutrient cycling and sediment sources in the upper Susquehanna River basin. Journal of Contemporary Water Research & Education, 7–14 (2008)
12. Steinhart, G., Lowe, B.J.: Data curation and distribution in support of Cornell University's Upper Susquehanna Agricultural Ecology Program. In: DigCCurr 2007, Chapel Hill, NC (April 2007)
13. Berkley, C., Jones, M., Bojilova, J., Higgins, D.: Metacat: a schema-independent XML database system. In: Thirteenth International Conference on Scientific and Statistical Database Management (SSDBM), pp. 171–179 (2001)

Structured Metadata for Representing and Managing Complex 'Narrative' Information

Gian Piero Zarri

University Paris-Est/Paris12 – LiSSi Laboratory
120-122, rue Paul Armangot – 94400 Vitry-sur-Seine, France
zarri@noos.fr, gian-piero.zarri@univ-paris12.fr

Abstract. In this paper, we evoke first the ubiquity and the importance of the so-called 'non-fictional narrative' information. We show then that the usual knowledge representation and 'ontological' techniques have difficulties in finding complete solutions for representing and using this type of information. We supply then some details about NKRL, a (complex metadata) representation language and a querying/inferencing environment especially created for an 'intelligent' exploitation of (non-fictional) narratives. The paper will be illustrated with some examples concerning recent concrete applications of this environment/language.

Keywords: Narrative information, narratology, artificial intelligence, metadata, knowledge representation, ontology of concepts, ontology of events, querying and inferencing.

1 Introduction

'Narrative' information concerns in general the account of some real-life or fictional story (a 'narrative') involving concrete or imaginary 'characters'. In this paper we will deal, essentially, with *(multimedia) non-fictional narratives of an economic interest*. This means, first, that we are only *partially concerned* with those sorts of *fictional narratives* that have principally an *entertainment value*, and represent a narrator's account of a story that happened in an imaginary world: a novel is a typical example of fictional narrative. Secondly, our 'non-fictional narratives' must have some sort of *economic value*: this means, from a practical point of view, that people *could be willing to pay for a system able to process in an 'intelligent' way this sort of information and/or for the results of the processing*. Narratives of this type are typically embodied into corporate memory documents, news stories, normative and legal texts, medical records, intelligence messages, surveillance videos, actuality photos for newspapers and magazines, multimedia material for eLearning, Cultural Heritage material, etc.

In this paper, we will first introduce the general background of the narrative domain. We will then describe a high-level metadata system, NKRL (Narrative Knowledge Representation Language) that tries to represent, without a too important loss of information, the *deep 'semantic meaning' of complex (non-fictional) narratives*.

F. Sartori, M.Á. Sicilia, and N. Manouselis (Eds.): MTSR 2009, CCIS 46, pp. 151–163, 2009.

2 General Background

2.1 Narratology and Related Disciplines

'Narratives' represent presently a very 'hot' domain. From a *theoretical point of view*, they constitute the object of a full discipline, narratology, whose aim can be defined in short as that of producing an in-depth description of the 'syntactic/semantic structures' of the narratives. This means that the narratologist is in charge of dissecting narratives into their component parts in order to establish their functions, their purposes and the relationships among them. A good introduction to the full domain is [1]. Several humanistic disciplines, from stylistic to literary techniques, are in a strict relationship with narratology, which is then particularly concerned with literary analysis (and, therefore, with 'fictional' narratives). But this sort of research is also directly related to our interests.

First of all, we can note immediately that at least two sub-disciplines that can be associated with the narratology's domain, 'storytelling' and 'eChronicles', are also of import from a general metadata, artificial intelligence and computer science point of view; moreover, they can deal as well with non-fictional data.

'Storytelling', see [2], concerns in general the study of the *different ways of conveying 'stories' and events in words, images and sounds in order to entertain, teach, explain etc.* 'Digital Storytelling' deals in particular with the ways of introducing characters and emotions in the interactive entertainment domain: it is of interest then for the videogames, the (massively) multiplayer online games, interactive TV, virtual reality etc., see [3]. An 'eChronicle' system [4, 5] can be defined as way of recording, organizing and then accessing *streams of multimedia events* captured making use of video, audio and any other possible type of sensors. These 'eChronicles' may concern any sort of 'narratives' like conference records, surveillance videos, football games, visitor logs, sales activities, 'lifelogs' obtained from wearable sensors, etc. The technical challenges concern mainly the ways of aggregating the events into coherent 'episodes' and of providing access to this sort of material at the required level of granularity. Note that 'exploration', and not 'normal querying', is the predominant way of interaction with the eChronicle repositories.

Secondly, some classic studies in the narratology domain, like *Mieke Bal's analysis of the narrative phenomena* [6], can be very useful to build up a solid, theoretical background for our research.

Bal sees narrative phenomena as structured into *three layers* called 'fabula' (a Latin word: fable, story, tale, play), 'story proper', and '(narrative) presentation'. The *fabula* layer concerns a series of logically and chronologically related events (in other terms, a '*stream of elementary events*') that describe the activities or the experiences of given characters. The second level, the *story*, consists in *a particular subset of the fabula contents rearranged in a new sequence according to given purposes*, for example, to inform a user about the background of a series of events. It is then possible to derive a number of different 'stories' starting from a given 'fabula'. The third layer, the level of the final *presentation*, is *how the 'story' is eventually expressed according to a given language or media form*, e.g., as a verbal exchange, a picture, a novel, a film etc.

NKRL takes all these three layers into account. The proper *knowledge representation* component of NKRL consists of a general model for representing in a machine understandable way the main characteristics of the above 'stream of elementary events': it corresponds well, therefore, to the *proper fabula level*. The *story layer* refers to the extraction of *specific subsets of the representations introduced at the fabula level* to answer some user queries, or to validate, e.g., specific hypotheses about the causal chaining of events/streams of events; it denotes then, in short, a querying/inferencing system. Given that NKRL is more interested in an 'intelligent information retrieval' form of exploitation of the fabula material than in the use of this material for 'generation' purposes, the *presentation* layer concerns only the usual forms of displaying the results of the *story* activities to the user.

It is then evident that representing correctly the 'fabula' aspects is of paramount importance for the success of the NKRL's effort. For simplicity's sake, *we will now assimilate a (non-fictional) NKRL 'narrative' with its 'fabula level'*. We will then state, accordingly, that *the NKRL understanding of a 'narrative' corresponds to a logically and temporally connected stream of (non-fictional) elementary events* – we can also add that the *verbalized expression* of a single elementary event includes at most an (implicit or explicit) '*generalized natural language (NL) predicate*', see [7: 7-13] for a discussion about this topic. Subsets pertaining to different narratives can be extracted at the 'story' and 'presentation' levels to satisfy specific requirements of the users. A 'narrative' as defined above is characterized, among other things, by the following general properties – independently, by the way, by any 'economically relevant' or not consideration:

- One of the features defining the 'connected' character of the elementary events of the stream concerns the fact that these events are *chronologically related*, i.e., narratives *extend over time*. This *diachronic* aspect of narratives (a narrative has a beginning, a development and an end) represents one of their most important characteristics.
- *Space* is also very important, given that the elementary events of the stream occur generally in well defined 'locations', real or imaginary ones. The connected events that make up a narrative are then both *temporally and spatially bounded*.
- A simple chronological successions of elementary events that take place in given locations cannot, however, be defined as a 'narrative' without some sort of '*semantic coherence*' and '*uniqueness of the theme*' that characterise the different events of the stream. *If this logical coherence is lacking, the events pertain to different narratives.*
- When the 'elementary events' of a narrative are verbalized in NL terms, their *coherence* is normally expressed through *syntactic constructions like causality, goal, indirect speech, co-ordination and subordination*, etc. In this paper, we will make use of the terms '*connectivity phenomena*' to denote this sort of clues, i.e., to denote what, in a stream of events, i) leads to a 'global meaning' that goes beyond the addition of the 'meanings' conveyed by the single elementary events; ii) defines the influence of the context where a particular event is used on the meaning of this event.
- Eventually, narratives concern *the behaviour or the condition of some 'actors'* (persons, characters, personages etc.). They try to attain a specific result, experience

particular situations, manipulate some (concrete or abstract) materials, send or receive messages, etc. In short, *they have a specific 'role' in the event* (in the stream of events representing the global narrative). Note that these actors or personages are not necessarily human beings; we can have narratives concerning, e.g., the vicissitudes in the journey of a nuclear submarine (the 'actor', 'personage' etc.) or the various avatars in the life of a commercial product.

2.2 Present (and Past) Solutions for Representing Narratives

An in-depth analysis of the existing, computer-oriented solutions that *could* be used to represent and manage (non-fictional) narratives is beyond the possibilities of this paper – see [7] in this context. We will limit ourselves, here, to some quick consideration.

We can note, first of all, that the now so popular Semantic Web (W3C) languages like RDF (Resource Description Framework), see [8], and OWL (Web Ontology Language), see [9] are unable to fit the bill because their core formalism is based on the classical '*attribute – value*' model. For these 'binary' languages then, a property can only be a *binary relationship*, linking two individuals or an individual and a value.

When these languages must represent simple (fabula level) 'narratives' like "John gives a book to Mary", several difficulties arise. In this example, "give" is an *n-ary (ternary) relationship* that, to be represented in a *complete and unambiguous way*, asks for a form of complex syntax where the arguments of the predicate, i.e., "John", "book" and "Mary", are introduced by '*conceptual roles*' such as, e.g., "agent of give", "object of give" and "beneficiary of give" respectively. As noted in the previous Section, the notion of '*role*' must, then, be necessarily added to the traditional (binary) 'generic/specific', 'property/value' etc. representational principles in order to specify the *exact function* of the different components of the events in the context of narrative documents. Note, moreover, that the argument often raised stating that an *n*-ary relation can always be converted to a set of binary ones *without loss of semantics* is incorrect with respect to the last part of this sentence. In fact, it is true that, from a pure formal point of view, any *n*-ary relationship with $n > 2$ can always be reduced to a set of binary relationships. However, this fact does not change the *intrinsic, 'semantic'* n*-ary nature* of a simple statement like "John gives a book to Mary" that, to be fully understood, requires that all the constituents of the *n*-ary representation – predicates, roles, arguments of the predicate etc. – must *necessarily be managed at the same time as a coherent block*, see [7: 14-17] for the formal details. The *impossibility of reducing* n*-ary to binary from a conceptual and semantic point of view* has, as a practical consequence, the need of using specific *n*-ary tools for reasoning and inference when dealing with narratives in a *non-restricted way*.

Several solutions for representing narratives in computer-usable ways according to some sort of *genuine* '*n*-ary model' have been described in the literature. They range from Silvio Ceccato's 'correlations' to the Roger Schank's 'conceptual dependency' for the oldest ones, to the well-known proposals represented by John Sowa's 'conceptual graphs' or the (very controversial) Lenat's CYC system, to the recent Topic Maps, etc. None of them, however, seems to be able to satisfy completely the (non-fictional) narratives requirements as expounded in the previous Section because, among other things, the existence of a series of *epistemological problems* concerning, e.g., the lack of agreement about the list of 'roles' (conceptual cases) or the divergences of opinion about the use of 'primitives' – a detailed discussion about these topics can be found in [7: 14-33]. Conceiving, however, a *'practical' Knowledge Representation tool* for dealing *concretely* with (non-fictional) narrative information is far from being impossible.

Returning then to the "John gives a book..." example above – and leaving provisionally aside all the additional problems linked, e.g., with the temporal representation or the 'connectivity phenomena' – it is not too difficult to see that a complete, *n*-ary representation that captures the '*essential meaning*' of this narrative amounts to:

- Define JOHN_, MARY_ and BOOK_1 as '*individuals*', instances of general *'concepts'* like human_being and information_support or of more specific concepts. Concepts and instances (individuals) must be, as usual, collected into a 'binary' ontology – built up using a standard tool like, e.g., Protégé [10].
- Define an *n*-ary structure organised around a *conceptual predicate* like, e.g., MOVE or PHYSICAL_TRANSFER, and associate the above individuals (the arguments) to the predicate through the use of *conceptual roles* that specify their 'function' within the global narrative. JOHN_ will then be introduced by an AGENT (or SUBJECT) role, BOOK_1 by an OBJECT (or PATIENT) role, MARY_ by a BENEFICIARY role.
- '*Reify*' the obtained *n*-ary structure by associating with it a *unique identifier* under the form of a 'semantic label', to assure both i) the logical-semantic coherence of the structure; ii) a rational and efficient way of storing and retrieving it.

Formally, this *n*-ary structure can be described as:

$$(L_i\,(P_j\,(R_1\;a_1)\;(R_2\;a_2)\;\ldots\;(R_n\;a_n))) \qquad (1)$$

where L_i is the symbolic label identifying the particular *n*-ary structure (e.g., the global structure corresponding to the representation of the "John gives a book..." example), P_j is the conceptual predicate, R_k is the generic role and a_k the corresponding argument (e.g., the individuals JOHN_, MARY_ etc.). Note that each of the $(R_i\;a_i)$ cells of Eq. 1, *taken individually*, represents a *binary relationship* in the W3C (OWL, RDF...) languages style. The main point here is, however, that *the whole conceptual structure represented by (1) must be considered globally.* As we will see more in detail in the following Sections, the solution represented formally by Eq. 1 is *at the core of the NKRL proposals for the representation and management of (non-fictional) narratives.*

3 A Short Description of NKRL

3.1 The Knowledge Representation Aspects

NKRL innovates with respect to the current *ontological paradigms*, both the 'traditional' ones and those inspired by the Semantic Web research, by adding to the usual '*ontologies of concepts*' an '*ontology of events*', i.e., a new sort of hierarchical organization *where the nodes correspond to* n-*ary structures (sorts of structured metadata) called 'templates'*. This hierarchy is called HTemp (hierarchy of templates) in NKRL.

Note that, in the NKRL environment, an 'ontology of concepts' (according to the traditional meaning of these terms) not only exists, but it represents in fact an essential component for assuring the correct functioning of the whole environment. This 'standard' ontology is called HClass (hierarchy of classes): structurally and functionally, HClass is *not fundamentally different* from one of the 'traditional' ontologies that can be built up by using tools in the Protégé style, see again [10]. A fragment of HClass is reproduced in Figure 1 – note that HClass includes presently (June 2009) more than 7,500 concepts. See [7: 123-138] for a discussion about concepts as non_sortal_concept (the specialisations of this concept, i.e., its subsumed concepts like substance_ cannot be endowed with direct instances), sortal_concept etc.

Where the data structures representing the nodes of HClass are essentially 'binary', the templates included in HTemp follow the *n*-ary format defined by Eq. 1 above. Predicates (P_j in Eq. 1) pertain to the set {BEHAVE, EXIST, EXPERIENCE, MOVE, OWN, PRODUCE, RECEIVE}, and roles (R_k) to the set {SUBJ(ect), OBJ(ect), SOURCE, BEN(e)F(iciary), MODAL(ity), TOPIC, CONTEXT}. An argument a_i of the predicate can consist of a simple 'concept' or of a structured association ('expansion') of several concepts. Templates can be conceived as the formal representation of generic classes of elementary events (at the fabula level) like "move a physical object", "be present in a place", "produce a service", "send/receive a message", etc.

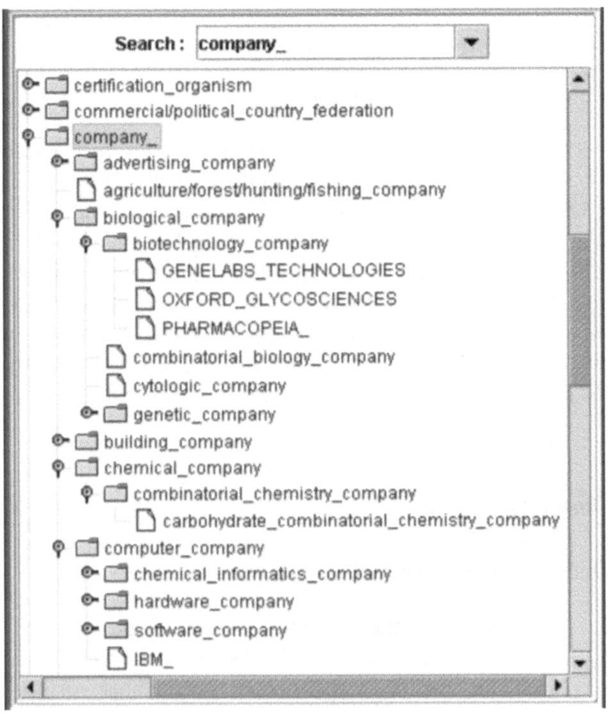

Fig. 1. Partial representation of HClass, the 'traditional' ontology of concepts

When a *particular elementary event* pertaining to one of these general classes must be represented, the corresponding template is *instantiated* to produce what, in the NKRL's jargon, is called a *'predicative occurrence'*– i.e., the formal, NKRL representation of this event. To represent then a simple narrative like: "British Telecom will offer its customers a pay-as-you-go (payg) Internet service in autumn 1998", we must select firstly in the HTemp hierarchy the template corresponding to 'supply a service to someone', represented in the upper part of Table 1. This template is a specialization of the particular MOVE template corresponding to 'transfer of resources to someone' – Figure 2 below reproduces a fragment of the 'external' organization of HTemp that includes, in particular, the offsprings of Move:TransferToSomeone.

In a template, the arguments of the predicate (the a_k terms in Eq. 1) are represented *by variables with associated constraints* – which are expressed as concepts or combinations of concepts, i.e., using the terms of the NKRL standard 'ontology of concepts' (HClass). When creating a *predicative occurrence (an instance of a template)* like c1 in the lower part of Table 1, the role fillers in this occurrence *must conform to the constraints of the father-template*. For example, in occurrence c1, BRITISH_TELECOM is an individual, instance of the concept company_: this last is, in turn, a specialization of human_being_or_social_body. payg_internet_service is a specialization of service_, a specific term of social_activity, etc. The meaning of the expression "BENF (SPECIF customer_ BRITISH_TELECOM)" in c1 is self-evident: the beneficiaries (role BENF) of

Table 1. Deriving a predicative occurrence from a template

```
name:  Move:TransferOfServiceToSomeone
father:  Move:TransferToSomeone
position:  4.11
natural language description: 'Transfer or Supply a Service to Someone'

MOVE   SUBJ          var1: [var2]
       OBJ           var3
       [SOURCE       var4: [var5]]
       BENF          var6: [var7]
       [MODAL        var8]
       [TOPIC        var9]
       [CONTEXT      var10]
       {[modulators]}

var1, var4, var6 = human_being_or_social_body
var3   =   service_
var8   =   process_, sector_specific_activity
var9   =   sortal_concept
var10 =    situation_
var2, var5, var7 = geographical_location

c1)    MOVE      SUBJ        BRITISH_TELECOM
                 OBJ         payg_internet_service
                 BENF        (SPECIF customer_ BRITISH_TELECOM)
                 date-1:     after-1-september-1998
                 date-2:
```

the service are the customers of – SPECIF(ication) – British Telecom. The 'attributive operator', SPECIF(ication), is one of the *four operators that make up the AECS sub-language*, used for the set up of the *structured arguments (expansions)*, see [7: 68-70]. In the occurrences, the two operators date-1, date-2 materialize *the temporal interval normally associated with narrative events*; a detailed description of the methodology for representing temporal data in NKRL can be found in [11].

More than 150 templates are permanently inserted into HTemp; HTemp, *the NKRL ontology of events*, corresponds then to a sort of *'catalogue' of narrative formal structures*, which are very easy to extend and customize.

To deal now with the 'connectivity phenomena', we make use of *second order structures created through reification of the conceptual labels of predicative occurrences*, see [7: 86-98, 11] for further details. For example, the '*binding occurrences*' consist of lists of symbolic labels (c_i in Eq. 1) of predicative occurrences: the lists are differentiated using binding operators like GOAL, CAUSE or COND(ition).

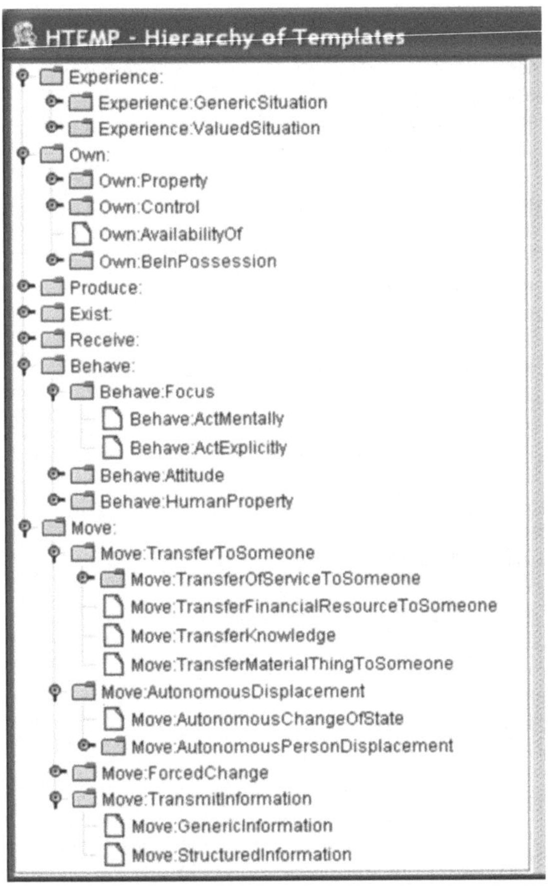

Fig. 2. 'MOVE' etc. branch of the HTemp hierarchy

Returning to the Table 1 example, let us suppose we would now state that: "We can note that, on March 2008, British Telecom *plans to offer* to its customers, in autumn 1998, a pay-as-you-go (payg) Internet service…", where the elementary event corresponding to the 'offer' is still represented by the predicative occurrence c1 in Table 1. To encode correctly the new information, we must introduce first an *additional predicative occurrence* c2 meaning that: "at the specific date associated with c2 (March 1998), it can be noticed that British Telecom is planning to act in some way". We will eventually add a *binding occurrence* c3, "c3) (GOAL c2 c1)", to link together the conceptual labels c2 (the planning activity) and c1 (the intended result). The global meaning of the c3 is then: "the activity described in c2 is focalized towards (GOAL) the realization of c1".

3.2 The Querying and Inferencing Aspects

Reasoning in NKRL ranges from the *direct questioning* of a knowledge base of narratives represented in NKRL format – by means of *search patterns* (formal queries) that unify information in the base thanks to the use of a *Filtering Unification Module* (*Fum*), see [7: 183-201] – to *high-level inference procedures*. These last make use of the richness of the representation to automatically establish 'interesting' relationships among the narrative items separately stored within the base; a detailed paper on this topic is [12].

The NKRL rules are characterised by the following general properties:

- All the NKRL high-level inference rules can be conceived as *implications* of the type:

$$X \text{ iff } Y_1 \text{ and } Y_2 \dots \text{and } Y_n \,. \quad (2)$$

- In Eq. 2, X corresponds either to a *predicative occurrence* c_j (formal representation of an elementary event) or to a *search pattern* p_i (formal representation of a simple query) and $Y_1 \dots Y_n$ – the NKRL translation of the '*reasoning steps*' that make up the rule – correspond to *partially instantiated templates*. They include then, see the upper part of Table 1, *explicit variables* of the form var_i.
- According to the conventions of logic/rule programming, the NKRL *InferenceEngine* understands each *implication* as a *procedure*. This allows us to reduce 'problems' of the form X to a succession of 'sub-problems' of the form $Y_1 \dots Y_n$.
- Each Y_i is interpreted in turn as a *procedure call* that tries to convert – using, in case, *backtracking procedures* – Y_i *into (at least) a successful search pattern* p_i. These last should be able to unify one or several of the occurrences c_j of the knowledge base.
- The success of the unification operations of the patterns p_i derived from Y_i means that the 'reasoning step' represented by Y_i has been validated. *InferenceEngine* continues then trying to validate the reasoning step corresponding to the sub-problem Y_{i+1}.
- In line with the presence of the operator 'and' in Eq. 2, the implication represented by this formula *is fully validated iff all the reasoning steps* $Y_1, Y_2 \dots Y_n$ *are validated.*

All the unification operations p_i/c_j make use only of the unification functions supplied by the Filtering Unification Module (*Fum*). Apart from being used for direct questioning, *Fum* constitutes as well, therefore, the 'inner core' of *InferenceEngine*.

From a concrete point of view, the NKRL high-level inference procedures concern *mainly* two classes of rules, 'transformations' and 'hypotheses', see [12].

Let us consider, e.g., the 'transformations'. These rules try to '*adapt*', from a *semantic* point of view, a search pattern p_i that '*failed*' (that was unable to find an unification within the knowledge base) to the *real contents* of this base making use of a sort of '*analogical reasoning*'. In a transformation context, the '*head*' X of Eq. 2 is then represented by a search pattern, p_i. The transformation rules try to *automatically 'transform'* p_i into one or more *different* p_1, p_2 ... p_n that *are not strictly 'equivalent' but only 'semantically close'* to the original one. Let us suppose that, in the context of a recent NKRL application about 'Southern Philippine terrorism', see [12], we ask: "Search for the existence of links between ObL (a well-known 'terrorist') and Abubakar Abdurajak Janjalani, the leader of the Abu Sayyaf' group (a separatist group in Southern Philippines)". In the absence of a direct answer, the corresponding search pattern can be *transformed* into: "Search for the attestation of the transfer of economic/financial items between the two". This could lead to retrieve: "During 1998/1999, Abubakar Abdurajak Janjalani has received an undetermined amount of money from ObL through an intermediate agent".

A transformation rule can be conceived as made up of a *left-hand side*, the '*antecedent*' – i.e. the formulation, in search pattern format, of the 'query' to be transformed – and of one or more *right-hand sides*, the '*consequent(s)*' – the NKRL representation(s) of one or more queries (search patterns) to be substituted for the given one. Denoting then with A the antecedent and with Cs all the possible consequents, these rules can be expressed as:

$$A(var_i) \Rightarrow Cs(var_j), \qquad var_i \subseteq var_j \tag{3}$$

With respect then to Eq. 2 above, X coincides now with A – operationally, a *search pattern* – while the reasoning steps Y_1, Y_2 ... Y_n are used to produce the *search pattern(s)* Cs *to be used in place of* A. The restriction $var_i \subseteq var_j$ – all the variables declared in the antecedent A *must also appear* in Cs accompanied, in case, by *additional variables* – has been introduced to assure the logical congruence of the rules.

The 'transformation arrow' of Eq. 3, '⇒', has a *double meaning*:

- *Operationally speaking, the arrow indicates the direction of the transformation.* The original pattern (a specialisation of the left-hand side A of the transformation rule) is then *removed* and *replaced* by one or several new search patterns obtained through the updating, using the parameters of the original pattern, of the right-hand side Cs.
- From a logical/semantic point of view, we assume that between the information retrieved through Cs and the information we wanted to obtain through an instantiation of A there is an implication relationship, that normally denotes solely a possible (a weak) implication.

More formal details are given in [7: 212-216]. A representation of the above 'financial transfer' transformation is reproduced in Table 2. Note that its left-hand side (antecedent) corresponds to a partial instantiation of the template Behave:FavourableConcreteMutual, routinely used in NKRL to represent a (positive) mutual behaviour among two or more entities. Many of the transformation rules are characterized by the format of Table 2, which implies then *only one 'consequent'*. This is not true in general: examples of '*multi-consequent transformations*' can be found in [7, 12] – and in Table 4 below.

Table 2. A simple example of 'transformation' rule

t1)	BEHAVE	SUBJ	(COORD1 *var1 var2*)	⇒	RECEIVE	SUBJ	*var2*
		OBJ	(COORD1 *var1 var2*)			OBJ	*var4*
						MODAL	*var3*
						SOURCE	*var1*

var1, var2 = human_being_or_social_body
var3 = business_agreement, mutual_relationship
var4 = economic/financial_entity

To verify the existence of a relationship or of a business agreement between two (or more) people, try to see if one of them has received a 'financial entity' (e.g., money) from the other.

With respect now to the *hypothesis rules*, these allow us to build up automatically a sort of '*causal explanation*' for a narrative information (a predicative occurrence c_j) retrieved within a NKRL knowledge base using *Fum* and a search-pattern in a querying-answering mode. In a hypothesis context, the 'head' X of Eq. 2 is then represented by a predicative occurrence, c_j. Accordingly, the 'reasoning steps' Y_i of Eq. 2 – called 'condition schemata' in a hypothesis context – *must all be satisfied* (for each of them, at least one of the corresponding search patterns p_i must find a unification within the base) *in order that the set of* $c_1, c_2 \ldots c_n$ *predicative occurrences retrieved in this way can be interpreted as a context/causal explanation of the original occurrence* c_j. For example, let us suppose we have retrieved, in a querying-answering mode, an information like: "Pharmacopeia, an USA biotechnology company, has received 64,000,000 dollars from the German company Schering in connection with an R&D activity" that corresponds then to c_j. We can then be able to automatically construct, using a 'hypothesis' rule, a sort of 'causal explanation' of this event by retrieving in the knowledge base information like: i) "Pharmacopeia and Schering have signed an agreement concerning the production by Pharmacopeia of a new compound" (c_1) and ii) "in the framework of the agreement previously mentioned, Pharmacopeia has actually produced the new compound" (c_2).

An interesting, recent development of NKRL concerns the possibility of making use of the two above modalities of inference *in an 'integrated' way* [12].

In Table 3, we supply the informal description of the reasoning steps ('condition schemata') that, making use of the hypothesis tools, must be validated to prove that a generic 'kidnapping' corresponds, in reality, to a more precise 'kidnapping for ransom'. When many reasoning steps must be *simultaneously* validated, as usual in a hypothesis context, a failure is always possible. To overcome this problem – and, at the same time, discover all the possible *implicit information* associated with the original data – 'transformation' and 'hypotheses' can be combined: in practice, *it is possible to make use of 'transformations' within a main 'hypothesis' inferencing environment.* This means that, whenever a 'search pattern' is derived from a 'condition schema' of a hypothesis to implement, using *Fum*, one of the steps of the reasoning process, we can use it 'as it is' – i.e., as originally coded when the inference rule has been built up – but also in *a 'transformed' form if the appropriate transformation rules exist within the system.*

Table 3. Inference steps for the 'kidnapping for ransom' hypothesis

(Cond1)	The kidnappers are part of a separatist movement or of a terrorist organization.
(Cond2)	This separatist movement or terrorist organization currently practices ransom kidnapping of particular categories of people.
(Cond3)	In particular, executives or assimilated categories are concerned.
(Cond4)	It can be proven that the kidnapped is really a businessperson or assimilated.

Table 4. 'Kidnapping' hypothesis in the presence of transformations

(`Cond1`) The kidnappers are part of a separatist movement or of a terrorist organization.

– (**Rule `T3, Consequent1`**) *Try to verify whether a given separatist movement or terrorist organization is in strict control of a specific sub-group and, in this case,*

– (**Rule `T3, Consequent2`**) *check if the kidnappers are members of this sub-group. We will then assimilate the kidnappers to 'members' of the movement or organization.*

(`Cond2`) The movement/organization performs ransom kidnapping of specific classes of people.

– (**Rule `T2, Consequent`**) *The family of the kidnapped has received a ransom request from the separatist movement or terrorist organization.*

– (**Rule `T4, Consequent1`**) *The family of the kidnapped has received a ransom request from a group or an individual person, and*

– (**Rule `T4, Consequent2`**) *this second group or individual person is part of the separatist movement or terrorist organization.*

– (**Rule `T5, Consequent1`**) *Try to verify if a particular sub-group of the separatist movement or terrorist organization exists, and*

– (**Rule `T5, Consequent2`**) *check whether this particular sub-group practices ransom kidnapping of particular categories of people.*

– …

(`Cond3`) In particular, executives or assimilated categories are concerned.

– (**Rule `T0, Consequent1`**) *In a 'ransom kidnapping' context, we can check whether the kidnapped person has a strict kinship relationship with a second person, and*

– (**Rule `T0, Consequent2`**) *(in the same context) check if this second person is a businessperson or assimilated.*

(`Cond4`) It can be proven that the kidnapped person is really an executive or assimilated.

– (**Rule `T6, Consequent`**) *In a 'ransom kidnapping' context, 'personalities' like physicians, journalists, artists etc. can be assimilated to businesspersons.*

Making use of transformations, the hypothesis represented in an informal way in Table 3 becomes then, in practice, *potentially equivalent* to the hypothesis of Table 4. For example, the proof that the kidnappers are part of a terrorist group or separatist organization (reasoning step Cond1 of Table 3) can be now obtained *indirectly*, transformation T3 of Table 4, *by checking whether they are members of a specific subset of this group/organization*. Note that transformations T2 and T6 of Table 4 imply only one step of reasoning, whereas all the residual transformations are '*multi-consequent*'.

4 Conclusion

In this paper, we have evoked first the *ubiquity* and the *importance* of the so-called 'non-fictional narratives'. We have then supplied some information about NKRL (Narrative Knowledge Representation Language), a *fully implemented* language/environment expressly designed to deal with the narratives domain. We have mentioned first the *knowledge representation aspects* of NKRL. In this case, the most important innovation consists in the addition of a an *ontology of events* to the 'classical' ontology of concepts. We have then shown that the NKRL's rules correspond to *high-level reasoning paradigms* like the search for causal relationships or the use of analogical techniques.

References

[1] Jahn, M.: Narratology: A Guide to the Theory of Narrative (version 1.8). English Department of the University, Cologne (2005), `http://www.uni-koeln.de/~ame02/pppn.htm`

[2] Soulier, E. (ed.): Le Storytelling, concepts, outils et applications. Lavoisier, Paris (2006)

[3] Handler Miller, C.: Digital Storytelling. In: A Creator's Guide to Interactive Entertainment, Focal Press, Burlington (2004)

[4] Güven, S., Podlaseck, M., Pingali, G.: PICASSO: Pervasive Information Chronicling, Access, Search, and Sharing for Organizations. In: Proc. of the IEEE 2005 Pervasive Computing Conference. IEEE Comp. Society Press, Los Alamitos (2005)

[5] Westermann, U., Jain, R.: A Generic Event Model for Event-Centric Multimedia Data Management in eChronicle Applications. In: Proc. of the 22nd Int. Conference on Data Engineering – eChronicles Workshop. IEEE Comp. Society Press, Los Alamitos (2006)

[6] Bal, M.: Narratology: Introduction to the Theory of Narrative, 2nd edn. University of Toronto Press, Toronto (1997)

[7] Zarri, G.P.: Representation and Management of Narrative Information – Theoretical Principles and Implementation. Springer, London (2009)

[8] Manola, F., Miller, E.: RDF Primer – W3C Recommendation 10 February 2004. W3C (2004), `http://www.w3.org/TR/rdf-primer/`

[9] McGuinness, D.L., van Harmelen, F.: OWL WEB Ontology Language Overview – W3C Recommendation 10 February 2004. W3C (2004), `http://www.w3.org/TR/owl-features/`

[10] Noy, F.N., Fergerson, R.W., Musen, M.A.: The Knowledge Model of Protégé-2000: Combining Interoperability and Flexibility. In: Dieng, R., Corby, O. (eds.) EKAW 2000. LNCS (LNAI), vol. 1937, pp. 17–32. Springer, Heidelberg (2000)

[11] Zarri, G.P.: Representation of Temporal Knowledge in Events: The Formalism, and Its Potential for Legal Narratives. Information & Communications Technology Law – Special Issue on Models of Time, Action, and Situations 7, 213–241 (1998)

[12] Zarri, G.P.: Integrating the Two Main Inference Modes of NKRL, Transformations and Hypotheses. Journal on Data Semantics (JoDS) 4, 304–340 (2005)

A Semantic Web Framework to Support Knowledge Management in Chronic Disease Healthcare*

Marut Buranarach[1], Thepchai Supnithi[1], Noppadol Chalortham[1], Vasuthep Khunthong[2], Patcharee Varasai[2], and Asanee Kawtrakul[1,2]

[1] National Electronics and Computer Technology Center (NECTEC)
112 Thailand Science Park, Phahon Yothin Rd.
Klong Luang, Pathumthani, Thailand 12120
{marut.bur,thepchai.sup,asanee.kaw}@nectec.or.th
[2] Department of Computer Engineering, Faculty of Engineering,
Kasetsart University, Bangkok, Thailand 10900
{vasutap,patcha_matsu}@hotmail.com

Abstract. Improving quality of healthcare for people with chronic conditions requires informed and knowledgeable healthcare providers and patients. Decision support and clinical information system are two of the main components to support improving chronic care. In this paper, we describe an ongoing initiative that emphasizes the need for healthcare knowledge management to support both components. Ontology-based knowledge acquisition and modeling based on knowledge engineering approach provides an effective mechanism in capturing expert opinion in form of clinical practice guidelines. The Semantic Web framework is adopted in building a knowledge management platform that allows integration between the knowledge with patient databases and supported publications. We discuss one of the challenges, which is to apply the healthcare knowledge into existing healthcare provider environments by focusing on augmenting decision making and improving quality of patient care services.

Keywords: Ontology-based Knowledge Management, Knowledge-based Decision Support, Clinical Information System.

1 Introduction

Chronic illness is typically defined as condition that requires ongoing activities from both the patient and care givers in its treatment. Chronic conditions, such as diabetes, heart diseases, hypertension, etc. are major public health problems in developing countries, as well as in developed countries. As reported in 2004, it was suggested that approximately 45 percents of the US population have chronic illness [1]. While current healthcare systems are designed primarily to treat acute conditions, specific focus is increasingly applied to people with chronic conditions [2]. Treatments of chronic conditions normally require planning and management to maintain the patients' health status and functioning.

* This work is collaboration with the Dept. of Medical Services, Ministry of Public Health.

F. Sartori, M.Á. Sicilia, and N. Manouselis (Eds.): MTSR 2009, CCIS 46, pp. 164–170, 2009.

Healthcare processes heavily depend on both information and knowledge [3]. Information systems are typically integrated into hospitals to support organization processes such patient record entry and management, result reporting, etc. Although medical databases and information management systems are common, healthcare knowledge, which is important for medical treatment, is rarely integrated in supporting healthcare processes. It has been recognized that integration of knowledge into institutional workflows can help to improve the quality and efficiency of healthcare delivery system [4].

This paper describes our ongoing project in applying knowledge management to augment diabetes healthcare processes. A healthcare knowledge management framework is designed to support two chronic care components: decision support and clinical information system. Ontology is used as a means for knowledge acquisition and modeling based on expert opinion in form of clinical practice guidelines. Ontology-based healthcare knowledge management framework is a core component that focuses on building a repository of knowledge resources to support knowledge activities, i.e. problem solving and decision making. The framework utilizes the Semantic Web technologies in providing a knowledge management platform that allows linking between the knowledge with patient databases and supported publications. Finally, we discuss one of the main challenges, which is to apply the knowledge resources into the existing healthcare provider environments by focusing on augmenting decision making and improving quality of patient care services.

2 Healthcare Knowledge Management for Chronic Disease

In the Diabetes Healthcare Knowledge Management project, we emphasize the need for healthcare knowledge management to support diabetes healthcare processes. The Chronic Care Model (CCM) [2] is a guide towards improving quality of healthcare for people with chronic conditions. The model aims at producing more informed and knowledgeable patients and healthcare providers that can result in higher quality of chronic care. Decision support and clinical information system are two of the main components for improving chronic care. These components must rely on relevant and reliable information and knowledge in order to assist healthcare providers to deliver higher-quality care service. Knowledge captured from clinical practice guideline (CPG) should be embedded into healthcare applications to assist healthcare providers' decision making. In addition, updates in the medical guideline knowledge are typically based on proven research studies and results, i.e. evidence-based medicine (EBM). As a result, convenient access to medical publication databases, such as PubMed[1], the Cochrane Library[2], etc. should be provided and linked with the guideline knowledge. The guideline knowledge can also be integrated with existing hospital databases, e.g. patient registries. For example, based on a patient's clinical data, a clinician may be automatically reminded about the routine examinations that the patient should receive based on the medical guideline recommendations. Together, they allow for knowledge-enabled chronic care components that provide support for the diabetes care processes.

[1] http://www.pubmed.gov/

[2] http://www.thecochranelibrary.com/

3 Ontology-Based Healthcare Knowledge Management Framework

3.1 Knowledge Resources

Ontology-based knowledge management [5], [6], [7] framework focuses on providing information and knowledge support for the knowledge-enabled chronic care services. In Fig. 1, the framework focuses on integration of three forms of knowledge resources: ontologies, patient registries and evidence-based healthcare resource repository. The Semantic Web technologies are adopted for building a knowledge management platform that allows various forms of data to be integrated and associated with the ontology-based knowledge structure [8]. Ontologies provide a means for knowledge acquisition and modeling of the relevant healthcare knowledge. Two types of ontologies are utilized. The first type is developed based on translation of existing clinical guideline documents. This ontology type is mainly used in providing structural schema for the data in patient registries, i.e. concept instantiation. It also contains sets of production rules that represent decision models defined in the guideline to support inferences. The second type is controlled medical vocabulary, i.e. Medical Subject Headings (MESH)[3], which is a standard set of controlled vocabulary arranged in hierarchical structure for indexing and retrieving medical publications. MESH terms and structure are utilized as lightweight ontology used for semantic-based indexing and access to the evidence-based resource repository. Vocabulary mapping is a process of translating CPG-based ontology concepts on into MESH terms in order to allow linking the different types of knowledge resources.

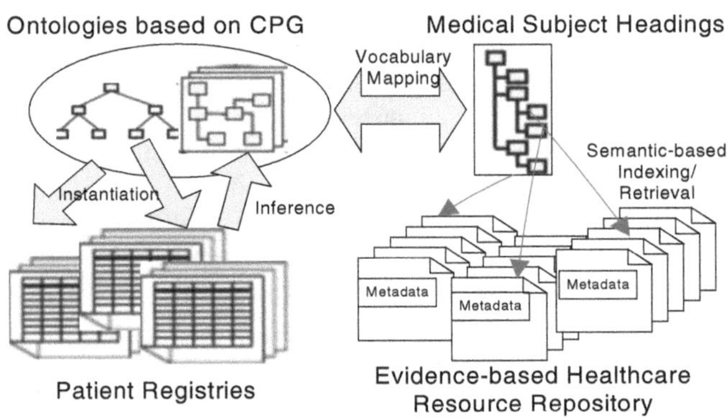

Fig. 1. Knowledge resources in ontology-based healthcare KM framework

3.2 Ontology Development

Ontology development in this project relies on expert opinions in form of clinical guidelines. Clinical guideline recommendations are normally provided based on the best available evidence. Thus, ontologies developed based on the guidelines typically

[3] http://www.nlm.nih.gov/mesh/

represent reliable knowledge and are agreeable in terms of expert opinions. In developing the ontologies, the clinical guideline for diabetes care issued by Thailand's Ministry of Public Health was translated from free text into a formal representation using the knowledge engineering approach.

The development can be divided into two major phases: knowledge acquisition and modeling and knowledge verification. In the knowledge acquisition and modeling phase, diabetes healthcare ontologies are designed and developed by a team of knowledge engineers and medical experts, i.e. medical doctors and public health specialists, using ontology development tools, which result in drafted ontologies. In the knowledge verification phase, a group of medical doctors who are diabetes specialists are invited to a public hearing session to verify and give additional comments on the drafted ontologies. The results are verified ontologies, which can be deployed in healthcare applications and services.

```
<owl:Class rdf:ID="DM_patient">
   <rdfs:subClassOf rdf:resource="#user"/>
</owl:Class>
<owl:DatatypeProperty rdf:ID="hasName">
   <rdfs:domain rdf:resource="#user"/>
   <rdfs:range rdf:resource="&xsd;string"/>
</owl:DatatypeProperty>
<owl:DatatypeProperty rdf:ID="hasAge">
   <rdfs:domain rdf:resource="#user"/>
   <rdfs:range rdf:resource="&xsd;int"/>
</owl:DatatypeProperty>
<owl:DatatypeProperty rdf:ID="hasFPG">
   <rdfs:domain rdf:resource="#user"/>
   <rdfs:range rdf:resource="&xsd;int"/>
</owl:DatatypeProperty>
<user rdf:ID="somchai">
   <hasAge rdf:datatype="&xsd;int">45</hasAge>
   <hasName rdf:datatype="&xsd;string">Somchai</hasName>
   <hasFPG rdf:datatype="&xsd;int">130</hasFPG>
</user>
```

Structural Knowledge

Instantiation

a) Structural Knowledge

```
<swrl:Imp rdf:ID="Rule-FPG1">
  <swrla:isRuleEnabled rdf:datatype="&xsd;boolean">true</swrla:isRuleEnabled>
  <swrl:head><rdf:List><rdf:first><rdf:Description>
     <rdf:type rdf:resource="&swrl;ClassAtom"/>
     <swrl:argument1 rdf:resource="#x"/>
     <swrl:classPredicate rdf:resource="#DM_patient"/>
  </rdf:Description></rdf:first><rdf:rest rdf:resource="&rdf;nil"/>
  </rdf:List></swrl:head>
  <swrl:body><rdf:List><rdf:first><rdf:Description>
     <rdf:type rdf:resource="&swrl;ClassAtom"/>
     <swrl:argument1 rdf:resource="#x"/>
     <swrl:classPredicate rdf:resource="#user"/>
  </rdf:Description></rdf:first><rdf:rest><rdf:List><rdf:first><rdf:Description>
     <rdf:type rdf:resource="&swrl;DatavaluedPropertyAtom"/>
     <swrl:argument2 rdf:resource="#y"/>
     <swrl:argument1 rdf:resource="#x"/>
     <swrl:propertyPredicate rdf:resource="#hasFPG"/>
  </rdf:Description></rdf:first><rdf:rest><rdf:List><rdf:first><rdf:Description>
     <rdf:type rdf:resource="&swrl;BuiltinAtom"/>
     <swrl:builtin rdf:resource="&swrlb;greaterThanOrEqual"/>
     <swrl:arguments><rdf:List>
        <rdf:first rdf:resource="#y"/>
     <rdf:rest> <rdf:List>
        <rdf:first rdf:datatype="&xsd;int">126</rdf:first>
     <rdf:rest rdf:resource="&rdf;nil"/>
     </rdf:List></rdf:rest></rdf:List></swrl:arguments>
</rdf:Description></rdf:first><rdf:rest rdf:resource="&rdf;nil"/></rdf:List></rdf:rest>
</rdf:List></rdf:rest></rdf:List></swrl:body></swrl:Imp>
```

b) Procedural Knowledge

Fig. 2. Diabetes healthcare knowledge modeling using Semantic Web standards

Two forms of knowledge are distinguished: structural and procedural knowledge.

1. Structural Knowledge. This knowledge type allows the computer to be able to make use of patient's clinical data. Thus, the knowledge provides structural information, i.e. schema, of patient's clinical data. This includes personal data, assessment and therapeutic data and history, which are critical for decision support and clinical information systems. OWL and RDF standards are utilized in defining structural knowledge and its instantiation respectively. Fig. 2a shows a simplified structural knowledge modeling and its instantiation using OWL/RDF syntax format.

2. Procedural Knowledge. This knowledge type represents the guideline recommendations that help to support decision making in medical diagnosis, treatment and planning processes. This process-oriented knowledge together with the patient's clinical data will assist the healthcare providers to make well-informed decisions. Numerous models have been developed to formally represent medical guideline knowledge, such as GLIF[4], DeGel[5], etc. We utilize the Semantic Web Rule Language (SWRL) [9] to construct decision models based on the clinical guideline knowledge. Fig. 2b shows a procedural knowledge modeling for a diabetes diagnosis process based on the defined structural knowledge using SWRL syntax format.

3.3 Evidence-Based Healthcare Resource Repository

Development and updates of medical guideline knowledge typically rely on proven research studies and results in medical literature, e.g. journals and publication databases. Guideline recommendations are usually provided with reference to publications. In our project, evidence-based healthcare resource repository is the component that provides a uniform and semantic-based access to medical literature. Similar to the federated medical search approach [10], [11], the repository consists of wrappers developed for each individual source for query translation and integration of search results. The medical subject headings (MESH) terms assigned to the retrieved resources are used as subject indexing terms that will allow for semantic-based search and navigation based on MESH hierarchical structure. To link the guideline knowledge with medical literature, vocabulary mapping between concept terms defined based on the medical guideline and MESH terms must be provided. This will enable the medical guideline knowledge to be linked with the supported evidence that can be accessed through a search interface as exemplified in Fig.3.

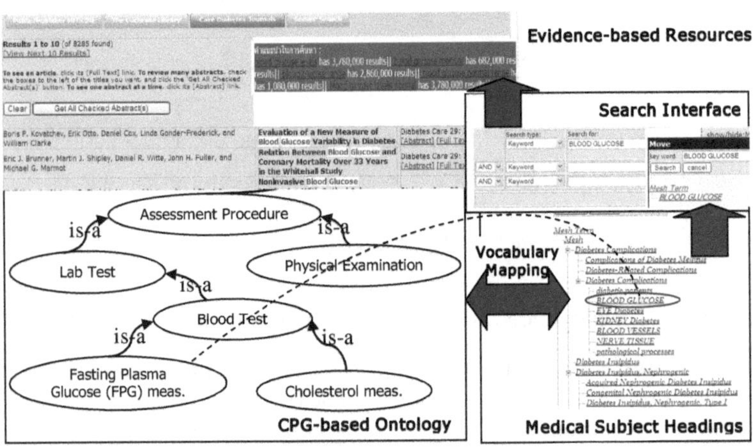

Fig. 3. Medical guideline knowledge linked with the evidence-based resources

[4] http://www.openclinical.org/gmm_glif.html
[5] http://www.openclinical.org/gmm_degel.html

4 Discussion

One of the challenges is to apply reliable knowledge into existing healthcare provider environment by focusing on augmenting decision making and improving quality of patient care services. The healthcare knowledge management approach [4] focuses on embedding knowledge into the clinical work environment that would not require the providers to explicitly request for, i.e. using automatic alerts and reminders. Medical errors and omissions in healthcare process may be minimized by means of detection and prevention. For example, based on medical knowledge from the guideline, an automatic reminder may be triggered when a patient has not received some recommended tests within some recommended periods. Alerts can be triggered to inform the provider when the patient's lab test data is above or below recommended values, which may affect the clinician's decision making. In addition, the knowledge based on medical guideline recommendations may be used to support monitoring of practices for assuring quality control.

In this paper, we have provided a healthcare knowledge management framework that is important for chronic disease care management. The framework is designed to support two chronic care components: decision support and clinical information system. The framework focuses on building of healthcare knowledge resources that link clinical guideline knowledge with patient registries and medical literature databases to support evidenced-based healthcare. The Semantic Web technologies provide an effective platform to support the knowledge management process. It supports modeling of ontologies and metadata in the standard formats that can enable semantic-based integration, processing and access of the knowledge resources. Our future work will focus on embedding and applying the knowledge to the existing healthcare applications and services and conducting user evaluations.

References

1. Anderson, G., Horvath, J.: The Growing Burden of Chronic Disease in America. Public Health Reports 119(3), 263–270 (2004)
2. Bodenheimer, T., Wagner, E.H., Grumbach, K.: Improving Primary Care for Patients with Chronic Illness. The Journal of the American Medical Association 288(14), 1775–1779 (2002)
3. Lenz, R., Reichert, M.: IT Support for Healthcare Processes - Premises, Challenges, Perspectives. Data & Knowledge Engineering 61(1), 39–58 (2007)
4. Abidi, S.: Healthcare Knowledge Management: the Art of the Possible. In: Riaño, D. (ed.) K4CARE 2007. LNCS (LNAI), vol. 4924, pp. 1–20. Springer, Heidelberg (2008)
5. Jurisica, I., Mylopoulos, J., Yu, E.: Ontologies for Knowledge Management: an Information Systems Perspective. Knowledge and Information Systems 6(4), 380–401 (2004)
6. Davies, J., Studer, R., Warren, R. (eds.): Semantic Web Technologies: Trends and Research in Ontology-based Systems. John Wiley & Sons, Chichester (2006)
7. Aldea, A., et al.: An Ontology-Based Knowledge Management Platform. In: Proc. of the IJCAI 2003 Workshop on Information Integration on the Web, IIWeb 2003 (2003)
8. Kozaki, K., et al.: Understanding Semantic Web Applications. In: Domingue, J., Anutariya, C. (eds.) ASWC 2008. LNCS, vol. 5367, pp. 524–539. Springer, Heidelberg (2008)

9. Horrocks, I., et al.: SWRL: A Semantic Web Rule Language Combining OWL and RuleML, W3C Member Submission (May 21, 2004), http://www.w3.org/Submission/SWRL/
10. Coiera, E., et al.: Architecture for Knowledge-based and Federated Search of Online Clinical Evidence. Journal of Medical Internet Research 7(5) (2005)
11. Bracke, P.J., et al.: Evidence-based Medicine Search: a Customizable Federated Search Engine. Journal of Medical Library Association 96(2), 108–113 (2008)

Ontological Enrichment of the *Genes-to-Systems Breast Cancer* Database

Federica Viti, Ettore Mosca, Ivan Merelli, Andrea Calabria, Roberta Alfieri, and Luciano Milanesi

Institute for Biomedical Technologies, National Research Council
Via Fratelli Cervi, 93, 20090 Segrate, Milan, Italy
{federica.viti,ettore.mosca,ivan.merelli,andrea.calabria, roberta.alfieri,luciano.milanesi}@itb.cnr.it

Abstract. Breast cancer research need the development of specific and suitable tools to appropriately manage biomolecular knowledge. The presented work deals with the integrative storage of breast cancer related biological data, in order to promote a system biology approach to this network disease. To increase data standardization and resource integration, annotations maintained in Genes-to-Systems Breast Cancer (G2SBC) database are associated to ontological terms, which provide a hierarchical structure to organize data enabling more effective queries, statistical analysis and semantic web searching. Exploited ontologies, which cover all levels of the molecular environment, from genes to systems, are among the most known and widely used bioinformatics resources. In G2SBC database ontology terms both provide a semantic layer to improve data storage, accessibility and analysis and represent a user friendly instrument to identify relations among biological components.

Keywords: Ontologies, database, breast cancer, data integration.

1 Introduction

Breast cancer is one of the most studied cancer types and there are many resources on the web that contain general as well as scientific information relevant to this pathology. Nevertheless the state-of-art lacks integrated and easily accessible databases related to breast cancer in a *systems biology* context, which means considering not only the biological components as standalone entities but even the systems they create, that are characterized by emergent properties. Cancer is considered a robust system disease [1], and often can be referred as a *network disease*, as reported in literature [2]: actually, genes do not work as standalone entities within the cell, but strongly interact among each other. Therefore it is important to face cancer study exploiting an integrated approach which includes a systemic view along with a reductionist one. Two limits must be highlighted in dealing with breast cancer system approach: the first concerns the spread in different resources of the available knowledge related to the functioning of these

F. Sartori, M.Á. Sicilia, and N. Manouselis (Eds.): MTSR 2009, CCIS 46, pp. 171–182, 2009.

molecular-systems [3]. The second concerns the difficulty of integrating information such as protein-protein interactions (PPIs), or post-translational modifications, which are protein changes that occur after protein synthesis. The same limit is present for molecular pathways data, that are accumulating in literature [4], which describe sets of biochemical processes. The integration of this information can lead to predictions (or at least suggestions) and hypothesis formulation for annotating new genes, discovering yet unknown biological pathways, finding more biological compliant interactions and reactions.

In order to promote and support resources a semantic integration approach to describe entities descriptions is fundamental and ontologies can play a crucial role in this context.

Belonging to the field of *knowledge representation*, an ontology is a collection of terms naming descriptors in a hierarchical structure that allows searching at various levels of specificity in a particular domain. Ontologies provide a formal representation of a set of concepts through the description of *individuals*, which are the basic objects, *classes*, that are the categories which objects belong to, *attributes*, which are the features the objects can have, and *relations*, that are the ways objects can be related one another.

Ontologies allow the direct and correct link of terms from the same domain, even when they belong to different sources, and the generation of concepts hierarchies, that facilitate creating relationships among diverse entities and their matching. In the context of systems biology an ontology based resource can not only improve data integration, but even simplify the information searching. A common problem concerns, for example, the generality of the term *cancer*. A direct query on that term will retrieve just the specific word in all the occurrences found into the screened resource. Employing a specialized ontology the output will be richer, including words such as *sarcoma* and *carcinoma* that will not be retrieved otherwise.

2 Background

Several public resources focused on cancer genes have been developed by scientists, some of which are general purpose, while other are specific for particular cancer subtypes. The most known sources have been analyzed and are listed here: Oncomine [5], developed for cancer gene expression analysis, The Tumor Gene Family of Databases [6], which contains information about genes that are targets for cancer-causing mutations, BreastCancerDatabase [7], which collects the molecular alterations associated with breast cancer, Breast Cancer Information Core Database [8], which stores mutations of main breast cancer genes. None of the existing resources provides an easily accessible databank dealing with breast cancer in a systems biology context and none of them make use of ontologies for data integration and mining. In this scientific context arose the development of the Genes-to-Systems Breast Cancer (G2SBC) Database [9], which provides a multi-level approach to breast cancer study. The resource, which collects information about breast cancer genes, proteins and mathematical models and provides a number of tools to analyze integrated data, is focused on the systems level other than on the components one, thus overcoming the limits of a classical data integration approach and enabling predictions and new hypothesis formulation other than data exploration. Most of the data about genes have been collected from NCBI *Entrez Gene* system [10] and their tissue expression pattern from NCBI *Unigene* [11], while data on proteins and protein domains have been retrieved from *Uniprot* [12] and *InterPro* [13] respectively. A primary gene association has been performed according to the *Gene Ontology* (GO) project. The published evidences of molecular alterations on breast cancer have been obtained from the Breast Cancer Database, pathway membership comes from *KEGG* [14] and *Reactome* [15]. Finally, protein-protein interactions (PPIs) have been taken from *BioGRID* [16] and cell cycle gene membership and mathematical models come from *Cell Cycle Database* [26]. Moreover, mathematical models related to breast cancer have been collected through a manual search in literature. The database relies on a relational database, managed by a MySQL server. A data warehousing approach has been chosen to develop the resource using a snowflake schema to organize the data. A set of Perl scripts are used to retrieve data from the external resources, to transform them according to the proposed schema, and to load them into the warehouse data model. Information is available to users through a User Interface (UI) which consists of two types of pages: (i) reports for genes, proteins, mathematical models and (ii) pages showing results of the available analysis tools. The later improvement of our resource is shown in this work and consists in the enrichment of G2SBC Database by embedding ontological information in order to better describe uploaded concepts and to provide suitable concept linkage. Literature presents many works concerning the use of ontologies in the biological field, especially pathology, cancer research and bioinformatics. Niepage et al. [17] developed an ontology based approach to improve data retrieval in telepathology, specifically related to human lung diseases. Always in the field of lung digital pathology Bontas et al. [18] describe a semantic web approach to perform a content-based retrieval system for text and image data. The ACGT project [19] presents a terminology source for transnational data exchange in oncology, which relies on an cancer ontology for research and management. Abidi [20] proposes a semantic web approach to develop a clinical decision support system to support family physicians while providing breast cancer follow-up care, based on a developed breast cancer ontology. Steichen et al. [21] built an ontology of morphological abnormalities in breast pathology to assist inter-observer consensus. In the bioinformatics field scientists often [22,23] exploit the power of ontologies, particularly Gene Ontology (GO) [24], in order to standardize their concepts and data, to enable statistical analysis and to facilitate information sharing by providing consistent descriptions of gene products. Even private companies such as The Mathworks implemented specific functions to handle GO annotation for data analysis (Matlab - Bioinformatics Toolbox 3.2) [25].

3 Methods

3.1 Ontologies

All the ontologies chosen to enrich the information maintained in the G2SBC database are *relational* ontologies, which means that the *relations* between entities (i.e. is_a, part_of) are ontologically more important than the entities themselves.

The most used biomedical ontology format is the Open Biomedical Ontologies (*obo*) format [27], which attempts to guarantee human readability, facility of parsing, extensibility, minimal redundancy. It relies in a list of *stanza*, a labeled section of the document which indicates that an object of a particular type is being described. Stanzas consist of a stanza name in square brackets ([Term]), followed by a series of new-line separated *tag:value* pairs (i.e. id: GO:00800381, name: ribosome assembly, is_a: GO:0034097, relationship: part_of GO:0045087).

To easily handle the *obo* format the open source Ontology Lookup Service (*OLS*) [28], developed by the European Bioinformatics Institute (EBI), has been integrated into G2SBC web site. The system provides a user-friendly web based single entry point to look into the ontologies for a single specific term, that can be queried using an useful auto-completion search engine. Otherwise it is possible to browse the complete ontology tree using AJAX library, querying the system through a standard SOAP web service described by a WSDL descriptor.

The whole OLS system has been installed on a local server to allow the interoperability, and customized by uploading the desired ontologies.

4 The Ontology Based G2SBC

4.1 Ontologies Description

The G2SBC consists of a web based graphical interface that allows to show the complete reports for genes, proteins and pathways, enriched by the ontological information more suitable for each level of biomolecular analysis. Whenever no *obo* version of the desired ontology was found, it has been generated starting from the most useful available resource. For example, the starting point for creating the KEGG ontology in *obo* format has been the KEGG Orthology Database [29].

During database population ontologies terms have been associated to entries, in order to map each database record with the most suitable ontological reference, providing a standardized and hierarchical organization of the knowledge in breast cancer genetic domain.

Exploited ontologies are: Brenda Tissue ontology (BTO), which supports the description of human tissues; Cell ontology (CL), which provides an exhaustive organization about cell type; Disease ontology (DOID), which focuses on the classification of breast cancer pathology compared to the other human diseases; Gene ontology (GO), which describes gene and gene products attributes, considering cellular components, biological processes and molecular functions classification; Protein ontology (PRO), which describes the protein evolutionary classes to delineate the multiple protein forms of a gene locus; Post-translational modification (PSI-MOD), which presents protein chemical modifications, classifying them either by the molecular structure of the modification, or by the amino acid residue that is modified; KEGG ontology (KO), which provides a pathway based annotation of the genes in all organisms; Systems Biology ontology (SBO), which is tailored especially for the context of computational modeling; MeSH thesaurus (MESH), which is a hierarchical controlled vocabulary able to index biomedical and health-related information.

4.2 The Ontology Layer Implementation

Each ontology has been punctually used to support specific information. BTO, CL, DOID supports the description of the pathology from the points of view of the tissue classification, the cell types and the disease identification. GO enriches genes information by linking the related biological concepts. PRO is used to define and organize knowledge in protein domain, even interconnecting information coming from other sources like Pfam [30] and Panther. PSI-MOD is used to define modifications that occur in a protein after its synthesis. KO is fundamental to describe the pathways where each considered protein intervenes. SBO is closely associated with the Systems Biology Markup Language (SBML) [31] and provides a complete hierarchical vocabulary to define and annotate molecular models. MeSH terms are inserted in every biological entity or phenomena annotation whenever its classification is provided in this thesaurus.

For what concerns GO and KEGG identifiers, ontological information has been obtained directly from the original resources. With regards to the other ontologies, Perl scripts have been implemented to automatically associate database core records to ontological terms.

The mapping of breast cancer related proteins on the PRO ontological terms was performed through the Protein Information Resource *SuperFamily* (PIRSF) [32], a database which, relying on the evolutionary relationships of whole proteins, allows annotation of both specific biological and generic biochemical functions and classifies proteins from superfamily to subfamily levels. PIRSF identifier is shown in PRO as an external reference (xref) for describing each protein at a family level. For this reason protein annotations within G2SBC database include their *InterPro* ids, which are directly related to PIRSF identifiers. In case PIRSF id is missing the same procedure was applied to Pfam identifier, a large collection of protein families, each represented by multiple sequence alignments and hidden Markov models (HMMs).

Post-translational information associated to proteins involved in breast cancer is retrieved from DBPTM [33], by *SwissProt* ids mediation. Each type of modification has been then mapped on PSI-MOD ontology.

Relation with SBO terms is directly performed during models implementation, by defining each SBML component (such as parameter, reaction and species) with the provided ontological identifiers.

Regarding BTO, CL, DOID no direct matching among database entries and ontological terms is needed: the developed system enables the availability of the whole ontology trees to define the relationships among different kinds of cells present in the breast, the types of tissues involved in the disease and the classification of all the subtypes of this pathology.

MeSH broad-spectrum ontology represents a good interface for accessing *PubMed* information, since the articles searching engine relies on this ontology for performing paper retrieval. This ontology is therefore handled in a different and separated way: the system directly queries the NCBI server system for each specific gene or protein term involved in the database and the link to MeSH is provided. Fig. 1 schematically presents the database ontological structure.

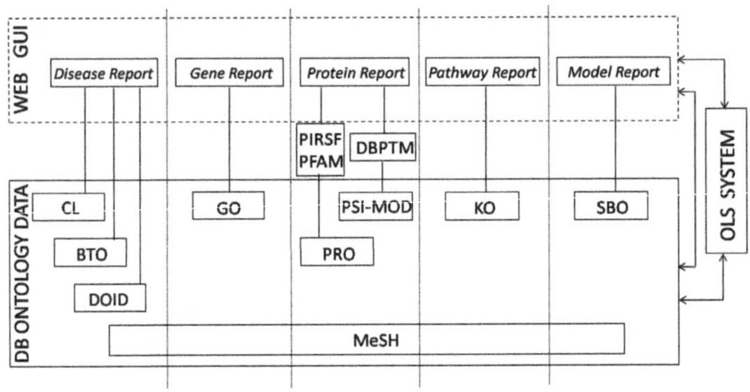

Fig. 1. Schema of the ontologies to support and complete traditional annotation data

5 Discussion

A header page, reachable from the side menu by clicking on *About breast cancer* link, introduces the context of breast cancer. The ontological description of the pathology concerning the disease and the related types of tissues and cells helps to focalize the hierarchical classification of the involved biological elements, providing an overview from different levels of organization.

The *Gene report* page is central in this database. For each gene it contains general information about its identity, tissue expression pattern, main gene products, associations with breast cancer, associations with cell cycle, related pathways and list of protein-protein interactions (PPI) of the gene products.

The previously described ontological annotation provides many advantages. First of all it allows performing more accurate queries on the studied genes. An exemplificative case is the searching for a pyruvate kinase (PKM2), which belongs to the glycolysis process, the biochemical pathway responsible for the conversion of the glucose to pyruvate with production of chemical energy molecules. Through its GO terms it can be easily associated to all the other genes which take part in the same process, i.e. the lactate dehydrogenase (LDHB) which can be identified even if it does not directly belong to the glycolysis process but intervenes in the anaerobic glycolysis, a child of the glycolysis in the hierarchy (see Fig. 2).

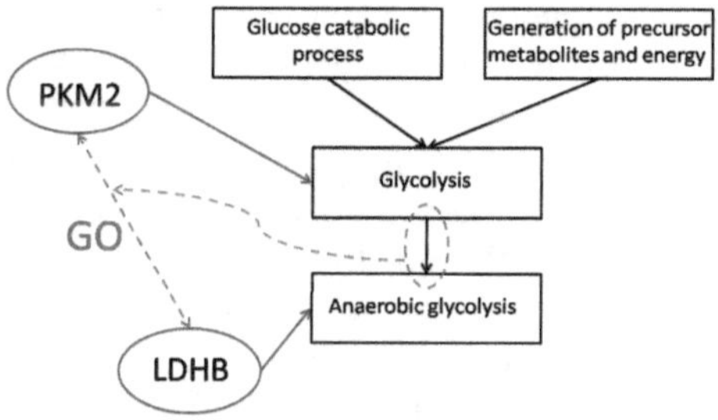

Fig. 2. An example of the exploitation of GO at the gene level in the breast cancer context. A relation can be determined among genes that belong to different but hierarchically connected terms. This case shows the pyruvate kinase PKM2 of the glycolysis pathway and the lactate dehydrogenase LDHB, which intervenes in the anaerobic glycolysis pathway.

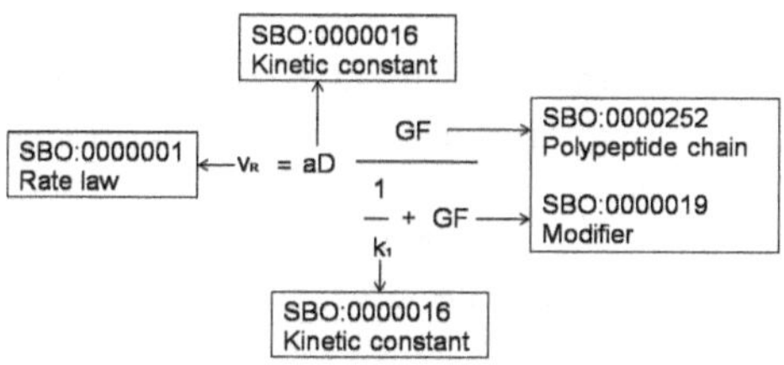

Fig. 3. The cyclin D synthesis kinetic equation. Each term has been associated to ontological identifiers which organize knowledge and allow performing queries over the mathematical models.

The GO support becomes crucial even in performing statistical analysis. For example its power has been exploited in the analysis of the *gene first neighborhood* in the PPI network, which is represented as a disconnected graph. This approach analyzes the set of annotations belonging to genes directly connected with a selected one in the PPI network, in order to suggest gene possible annotations. This strategy is particularly useful for not yet or poorly annotated genes, following the observation that the smaller the distance between two proteins in the PPI network, the higher the probability that they are involved in the same biological process.

Finally GO improves standardization, since genes annotations in the database are common to those available into most of the bioinformatics resources.

Through the system embedded OLS service, each GO term can be visually localized within the GO tree, both in the list and description of its children and in the visualization of the upper relations until the achievement of the GO roots.

Similar improvements are guaranteed for proteins, by using PRO. The *Protein report* contains general information about their identity and their domains and provides access to the OLS-system for the protein ontological classification. As previously said, this ontology is mainly devoted to delineate the multiple protein forms of a gene locus. A typical example is represented by the connection that is possible to create among isoforms *eta* and *zeta* of CD3 protein, which are both translation products of the CD247 gene but that would not be joined together without relying on a suitable ontology. Post-translational information has been annotated by exploiting PSI-MOD ontology.

For what concerns the evaluation of the molecular pathways associated to the genes involved in the breast cancer, information is provided in the *Pathway report*. Gene data have been associated to KEGG Pathway database information in order to provide the list of pathways in which each gene product is involved. Ontology terms can be even graphically shown by the creation and embedding of the *obo* version of KO into the OLS system. The ontological approach in this level of molecular biology is very interesting, particularly in query execution. For example it can be considered the interest in studying the genes and proteins involved in the apoptosis, the process by which cells undergo a controlled death. Without an ontology base, the output would concern just the terms directly associated to that process. On the contrary, the use of the ontology allows the retrieving of all the truly involved biological components, i.e. even those annotated as belonging to anoikis, which is an apoptosis process triggered by loss of contact with the extracellular matrix.

The G2SBC database even contains a *Model report* section, which shows the mathematical models involved in the breast cancer onset. In this context the ontological approach allows the performance of queries over the developed mathematical models, providing natural language descriptions that support information searching. An example is reported in Fig. 3, showing the kinetic equation of the biochemical reaction that describes the cyclin D synthesis from aminoacids, included in a cell cycle model. The ontology exploitation allows to define, classify and organize each equation term, thus providing a searchable field to records that could not be queried otherwise.

Finally the web site provides the automatic link to MeSH ontology, thus promoting the related semantic web search over articles and other literature resources maintained in PubMed server.

5.1 Improvement Related to Ontologies Exploitation

To better highlight the usefulness of an ontological layer to support integration and mining, here we propose a quantitative approach, aimed to compare the effectiveness of data analysis performed through ontological and a not ontological systems. The exploited approach is formalized below.

Let us consider the set of properties $\mathcal{P} = \{P_1, \ldots, P_n\}$, that are useful to represent the knowledge available for the genes $G = \{g_1, \ldots, g_m\}$. Each property P_i represents a particular type of information, such as the tissue expression or the molecular pathways in which the gene is involved, and is structured as a graph $P_i = (T_i, R_i)$, where T_i is the set of strings representing concepts relative to the considered property and R_i is the set of the relations among concepts. According to this formalization, we can define the annotation of a single gene g_j

Table 1. Similarity scores (s) between couples of genes using annotations $(\mathcal{A}_j, \mathcal{A}_k)$ (first and second rows) and annotations $(\mathcal{A'}_j, \mathcal{A'}_k)$ (third and fourth rows)

Gene Names	BARD1	CHEK2	DMP1	ELMOD3
BRCA1	0.263	0.080	0.030	0.000
BRCA2	0.100	0.080	0.073	0.000
BRCA1	0.340	0.230	0.117	0.000
BRCA2	0.323	0.280	0.120	0.010

as the set of sets $\mathcal{A}_j = \{A_{j,1}, \ldots, A_{j,n}\}$, where $A_{j,i} \subseteq T_i$. The similarity between the annotations $(\mathcal{A}_j, \mathcal{A}_k)$ of genes (g_j, g_k) is calculated through the score function $s : \mathcal{T} \times \mathcal{T} \rightarrow [0,1] \in \mathcal{Q}$, where

$$s(\mathcal{A}_j, \mathcal{A}_k) = \frac{1}{n} \sum_{i=1}^{n} \frac{|A_{j,i} \cap A_{k,i}|}{|A_{j,i} \cup A_{k,i}|}.$$

The function maps couples of concepts sets $\mathcal{T} = \{T_1, \ldots, T_n\}$ to the interval $[0,1]$ of rationals $\mathcal{Q}$ where 0 indicates that there are not concepts in common between the annotations and 1 means that all concepts are the same.

The availability of the relations R_i allows to expand $A_{j,i}$ by including all the neighborhoods of the concepts in $A_{j,i}$ that rely at a given distance d on the graph P_i, thus obtaining the new set $A'_{j,i}$. This operation is justified by the fact that concepts that are close on P_i are supposed to be semantically similar.

The previous formalization has been used to test the improvement provided by the exploitation of the relations existing between concepts. First, we considered a couple of reference genes, which are well known to be involved in breast cancer (BRCA1 and BRCA2), and a list of target genes composed by two genes that present a similar set of annotations whit respect to the reference genes and that are involved in breast cancer (BARD1, CHEK2) and other two genes that have a weaker annotations similarity in relation to the reference and that are not involved in breast cancer (DMP1, ELMOD3). Properties chosen to describe and compare those genes are: tissue expression, which includes the list of tissues where gene expression has been evidently localized; protein-protein interactions, which presents the set of proteins that are proved to interact physically with the considered gene product; Gene Ontologies Project Biological Functions, Biological Processes and Biological components.

We calculated the scores obtained by comparing each target with each reference. The score has been computed into two cases: by exploiting just gene annotations, $(\mathcal{A}_j, \mathcal{A}_k)$, and by considering gene annotations enriched by the inclusion of the concepts at $d = 1$ from the original concepts on the graph $P_i = (T_i, R_i)$, $(\mathcal{A'}_j, \mathcal{A'}_k)$. The most discriminant parameters, which determine the main differences between the two conditions (with or without the ontology graph support), are the ontology based terms, in particular the GO ones. The results, Table 1 and Fig. 4, focalize on the similarity scores based on those terms and highlight that the use of an ontological layer increases the score of similarity especially

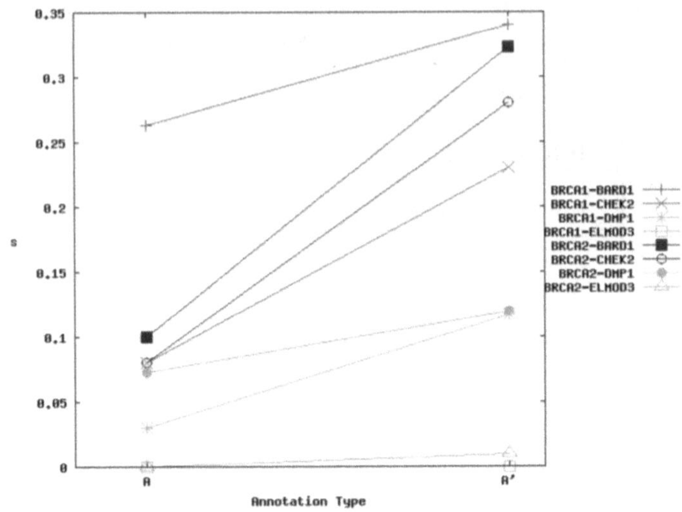

Fig. 4. Similarity scores (s) calculated considering the two annotation types where A= $(\mathcal{A}_j, \mathcal{A}_k)$ and A'= $(\mathcal{A'}_j, \mathcal{A'}_k)$

when considering similar genes. Indeed, the trend is that the gain is greater for genes BARD1 and CHECK2 then for DMP1 and ELMOD3.

The whole test thus shows that an ontological enrichment is crucial to go deeper in defining the objects relations, and that it represents an important support in highlighting hidden connections, maintaining correct and biologically compliant relations among the considered entities.

6 Conclusions

The presented work describes the enrichment of the G2SBC database through the use of ontologies. Suitable ontological terms accompany the annotations of biological components and related molecular systems, to better characterize function, localization, structure and modification of the breast cancer involved genes and proteins. Hierarchical structured terms intervene in all the main level of analysis of bioinfomatics field, from genomics to proteomics, metabolomics and system biology, allowing a better organization and linkage of the inter and intra levels information. Ontology approach represents a powerful instrument, whose effectiveness has been exploited in the presented work and particularly arises in performing efficient queries, solid statistics, and even reliable semantic web searching. To provide a user friendly management an EBI developed system has been locally installed and embedded within the application interface, gaining the possibilities either to visualize and browse the whole ontology trees or to look for specific ontological elements by their name of identifier.

Acknowledgments. This work has been supported by the Net2Drug, EGEE3, BBMRI, EDGE European projects and by the MIUR LITBIO (RBLA0332RH), ItalBioNet (RBPR05ZK2Z), BioPopGen (RBIN064YAT), CNR-Bioinformatics national initiatives.

References

1. Kitano, H.: Cancer as a robust system: implications for anticancer therapy. Nat. Rev. Cancer. 4, 227–235 (2004)
2. Hornberg, J.J., et al.: Systems biology-based drug design: Control of oncogenic signaling, Cellular and Molecular Biology 21: Computational Biology and Bioinformatics. In: Proc. Amer Assoc. Cancer Res. 47 (2006)
3. Regev, A., Shapiro, E.: Cellular abstractions: Cells as computation. Nature 419, 343 (2002)
4. Mathivanan, S., et al.: An evaluation of human protein-protein interaction data in the public domain. BMC Bioinformatics 7(suppl. 5) (2006)
5. Rhodes, D.R., et al.: Oncomine 3.0: genes, pathways, and networks in a collection of 18,000 cancer gene expression profiles. Neoplasia 9, 166–180 (2007)
6. Tumor Gene Family Databases, NAR Molecular Biology Database Collection entry number 155, `http://www.tumor-gene.org/tgdf.html`
7. Pendey Lab & Institute of Bioinformatics, `http://www.breastcancerdatabase.org`
8. Szabo, C., et al.: The Breast Cancer Information Core: Database design, structure, and scope. Human Mutation 16(2), 123–131 (2000)
9. Mosca, E., et al.: The BreastCancerDB: a data integration approach for breast cancer research oriented to systems biology. In: NETTAB 2008, Varenna (LC), Italy (2008), `http://www.itb.cnr.it/breastcancer`
10. Maglott, D., et al.: Entrez Gene: gene centered information at NCBI. Nucleic Acids Res 31, D26–D31 (2007)
11. Miller, G., et al.: IMAGE cDNA clones, UniGene clustering, and ACeDB: an integrated resource for expressed sequence information. Genome Res. 7, 1027–1032 (1997)
12. Bairoch, A., et al.: The Universal Protein Resource (UniProt). Nucleic Acids Res. 33, D154–D159 (2005)
13. Mulder, N.J., Apweiler, R.: The InterPro database and tools for protein domain analysis. In: Curr. Protoc. Bioinformatics, Chapter 2: Unit 2.7 (March 2008)
14. Kanehisa, M., et al.: KEGG for linking genomes to life and the environment. Nucleic Acids Res. 36, D480–D484 (2008)
15. Vastrik, I., et al.: Reactome: a knowledge base of biologic pathways and processes. Genome Biol. 8(3), R39 (2007)
16. Stark, C., et al.: BioGRID: a general repository for interaction datasets. Nucleic Acids Res. 1(34), D535–D539 (2006)
17. Niepage, S., et al.: Building Domain Ontologies in a retrieval system for lung Pathology. In: EUROCON 2005, Serbia-Montenegro, Belgrade (November 2005)
18. Paslaru Bontas, E., et al.: Ontology-based Knowledge Organization in a Semantic Web for Pathology. In: Inquiring Knowledge Networks on the Web Conference, IKNOW 2004 (2004)
19. Brochhausen, M., et al.: The ACGT Master Ontology on Cancer? A New Terminology Source for Oncological Practice. In: 21st IEEE International Symposium on Computer-Based Medical Systems (CBMS 2008), pp. 324–329 (2008)

20. Abidi, S.R.: Ontology-Based Modeling of Breast Cancer Follow-up Clinical Practice Guideline for Providing Clinical Decision Support. In: 20th IEEE International Symposium on Computer-Based Medical Systems (CBMS 2007), June 2007, pp. 542–547 (2007)
21. Steichen, O., et al.: Computation of semantic similarity within an ontology of breast pathology to assist inter-observer consensus. Computers in Biology and Medicine - Special Issue on Medical Ontologies 36(7-8), 768–788 (2006)
22. Viti, F., et al.: Ontology-based, Tissue MicroArray oriented, image centred tissue bank. BMC Bioinformatics 9(suppl. 4), S4 (2008)
23. Hu, Y., Kasabov, N.: Ontology-Based Framework for Personalized Diagnosis and Prognosis of Cancer Based on Gene Expression Data. In: Ishikawa, M., Doya, K., Miyamoto, H., Yamakawa, T. (eds.) ICONIP 2007, Part II. LNCS, vol. 4985, pp. 846–855. Springer, Heidelberg (2008)
24. Ashburner, M., et al.: Gene ontology: tool for the unification of biology. The Gene Ontology Consortium. Nat Genet. 25(1), 25–29 (2000)
25. `http://www.mathworks.com/products/bioinfo/description1.html`
26. Alfieri, R., et al.: The cell cycle DB: a systems biology approach to cell cycle analysis. Nucleic Acid Res. 36, D641–D645 (2008)
27. Smith, B., et al.: Relations in Biomedical Ontologies. Genome Biology 6, R46 (2005)
28. Cote, R.G., et al.: The ontology lookup service, a lightweight cross-platform tool for controlled vocabulary queries. BMC Bioinformatics 7(1), 97 (2006)
29. KEGG Orthology, `http://www.genome.jp/kegg/ko.html`
30. Finn, R.D., et al.: The Pfam protein families database. Nucleic Acids Research Database Issue 36, D281–D288 (2008)
31. Hucka, M., et al.: The Systems Biology Markup Language (SBML): A Medium for Representation and Exchange of Biochemical Network Models. Bioinformatics 9(4), 524–531 (2003)
32. Wu, C.H., et al.: PIRSF: family classification system at the Protein Information Resource. Nucleic Acids Res. 32(Database issue), D112–D114 (2004)
33. Lee, T.Y., et al.: dbPTM: an information repository of protein post-translational modification. Nucleic Acids Res. 1;34(Database issue), D622–D627 (2006)

An Ontology Based Approach to Information Security

Teresa Pereira[1] and Henrique Santos[2]

[1] Polytechnic Institute of Viana do Castelo
Superior School of Business Studies
Valença, Portugal
[2] University of Minho
School of Engineering
Information System Department
Guimarães, Portugal
tpereira@esce.ipvc.pt,
hsantos@dsi.uminho.pt
http://www.esce.ipvc.pt,
http://www.dsi.uminho.pt

Abstract. The semantically structure of knowledge, based on ontology approaches have been increasingly adopted by several expertise from diverse domains. Recently ontologies have been moved from the philosophical and metaphysics disciplines to be used in the construction of models to describe a specific theory of a domain. The development and the use of ontologies promote the creation of a unique standard to represent concepts within a specific knowledge domain. In the scope of information security systems the use of an ontology to formalize and represent the concepts of security information challenge the mechanisms and techniques currently used. This paper intends to present a conceptual implementation model of an ontology defined in the security domain. The model presented contains the semantic concepts based on the information security standard `ISO/IEC_JTC1`, and their relationships to other concepts, defined in a subset of the information security domain.

Keywords: Ontology, Information Security, Security Information Systems, Security Information Management.

1 Introduction

Tim Berners-Lee – the creator of the Web – considers ontologies to be a critical part of his latest work on the Semantic Web, envisioning the Semantic Web as being machine processable, leading to a better understanding of the content of Web pages by machines [1]. The proliferation of Web markup languages are supported by the growing needs of marking up information about the contents and services, instead of just presenting information. Assigning meaning to the

F. Sartori, M.Á. Sicilia, and N. Manouselis (Eds.): MTSR 2009, CCIS 46, pp. 183–192, 2009.

contents is actually the main concern of the information management experts. Further, ontologies have an important role to support browsing and search semantic contents, and in promoting interoperability for facilitation of knowledge management and configuration. The use of ontologies is not restricted to a specific domain. They practically are used to construct a model or a theory of a specific domain. In the context of information security the use of ontologies contribute to unify the terminology involved in classification and storage of security data. The tragic events of 09/11, as well as the ones that followed, forced many countries to review the efficiency and the efficacy of their information systems security [10]. The information management has become a main concern for the national security organizations, in addition to the interoperability between diverse information systems, in order to promote the exchange of security information. Actually, security organizations daily collect a large amount of data from different information sources, resulting in huge databases. On these databases an intensive analysis is performed, through the use of sophisticated data mining technologies with advanced statistical techniques to find important patterns, in order to be able to anticipate and prevent terrorist attacks [8]. However the results retrieved by data mining systems bring several questions regarding de false positives resulted from casual information associations, and meaningless positives. These errors can have potential negative side effects, for instance, leading innocent citizen to the confrontation of law enforcement services [2]. Actually the efficiency of the data mining technology used to foreseeing terrorism activities has not been proven in the academic literature [8]. The main problems are the amount, the heterogeneity and dynamic nature of data, and it becomes absolutely necessary to structure and organize it for knowledge retrieval. Actually, it is very difficult to incorporate knowledge or concepts abstracted from the low level data, into statistical analyses.The adoption of knowledge–based mechanisms seems to be an appropriate strategy in order to enable a better interpretation of data and therefore a better identification of the main features of information security threats and attacks. In this context, the knowledge organized according to the ontology under proposel, intends to help to organize and structure the terminology and concepts involved in this domain, based on the standards `ISO/IEC_JTC1`[5]. Furthermore, it enables a better interoperability among different security systems. In this paper we present an ontological semantic approach for information security and propose an implementation model of the ontology. The paper is structured as follows: in the section 2 it will be presented an overview about information security and the associated technologies currently used to perform data analysis. In section 3 we briefly present related works based on an ontology approach, in the knowledge management area. In section 4 it is presented the ontological needs in information security. In section 5 it is presented the implementation model of the ontology, which contains the semantic concepts specified in the information security scope, and the relationships to other concepts. Lastly, some conclusions are presented in section 6.

2 Overview of Information Security

Over the past decades, governments were especially concerned with borders control and security, and with the illegal immigration. The terrorist attacks on September 11, 2001, as well as all disturb that followed, forced governments and the national security organizations, all over the world, to review the efficiency and the efficacy of their information systems security. The Schengen Information System (SIS) is an extensive database that stores information of millions of objects and individuals data that is shared by the 15 European countries, for different purposes [3]. Currently efforts have being developed to extend the SIS to the 25 Schengen countries. The contents collected and stored in the SIS are not exclusively used in Europe. Additional information systems such as those maintained by Eurodac and Europol collect and share information to control immigration and safeguarding security [3]. In fact the primarily use of SIS was to control illegal immigration. However, the dramatic proportions of the terrorist threats, have been promoting the discussion to extend the use of the SIS to different purposes, namely the establishment of a new Visa Information System and the use and storage of biometrical data. These databases store, in a daily bases, an extensive amount of information, becoming difficult to perform manually assess on these data. The analyses of these massive and complex data are extremely difficult. Among the efforts was considered the use of data mining to uncover plans of terrorist actions. Even though the key goal is to produce more accurate and useful information, in order to enable the appropriate analysis and interpretation of data, in a given context [6]. Some security specialists consider predictive data mining as counterproductive, in the scope of national security [6]. Although the use of data mining technologies has proven that they are well suited to certain endeavors, particularly in the area of consumer direct marketing, or for example to identify credit card frauds, which relies in models constructed using thousands of known examples of fraud per year, terrorism has no similar evidence [6]. In fact, terrorist incidents occur a couple of times per year and they are typically distinct in terms of planning and execution, becoming extremely difficult to get a meaningful pattern and therefore enabling the definition of a standard bad behavior, which indicates a plan or a preparation for a terrorism attack. Unlikely consumers shopping habits and financial fraud, terrorism attacks does not occur with enough frequency to enable the definition of valid predictive models. Moreover, the risks to privacy and other civil liberties concerns several communities and raise important issues, as the likelihood of the false positives [2]. Predictive data mining requires a considerably amount of data. The aggregation of all the required data in a central system introduces a number of significant problems, including the difficulty of protecting so much sensitive data from misuse. Predictive data mining usually provides a considered amount of information, but useful knowledge comes from the context. Therefore the use of ontology to semantically structure knowledge stored in the security information systems, introduces a new perspective to the data analysis previously presented, since the ontology enables the description of the semantics content of the data. Knowledge based methods, such as ontologies, includes the description of the semantics content of the data, promoting a proper data analyze and consequently improving the performance of the security information services. In the following sections it is presented the related work that use ontology structure to express security related information for different types of resources, as well an overview of ontologies, followed by the presentation of a proposed implementation model defined in the context of information security.

3 Ontology Based Applications in the Knowledge Management

The World Wide Web Consortium (W3C) is developing efforts in a language to encode knowledge on the Web pages, in order to make it understandable to electronic agents searching for information – this language is called Resource Description Framework (RDF). The Defense Advanced Research Projects Agency (DARPA) and W3C are working together in the development of a DARPA Agent Markup Language (DAML) by extending RDF with more expressive structure in order to promote agent interaction on the Web [9]. In several areas researchers are now trying to develop standardized ontologies towards a common objective: to share and annotate information in their knowledge fields. Some relevant examples are presented in the area of the Medicine. In this domain standardized and large structured vocabularies have been developed, such as SNOMED (`http://www.snomed.org/`) and the semantic network of the United Medical Language System (UMLS – `http://www.nlm.nih.gov/research/umls/`). In public health domain several systems have been developed in order to detect disease-outbreak patterns and also with administrative and business purposes. One example is the billing and pharmaceutical sales records, collected for inventory and marketing purposes. Other example is the Realtime Outbreak and Disease Surveillance (RODS) project developed in the University of Pittsburgh to detect earlier outbreak of a disease. It is recognized the increase use of ontology based applications in the knowledge management of data analysis, in particularly in bioterrorism surveillance, in order to early detect and characterize an epidemic threat resulting from bioterrorism act. According to Buckeridge, an effective intervention depends on how quickly an epidemic can be detected, how well it can be characterized and how rapidly a response is initiated [4]. An experimental system named BioSTORM (Biological Spatio–Temporal Outbreak Reasoning Module) is a knowledge–based framework for real time epidemic surveillance [4]. In fact, the use of ontologies to model and annotate information and knowledge involved in syndrome and epidemic surveillance is the main feature of the BioSTORM system approach. Another relevant work is conducted by Raskin et al. [12]. They propose the use of natural language to define, in a unique way, the meaning of the main concepts about security incident information. Basically, two major components compose the ontology: a set of high–level incident related concepts and a method of classifying incident information [12]. Further they establish that the two components are related. The hierarchical representation of the concepts provides a structure that presents the concepts and their relations, improving the ability to: (1) gather, organize and record incident information; (2) extract data from incident information; (3) promote incident information interoperability, enabling the sharing and comparison of incident information; (4) use of incident information to evaluate and decide on proper courses of action; (5) use incident information to determine effects of actions over time [12]. The items specified clearly present what an ontology for the domain of information security can do. Finally, the ontology developed by Moreira et al. [11], Ontologies for Information Security Management and Governance presents a vocabulary of concepts and relations to represent information related to security incidents, to be used and understood at higher levels, such as security governance tools and people. This ontology is distinct from the ontology proposed in this project and presented in this paper due to the fact that it uses the security standard `ISO/IEC_JTCI1`[5] to represent concepts and relations in the information security domain, presenting a new and different structure of the concepts and relationships between the concepts.

4 Ontological Needs in Information Security

The General Accounting Office (GAO) recommends the establishment of common metadata standards for electronic information as a strategy to integrate and manage homeland security functions, including new procedure for data sharing across government [8]. The definition of metadata standards in the scope of security information will support the integration of heterogeneous data collected, enabling a uniform analytic and interpretation process of the data resource. It is recognized the fact that the attackers are smarter in creating more sophisticated security attacks, especially distributed attacks. In order to detect and withstand such attacks, security information systems should collaborate and communicate with each other by sharing a common vocabulary. A vocabulary based on ontologies is a powerful solution to achieve the above goals. The ontology–based approach enable to define the security concepts and their dependencies in a comprehensible way to both humans and software agents.

In fact the use of ontologies in the domain of security information management is just a proposal solution and needs further studies. However, the novelty of this solution regards the use of ontology to enhance the abstract metadata rich view on data semantics resources.

In summary the reasons that support a proposal ontological approach in the scope of information security management are provided as follows:

- Ontologies enable to specify semantic relationships between diverse concepts;
- Ontologies share a common understanding of structured information among different parties such as humans or software agents, which enables to be reasoned and analyzed automatically;
- Ontologies are reusable and able to evolve over time;
- Ontologies are shared among different agents to solve interoperability problems.

These reasons justify the popularity of the ontological approach has a theoretical foundation and has a methodological tool. In fact this is a new and an ambitious proposition in the information security domain in order to improve the current mechanisms used. Therefore we hope this topic generates discussion among the researcher community, in order to enrich and specify this view.

5 IS Ontology Conceptual Model

The architecture of the presented system has four layers: Data Resources, Conceptual Layer, Management Layer and Strategic Layer. The Data Resources Layer is composed of distributed data repositories, that contain security data provided by different and heterogeneous information sources, such as blogs, documents, reports of security events, et cetera. The data retrieved from the data resources will be mapped into the concepts of the ontology defined in the conceptual layer, enabling a better management of the security information, in the upper layer. The definition and adoption of a common terminology of the concepts, in the security information domain will help the security administrators to deal with security events more efficiently and therefore the implementation of security policies by the security information experts. Moreover, accurate information will promote the implementation of strategic security policies. The Figure 1 illustrates the information flow in the four layers, defined in the presented architecture.

The methodology used to develop the proposed ontology was the one presented by Noy and McGuiness [9]. This methodology was used in order to provide the necessary knowledge to develop the conceptualization phase. The implementation model of the proposed ontology for information security presented in the figure 1 comprises a set of concepts and their relations involved in the area, which are derived from established standards `ISO/IEC_JTC1`[5]. After defining

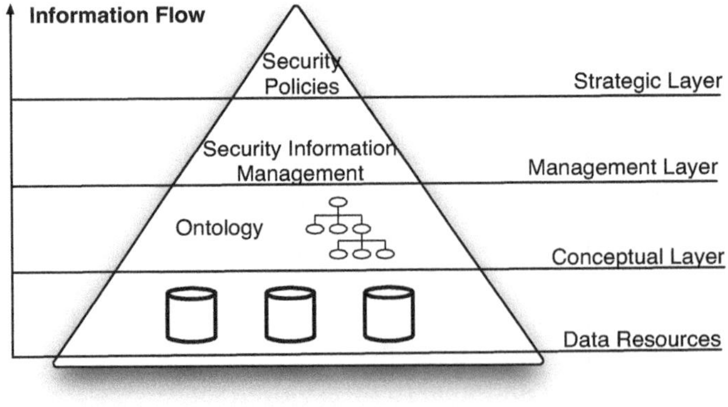

Fig. 1. System Architecture (adapted from [7])

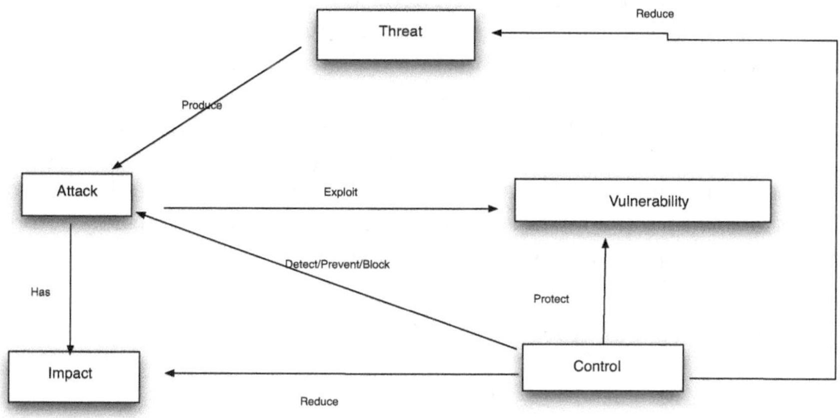

Fig. 2. Concepts and relationships of the Ontology

the concepts and the relationships to other concepts, the ontology for information security was formalized through the use of the W3C standard language for modeling ontologies Web Ontology Language (OWL). This web language has been developed by the Web Ontology Working Group as a part of the W3C Semantic Web Activity [13]. In spite of OWL has not been designed to specifically express security issues, it was selected because it is a W3C recommendation since February of 2004 and due to its expressiveness with superior machine interpretability. The OWL is build upon Resource Description Framework (RDF) and Resource Description Framework Schema (RDFS). In fact the OWL vocabulary is an extension of RDF and uses RDF/XML syntax.

Formally, an ontology is a tangled hierarchy of concepts related with properties. Figure 2 presents the main concepts related to the information security domain and the relationships among them. In this ontology were defined 5 main concepts and seven relationships. These concepts are described as following:

Threat – This concept represent the types of dangers against a given set of properties (security properties).
Attack – This concept represent the security incidents caused by some agent.
Impact – This concept represent the effects that a security incident can imply.
Control – This concept represent the mechanisms used to reduce or avoid the effects of an incident or to protect a vulnerability.
Vulnerability – This concept represent the weaknesses of the system.

The rational behind the ontology is structured as following: a threat *produce* an attack that may *has* impact. Attacks *exploit* one or more vulnerabilities and require a method, the opportunity and a given set of tools. By other side, the implementation of controls mechanisms aim to *reduce* the impacts of an attack, aim to *detect/prevent/block* an attack, aim to *protect* vulnerabilities and to *reduce* threats. The threat concept was included because it is important to correlate different attacks. The correlation can, for instance, help to establish which attacks succeed a threat.

In OWL description, the concepts correspond to classes and relations to properties. According to Smith et al., much of the power of ontologies comes from class–based reasoning [13]. In the proposed model the concepts defined such as threat, attack, impact, control and vulnerability, correspond to the root classes. Thus, the OWL representation of these classes is the following:

```
<owl:Class rdf:ID="Threat"/>

<owl:Class rdf:ID="Attack"/>

<owl:Class rdf:ID="Impact"/>

<owl:Class rdf:ID="Control"/>

<owl:Class rdf:ID="Vulnerability"/>
```

The properties enable to assert general facts about the classes. The following OWL sample presents the relations defined for the attack class.

```
<owl:ObjectProperty rdf:about="#has_Impact">
<rdfs:domain rdf:resource="#Attack"/>
<rdfs:range rdf:resource="#Impact"/>
</owl:ObjectProperty>

<owl:ObjectProperty rdf:about="#exploit_Vulnerability">
<rdfs:domain rdf:resource="#Attack"/>
<rdfs:range rdf:resource="#Vulnerability"/>
</owl:ObjectProperty>
```

The property `has_impact` is a relation between the class attack and the class impact, according to the model depicted in the Figure 2. The domain and range properties relate instances of the class attack to instances of class impact.

The structure of the presented concepts and their relations is the preliminary developing of the implementation model. However it needs further analysis and studies to complete the proposed ontology for information security.

6 Conclusions and Future Work

The tragic terrorism attacks occurred and their proportions forced many national agencies and governments to review the procedures used to manage information security. An astounding number of information security events is daily collected from distributed information sources, and stored. New approaches have being used to perform analysis on these data, such as Data Mining as well as sophisticated statistical techniques. However the efficiency of these techniques to predict attacks has been highly questionable, due to the fact that it is extremely difficult to establish a common pattern that fits a completely behavior, and because it can lead to false positives that can be generated and bringing potential negative side effects, for instance, leading innocent citizens to the confrontation of law enforcement services. The use of data mining systems for national security needs to be evaluated not only against the citizen privacy being subject of abuse, but also the likelihood of goal success. The ontology-based approach introduces a new perspective to model information in security domain. It allows the description of the data semantics in a machine-accessible way. In this paper we proposed an ontology–based approach to firm up and unify the concepts and terminology in the security information domain, based on the relevant `ISO/IEC_JTC1` standards. Adopting ontological approach as a theoretical foundation and a methodological tool is a promising new solution on the information security domain, and should be discussed by the research community.

The next steps of the ontology development process are: (1) completion of the proposed ontology, according to the points focused by the research community, (2) ontology evaluation, which includes the mapping security data into the ontology, and the development of the necessary applications to query and infer information security from this ontology.

References

1. Berners-Lee, T.: Semantic Web on XML. Presentation from XML (2000)
2. Anderson, S.R.: Total Information Awareness and Beyond. The Dangers of Using Data Mining Technology to Prevent Terrorism. Technical report, BORDC Bill *of* Rights Defense Committee (2007), `http://www.bordc.org/threats/data-mining.pdf`
3. Brouwer, E.: Data Surveillance and border control in the EU: Balancing efficiency and legal protection of third country nationals. Technical report (2005), `http://www.libertysecurity.org/article289.html?var_recherche=Data%20Surveillance`
4. Buckeridge, D.L., Graham, J., OÕConnor, M.J., Choy, M.K., Tu, S.W., Musen, M.A.: Knowledge-Based Bioterrorism Surveillance. In: AMIA Annual Symposium, San Antonio, TX (2002), `http://bmir.stanford.edu/file_asset/index.php/1147/SMI-2002-0946.pdf`
5. ISO/IEC FDIS 27001 Information technology – Security techniques – Information security management systems– Requirements, ISO copyright office. Geneva, Switzerland (2005)
6. Jonas, J., Harper, J.: Effective Counterterrorism and the Limited Role of Predictive Data Mining. Policy Analysis no 584, CATO Institute, December 11 (2006), `http://www.cato.org/pub_display.php?pub_id=6784`
7. Martimiano, L., Moreira, E.: The evaluation process of a computer security incident ontology. In: 2nd Workshop on Ontologies and their Applications (WONTO 2006), São Paulo, Brazil (2006)

8. Maxwell, T.A.: Information Policy, Data Mining, and National Security: False Positives and Unidentified Negatives. In: Proceedings of the 38th Hawaii International Conference on System Sciences, Hawaii (2005)
9. Noy, N.F., McGuinness, D.L.: Ontology Development 101: A Guide to Creating Your First Ontology. Technical Report SMI-2001-0880, Stanford Knowledge Systems Laboratory Technical Report KSL-01-05 and Stanford Medical Informatics (2001), `http://www-ksl.stanford.edu/people/dlm/papers/ontology-tutorial-noy-mcguinness-abstract.html`
10. Miller, R.R.: Information Management in the Aftermath of 9/11. Communications of the ACM 45(9) (2002)
11. Moreira, E.S., Martimiano, L.A., Brandão, A.J.d.S., Bernardes, M.C.: Ontologies for information security management and governance. Information Management & Security 16(2), 150–165 (2008)
12. Raskin, V., Hempelmann, C.F., Triezenberg, K.E., Nirenburg, S.: Ontology in information security: a useful theoretical foundation and methodological tool. In: Proceedings of the 2001 Workshop on New Security Paradigms. NSPW 2001, pp. 53–59. ACM, New York (2001)
13. Smith, M.K., Welty, C., McGuinness, D.L.: OWL Web Ontology Language Guide. W3C Recommendation February 10, 2004. Technical report, W3C (2004), `http://www.w3.org/TR/owl-guide/`

Reusability Evaluation of Learning Objects Stored in Open Repositories Based on Their Metadata

Javier Sanz[1], Salvador Sánchez-Alonso[2], and Juan Manuel Dodero[3]

[1] Universidad Carlos III, Madrid, Spain
jsrodrig@inf.uc3m.es
[2] Universidad de Alcala, Alcalá de Henares, Spain
salvador.sanchez@uah.es
[3] Universidad de Cádiz, Cádiz, Spain
juanma.dodero@uca.es

Abstract. Reusability is considered to be the key property of learning objects residing in open repositories. In consecuence, measurement instruments for learning object reusability should be developed. In this preliminary research we propose to evaluate the reusability of learning objects by a priori reusability analysis based on their metadata records. A set of reusability metrics extracted from metadata records are defined and a quality assessment of the metadata application profiles defined in repositories eLera and Merlot is exposed.

1 Introduction

Reusability can be defined as the extent to which a Learning Object can operate effectively for a variety of users in a variety of digital environments and a variety of educational contexts over time [10], bearing in mind that there are technical, educational and social factors affecting reusability and in most cases a degree of adaptation will be required for reuse.

This concept of reusability is a key issue on e-learning contents and systems. Providing reusable learning objects can facilitate its further development and adaptation, augment learning object development productivity, reduce development costs and improve quality of e-learning systems. Reusability is in fact an intrinsic characteristic of a learning object that can a priori provide a measure of its quality [14]. Besides the level of reuse of learning objects is empiric and observable and thus can be compared with specific metrics by a posteriori analysis of the data compiled from actual use. Nevertheless, studies on reusability indicators and design criteria that guarantee reusability are scarce [12]. Learning objects are described by metadata (ideally in a standard form), and given that this metadata information can be used to search learning objects in repositories, it is regarded as a vehicle for sharing and reuse [13].

The objective of this work is to develop a learning object reusability evaluation method using those elements of IEEE LOM [1] that have an impact on

F. Sartori, M.Á. Sicilia, and N. Manouselis (Eds.): MTSR 2009, CCIS 46, pp. 193–202, 2009.

reusability. To this aim, we will analyze the use of metadata in two different public, open learning object repositories: Merlot (http://www.merlot.org) and eLera (http://www.elera.net).

The structure of this paper is the following: Section 2 analyzes different approaches of learning object evaluation. In Section 3 we propose our set of reusability metrics extracted from metadata records. In Section 4, a quality assessment of the metadata application profiles used in repositories eLera and Merlot is exposed, and finally the paper concludes with acknowledgements and some conclusions about related work.

2 Learning Object Evaluation

Several initiatives have approached the evaluation of learning objects to provide an estimation of the guaranteed quality. Merlot classifies objects in seven discipline categories and compiles experts and users evaluations on three dimensions (i.e. content quality, usability and effectiveness as a learning tool) [16]. ELera extends this evaluation scheme by the LORI (Learning Object Review Instrument) tool [7], which evaluates aspects such as content quality, objective fulfillment, feedback and adaptation capability, motivation, presentation, usability, accessibility, reusability and standards compliance.

Summing up, usual learning object evaluation methods are based upon compiling opinions from users and experts about different aspects of a learning object. However, there exist a few exceptions such as the approach proposed by Ochoa and Duval [9], where a set of metrics to estimate the relevance of learning objects are calculated mainly from usage and contextual information, and the proposal of Zimmerman [20] that ranks learning resources according to adaptation effort needed for reuse.

In contrast to these iniciatives, we propose an aprioristic approach to the evaluation of learning objects reusability based upon the standardized descriptions of metadata that describe learning objects. We must note that metadadata in standardized format LOM it is a prerequisite to apply our reusability metrics to learning objects.

3 Learning Object Reusability Metrics Based on Metadata Records

We have based on evaluation methodology used to measure reusability of object-oriented software [6], adapted in the following steps:

1. Study and identify those learning object aspects and factors that might have an influence on the capability of reusing.
2. Define metrics to measure reusability factors that have been identified, based upon analysis of IEEE Learning Object Metadata (LOM) standards.
3. Develop a quality assessment of the metadata application profiles defined in the repositories eLera and Merlot.

4. In subsequent studies these metrics will be validated with the reusability data obtained from expert evaluations of eLera and Merlot for a significant set of learning objects of both repositories.

Common software metrics have been source of ideas to provide reusability metrics for learning objects, based upon the reusability factors [3]. Traditionally, software engineering based upon a design principle to strive for strong cohesion and loose coupling [2]. These two principles head for building maintainable software that easily adapt to new requirements. Since learning objects are designed for reuse, we analyzed how these principles apply to determine learning objects reusability. Although in the field of software engineering reusability metrics are mainly related to cohesion and coupling, the particular nature of learning objects lead us to analyze standardized IEEE LOM conformant metadata elements to evaluate other reusability factors, such as portability, size, complexity and difficulty of comprehension.

3.1 Cohesion

Cohesion analyzes the kind of relationships among different modules. A module, that can be different things depending on the language - a class, package, etc. - must realize a single task to be maximally cohesive [15]. Greater cohesion usually implies greater reusability [17]. Cohesion is a software quality indicator that, applied to learning objects, is fulfilled by the following elements:

- A learning object involves a number of concepts (LOM 9 Classification category). The less concepts, the greater the module cohesion [17].
- A learning object should have an only and clear learning objective [12]. The more learning objectives it has, the less cohesive it will be considered. The information about learning objectives. It is covered by educational objective in LOM 9.1 Purpose.
- The semantic density of a learning object (LOM 5.4 Educational category) indicates how concise it is. The more conciseness, the more cohesion for the learning object.
- A learning object must be self-contained to be highly cohesive [17]. LOM 7 Relation category defines as many instances as relationships the learning object has (notably is-version-of, has-version, is-format-of, has-format, references, is-referenced-by, is-based-on, is-basis-for, is-required-by, requires, is-part-of and has-part). For some types of relationships like references or requires we can say: the more relationship instances a learning object has, the less self-contained and, therefore, less cohesive. Moreover, LOM 1.8 Aggregation level element summarizes the level of aggregation of a learning object as ranging from 1 for single resources to 4 for a set of related courses. The lower level of aggregation, the more cohesion.
- Structure indicates the organizational structure of a learning object. It can be: Atomic, Collection, Networked, Hierarchical or Linear. We observed that there is a relationship between the level of aggregation of an object and its structure, e.g. an object with atomic structure will add a level of 1, whereas the other types of structures have values ranging from 2-4 [1].

We can conclude that learning object cohesion depends on semantic density, the number of relationships, aggregation level, number of concepts dealt with, and number of learning objectives covered. These metadata elements can be source for a valid estimation of the reusability of a learning object.

3.2 Size and Complexity

Software size and complexity can be measured through several methods, e.g. lines of code, McCabe's software complexity, Halstead's difficulty, etc. The size of a learning object indicates its granularity, and in general terms, granularity provides clear information on learning object reusability, since fine-grained objects are more easily reusable [18]. Learning object granularity depends on the following LOM elements:

- Size: the number of bytes of a learning object. These data should be weighted depending on the learning object format, as it can be interpretated differently depending on the type of content:while a 2MB plain text is a huge text, the same size for a video could be considered small according to today's quality and image resolution.
- Duration: the estimated time to run the learning object.
- Typical Learning Time: the estimated time required to complete the learning object. This is a reliable source of information to estimate the size and complexity of a learning object.
- Resource type: Specific kind of learning object, exercise, simulation, etc.

3.3 Portability

In the field of portability, metrics measure the ability to transfer software from one system to another. These metrics are based on the analysis of modularity and hardware/software context independence [11]. Learning objects portability can be measured as the context dependence at technological and socio-educational levels. The few dependendencies found, the more portable the learning object.

Technical Portability. The following LOM values can be analyzed when considering portability at a technical level:

- Format: it determines the learning object components delivery format, such as video/mpeg, application/x-toolbook, text/html, etc. Some formats are more readily portable (e.g. text/html is more widespread than application/x-toolbook).
- Requirements: it involves the hardware and software required to run the object. The higher the complexity of the requirements, the less portable the object is.

Educational Portability. Regarding educational portability, we can deal with vertical or horizontal portability [4]. Vertical portability means the possibility for a learning object to be used and reused on different educational levels. In contrast, horizontal portability determines the inter-disciplinarity of the object. We have considered the following IEEE LOM elements of metadata:

- Context: potential educational contexts in which a learning object can be used (i.e. school, high school, higher education, professional training, etc.) Educational portability is greater for those objects that can be used and reused in more different educational contexts.
- Typical age range: potential age ranges of the users which could benefit from using the object. Educational portability increases as the number of ranges grows.
- Language: the human languages supported by the object. An object is more reusable if it is available in more usual languages.
- Classification: information used to classify a learning object within the discipline it belongs or is related to. The more specific the classification scheme, the less reusable the learning object can be.

This section is summarized in Figure 1.

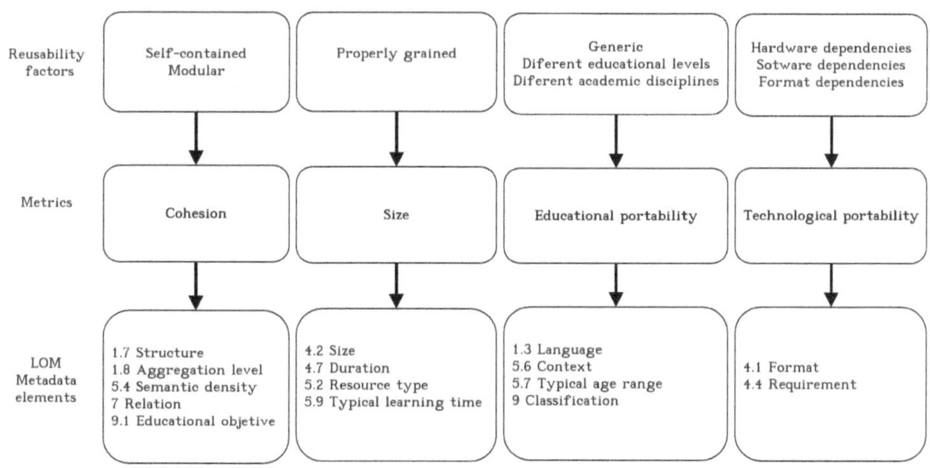

Fig. 1. Relationships between reusability factors, metrics and LOM metadata elements

4 Quality Evaluation in the Use of Metadata in Repositories eLera and Merlot

The quality of metadata records is a critical point for search and locate learning objects in repositories. Specifically, for our empirical study we needed a set of LOM metadata elements, sumarized in Figure 1, correctly filled to properly use our reusability metrics. To develop our research a set of learning objects were selected from both repositories. We selected learning objects with evaluations for in a later stage, be able to evaluate our metrics using this meaningful information.

From eLera were initially selected all objects with at least one evaluation (120 objects). We first preprocessed and examined in detail this set of 120 objects. As a result, 17 objects had to be discarded because they were not available at the time of our study, leaving the final study population in 103 objects. From Merlot, only those objects added to the repository from 2005 to 2008 including peer review evaluation and comments were selected. Thus, we analyzed a final set of 91 objects.

Unfortunately, metadata obtained in our study from eLera and Merlot repositories had different problems that describe the current use of metadata in open learning object repositories.

4.1 Source of Metadata

Concerning the source of metadata, we found significant differences between the repositories analysed:

- In eLera, authors fill metadata elements in when the object is added to the repository, and anyone can add comments to evaluate the objects. These comments are classified according to the nine areas of evaluation.
- In Merlot we have found three different levels of metadata: Merlot Material Detail (metadata provided by the author), Merlot Peer Review (Metadata provided by reviewers) and Merlot Comments (information provided by users). We discarded these unstructured comments because it is difficult to obtain meaningful information of them for our study.

In general, We conclude that metadata provided by the reviewers are often more accurate than those supplied by other sources.

4.2 Correspondence between eLera, Merlot and LOM Metadata Elements

Merlot and eLera adapt LOM defining application profiles that enable increased semantic interoperability in their communities, in a way that preserves ful compatibility with the larger LOM context [5]. The fundamental techniques for the definition of application profiles include:

- Giving elements a mandatory status.
- Restricting the value space of data elements.
- Imposing relationships between elements.
- Not including some LOM elements.
- Identifying taxonomies and classifications schemas.

These application profiles define a set of metadata elements that only cover a portion of metadata elements described in IEEE LOM. Table 1 summarizes LOM metadata elements needed for calculate our reusability metrics and compares them to the metadata elements found in Merlot and eLera application profiles. We remark those cases that require a human inspection of the learning object

Table 1. Metadata elements from eLera, Merlot and LOM needed for each metric

	LOM	eLera	Merlot Peer Review	Merlot Material Detail
Cohesion	Semantic density Relation Aggregation level Educational objective Structure	Resource type Description Inspection	Learning goals Overview Inspection	Description Inspection
Size	Size Duration Typical learning time	Resource type Description Inspection	Type of material Overview Inspection	Material type Description Inspection
Educational portability	Context Typical age range Language Classification	Educational context Description Language Subject	Target student population Overview Recommended use Prerequisite (knowledge or skills)	Primary audience Description Language Categories
Technical portability	Requirement Format	Description Inspection	Technical requirements	Technical requirements Technical Format

to complete the information provided by LOM metadata elements and needed to calculate the metrics.

We can observe that only the metadata required to calculate education portability are covered by the eLera metadata, for the rest of our metrics of reusability it will be necessary to inspect the object to complete the required IEEE LOM metadata elements.

Melot makes available some metadata elements needed to calculate technology portability and education portability, but if we want to estimate the size and cohesion it will be necessary to "manually" inspect the learning object, which hinders any automated calculation of the reusability.

4.3 Unstructured Metadata

We have observed that much of the information describing the learning object is in unstructured format. Although these free-text metadata elements are useful, they involve various problems: Increase errors in the completion of the metadata and hinder any possible automatic calculation of the reusability. Table 2 shows structured and unstructured metadata found in metadata elements selected from IEEE LOM, eLera and Merlot.

Table 2. Structure of Metadata values

	LOM	eLera	Merlot Peer Review	Merlot Material Detail
Structured	Semantic density Relation Aggregation level Size Context Language Structure	Subject Language Educational context Resource type		Material type Categories Primary audience Language Technical format
Unstructured	Format Requirement Typical age range Educational objective Duration Typical learning time Classification	Description	Overview Learning goals Target student population Recommended use Technical requirements Type of material Prerequisite (knowledge or skills)	Technical requirements Description

4.4 Different Value Spaces

Even though some metadata elements share the same value spaces in IEEE LOM and eLera, e.g. *Type resource* and *Context*, most use different value spaces and different ways of structuring information. It would be required to search and locate reusable learning objects that metadata elements from different repositories that represent the same concept share the same value space.

4.5 Metadata Completeness

To calculate our reusability metrics it is required that all metadata elements presented in Table 1 are complete. However, we have observed that some of them are optional. Specifically in eLera the following metadata elements are not mandatory to fill: *Description, Resource type and Educational context.* Whereas in Merlot, in Material Detail are not mandatory to fill: *Technical Requirements and Technical Format.* Table 3 presents the results of analyzing the presence of optional elements in the sets of learning objects both from Merlot and from eLera.

One problem encountered, for our study of reusability, is the presence of default values for some elements. For instance, in eLera's Educational Context, Higher education is the default option, whereas for the equivalent element in Merlot, Primary audience, is College General Ed. This may affect the proper description of the objects, because when metadata is filling, it is always easier to keep the suggested default value instead of entering the correct value.

We may also cite the presence of inadequate values for some metadata elements. By example in *Technical Format* of Merlot material details, we have found 13 learning objects with the value set to other, which doesn't add any information to the proper description of the learning object.

All these problems identified in the metadata elements constitute a limitation on the reusability of learning objects.

Table 3. Presence of optional elements

	Optional	Completion %
eLera	Resource type	67%
	Description	75%
	Educational Context	100%
Merlot Material Detail	Technical Requirements	46%
	Technical Format	67%
Merlot Peer Review	Overview	100%
	Learning Goals	100%
	Target Student Population	100%
	Prerequisite Knowledge or Skills	98%
	Technical Requirements	96%
	Type of Material	98%
	Recommended use	99%

5 Conclusions

The metrics based on metadata proposed in this paper could be useful to identify reusable learning objects, but quality and completeness of metadata elements would be helpful to get a proper result.

Metadata relating to education portability seem to be more properly filled in what we could propose a further investigation to automate a calculation of this metric.

For a full and proper development of our research in the calculation of reusability following conditions would be necessary:

- Every metadata elements of LOM should have their corresponding element in metadata profiles from eLera and Merlot.
- Each metadata element that represents the same concept in each repository should shares the same value space.
- Every metadata elements should be mandatory for avoid lack of information needed for reuse.
- Every metadata elements should present structured values to avoid errors in the filling process and to facilitate the automated processing of metadata.

In future works these metrics will be validated with the reusability data obtained from expert evaluations of learning objects available in eLera and Merlot repositories.

Acknowledgements

This work has been supported by project MARIA (reference code CCG08-UAH/TIC-4178), funded by the Comunidad de Madrid and the University of Alcala.

References

1. IEEE Learning Technology Standards Committee (LTSC): Learning Object Metadata (LOM), Final Draft Standard, IEEE 1484.12.1-2002
2. Boyle, T.: Design principles for authoring dynamic, reusable learning objects. Australian Journal of Educational Technology, 46–58 (2003)
3. Cervera, J.F., López, M.G., Fernández, C., Sánchez-Alosno, S.: Quality Metrics in Learning Objects. In: Proceeding of MTSR 2007 Conference, Corfu, Greece (2007)
4. Currier, S., Campbell, L.: Evaluating learning resources for reusability: the dner and learning objects study. In: Proceeding of The Australasian Society for Computers in Learning in Tertiary Education (ASCILITE 2002), Auckland, New Zealand (2002)
5. Duval, E., Hodgins, W.: A LOM Research Agenda. In: WWW 2003 Conference, Budapest, Hungria (2003)
6. Etzkorn, L.H., Hughes, W.E., Davis, C.G.: Automated reusability quality analysis. Information and Software Technology 43, 295–308 (2001)

7. Nesbit, J., Belfer, K., Leacock, T.: Learning Object Review Instrument (LORI) User Manual, www.elera.net
8. Neven, F., Duval, E.: Reusable learning objects: a survey of lom-based repositories. In: Proceedings of ACM Multimedia. ACM, New York (2002)
9. Ochoa, X., Duval, E.: Relevance ranking metrics for learning objects. In: Duval, E., Klamma, R., Wolpers, M. (eds.) EC-TEL 2007. LNCS, vol. 4753, pp. 262–276. Springer, Heidelberg (2007)
10. Palmer, K., Richardson, P.: Learning Object Reusability - Motivation, Production and Use. In: 11th International Conference of the Association for Learning Technology (ALT). University of Exeter, Devon (2004)
11. Poulin, J.: Measuring Software Reusability. In: Third International Conference on Software Reuse, Rio de Janeiro, Brasil, November 1-4, pp. 126–138 (1994)
12. Sánchez-Alonso, S., Sicilia, M.A.: Normative specifications of learning objects and processes. En International Journal of Instructional Technology and Distance Learning 2(3), 3–12 (2005)
13. Sicilia, M.A., Garcia, E., Pages, C., Martinez, J.J.: Complete metadata records in learning object repositories: some evidence and requirements. International Journal of Learning Technology 1(4), 411–424 (2005)
14. Sicilia, M.A., Garcia, E.: On the Concepts of Usability and Reusability of Learning Objects. International Review of Research in Open and Distance Learning (October 2003)
15. Sommerville, I.: Software engineering, 6th edn. Addison-Wesley, Reading (2000)
16. Vargo, J., Nesbit, J.C., Belfer, K., Archambault, A.: Learning Object Evaluation: Computer-Mediated Collaboration and Inter-rated Reliability. International Journal of Computers and Applications 25(3) (2003)
17. Vinoski, S.: Old Measures for New Services. IEEE Internet Computing, 72–74 (November-December 2005)
18. Wiley, D.A.: Connecting learning objects to instructional design theory: A definition, a metaphor, and a taxonomy. In: Wiley, D.A. (ed.) The Instructional Use of Learning Objects: Online Version, http://reusability.org/read/chapters/wiley.doc
19. Yang, D., Yang, Q.: Customizable Distance Learning: Criteria for Developing Learning Objects and Learning Model Templates. In: Proceedings of the 7th international conference on Electronic commerce (ICEC, ACM International Conference Proceeding Series, August, Xi'an (China), pp. 765–770. ACM, New York (2005)
20. Zimmermann, B., Meyer, M., Rensing, C., Steinmetz, R.: Improving retrieval of re-usable learning resources by estimating adaptation effort backhouse. In: First International Workshop on Learning Object Discovery and Exchange (2009), http://fire.eun.org/lode2007/lode07.pdf (Retrieved May 20, 2009)

A Comparison of Methods and Techniques for Ontological Query Expansion

Fabio Sartori

Department of Computer Science, Systems and Communication (DISCo)
University of Milan - Bicocca
viale Sarca, 336
20126 - Milan, Italy
Tel.: +39 02 64487913; Fax: +39 02 64487839
sartori@disco.unimib.it

Abstract. This paper presents an ongoing research on the comparison of ontological query expansion methods. Query Expansion is a technique that aims to enhance the results of a search by adding terms to the search query; today, it is a very important research topic in the semantic web and information retrieval areas. Although many efforts have been form the theoretical point of view to implements effective and general methods for expanding queries, based on both statistical and ontological approaches, the practical applicability of is nowadays restricted to few and very specific domains. The aim of this paper is the definition of a platform for the implementation of a subset of such methods, in order to make comparisons among them and try to define how and when use ontological QE. This work is part of JUMAS, a research project funded by European Community where query expansion is used to support the retrieval of signifiant information from audio–video transcriptions in the legal domain.

Keywords: Ontological Query Expansion.

1 Introduction

The purpose of this work was to explore and compare methods of query expansion based on ontologies. *Query Expansion* (QE) is a technique that aims to enhance the results of a search by adding terms to the search query. It is useful because it can deal with the ineherent ambiguity of natural language, indeed every language has problems of synonymy (more terms for a concept) and polysemy (more concepts for a term) that cause a decay of performance: the search engine doesn't return relevant documents that are present in the domain (low recall) and returns documents that are not relevant (low precision). For example if a user search for *bicycle* a traditional search engine won't return the documents that contain only *bike*. And if a user search for *jaguar* the search engine will return the documents on both the animal and the carmaker, although the user meant only one of the two. Query expansion aims to increase the number of relevant documents retrieved.

F. Sartori, M.Á. Sicilia, and N. Manouselis (Eds.): MTSR 2009, CCIS 46, pp. 203–214, 2009.

There are two main approaches to accomplish this task [1]: probabilistic query expansion and ontological query expansion. The first is based on statistic data that indicates the frequency of terms in a collection of supposed relevant documents, or the frequency of co-occurence of terms. The second is based on knowledge models, in particular ontologies, in which the method search for the terms that have to be added to the query.

Although probabilistic query expansion is the dominant approach, it has shown limits in different fields and tasks: Carpineto et al. [2] highlight how this approach weights the terms on the basis of their relevance within the text rather than the real benefits for the user; moreover, it has been pointed out [3][4] that probabilistic methods are very influenced by the corpus of documents and their relevance to the query.

On the contrary, ontological approaches are less developed and studied, but they virtually have a lot of undiscovered possibilities to semantically improve queries: being *corpus-independent*, they are more precise than probabilistic methods in the text disambiguation within a given domain. Moreover, they are particularly suitable to treat short queries. Anyway [5] [6], ontological methods have some important drawbacks: ontologies are typically difficult to create and maintain in order to guarantee the necessary level of precision to avoid the decrease of performance. The so called *query drift* (i.e. the choice of an expansion direction that is out of user scope) phenomenon is more probable with ontological approaches than probabilistic ones.

Due to the relative scarce knowledge on ontological query expansion, our work has been focused on them: in particular, we noticed how existing literature on ontological query expansion lacks of large comparisons of methods and analitic researches on which method or which technique is suitable for specific scenarios. Usually the comparisons are made between more versions of the same method (for example with different settings) and on the same domain with the same ontology.

Thus, our goal was to compare query expansion methods in an analitic way, focusing on specific characteristics of each method with the aim of seeing in which case they are useful and when they are not. To do so, since we have not found implemented methods available, we implemented some selected method ourselves. Moreover, we implemented a generic testing platform that allows to easily create, test and compare methods with different ontologies and on different domains.

The paper is organized as follows: Section 2 concerns a brief review of literature about query expansion and ontological methods for making it; then, in Section 3 the main choices adopted in the platform design and implementation are motivated, focusing on thesauri as target ontology, SKOS as the language for thesauri representation, and Java–based technology for the implementation of chosen QE methods. An example of platform application to an existent thesaurus is briefly introduced in Section 4, where some preliminary considerations about the effectiveness of our platform are also presented. Finally, conclusions and future work are briefly pointed out in Section 5.

2 State of the Art

2.1 A Definition of Query Expansion

In the analysis of literature we first explored the various definition of query expansion[7][8][9][10] and we tried to merge them to obtain a possible common definition that contains the main characteristic of each vision. Indeed, there are several different definitions because the problem of query expansion has been approached by different ways and different points of view. However the goals and basic mechanisms of expansion are common: to improve the effectiveness of a search, that is finding more relevant documents and less irrelevant documents, by adding new terms to the original search query.

We can consider query expansion as a system that accepts in input a set of words Q returning a new set Q' as the union between Q the set of expanded words E:

$$Q \Longrightarrow \boxed{\text{QEXP}} \Longrightarrow Q' = Q \cup E$$

Words belonging to E are functions of a subset of Q, i.e. each element of E has been obtained starting from an element of Q.

$$E = \{\ \omega_e\ :\ \omega_e\ = f(\ W_e \subseteq Q\ )\ \}$$

More ore less, all the existing approaches perform query expansion on the basis of one and only one word of Q:

$$E = \{\ \omega_e\ :\ \omega_e\ = f(\ \omega_i \in Q\ )\ \}$$

The general model of QE can be further developed according to the following constraints:

- the Q and Q' set can be *ordered* or *not ordered.* In the first case, the order according to which the different words are arranged must also be considered.;
- the Q' set can be a collection of pairs (w, p), where w is the word to expand and p is a *weight* related to it. This kind of output is useful when the result of query expansion must be returned to a research engine accepting weighted terms in input. The output would become a set Q'' as follows:

$$Q'' = \{\ (\ \omega, p\ )\ :\ \omega \in Q',\ \ p \in \{\,0, 1\}\ \}$$

- the output can be the result of a rewriting of Q' according a given rule r. For example, if the research engine supports operators like $\wedge(AND)$ e $\vee(OR)$, we could obtain a rewriting similar to the following one:

$$(\ \omega_1\ \vee\ \omega_{1,1}\ \vee\ \omega_{1,2}\ \vee\ ...\ )\ \wedge\ (\ \omega_2\ \vee\ \omega_{2,1}\ \vee\ \omega_{2,2}\ \vee\ ...\ )\ \wedge\ ...$$

where $\omega_{i,e}$ are all the expanded terms of the word ω_i.

2.2 Query Expansion Methods

Given the definition and main characteristics of QE in the previous section, in this paragraph we introduce some of the most known and used approaches, with the aim to identify among them the list of methods which will be considered for the implementation.

Since we have decided to focus on ontological methods, the proposed analysis will concern them[1].

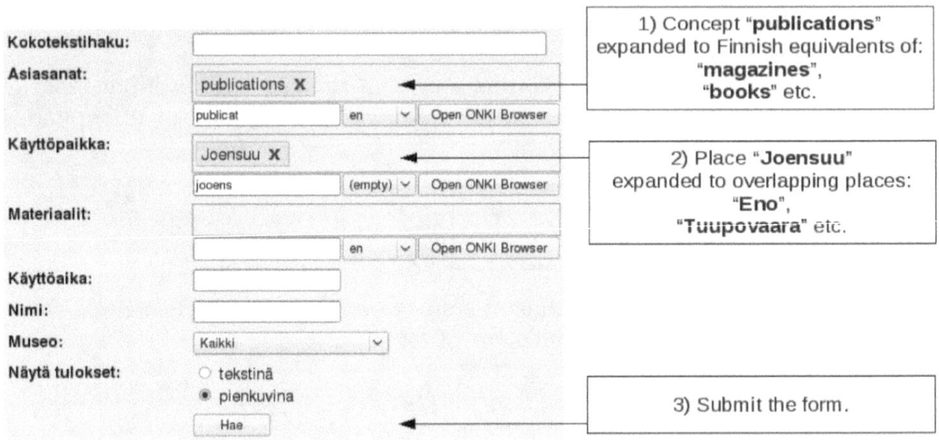

Fig. 1. The *Kantapuu.fi* system wiht the integrated l'*ONKI* widget

Voorhees 1994 [3] exploits a thesaurus to accomplish the expansion, by adding synonyms and any other kinds of term linked directly to the original;

ONKI [1] (see Figure 1) allows the user to select interactively by means of a thesaurus the concepts to search for. Then, it expands automatically the keywords exploiting the same thesaurus. The method is also interesting because it is *multi–ontology*: the *Finnish Collaborative Holistic Ontology (KOKO)* is used for expanding the concepts and the *Finnish Spatio-temporal Ontology (SAPO)* for expanding the places;

ATJ [11][12] is an approach which expands a query on every direction within the ontology, assigning different weights according to the nature of the followed relationship;

Navigli&Velardi (Note) [5] are two methods that, different from others, exploit ontologies to extract the concept descriptions. While in the first mode the QE is based on the terms present in the description, in the second case the similarity between two concepts is evaluated on the basis of the similarity value between their descriptions;

[1] In the following, some names of the QE methods derive from the names of the paper authors where they are described.

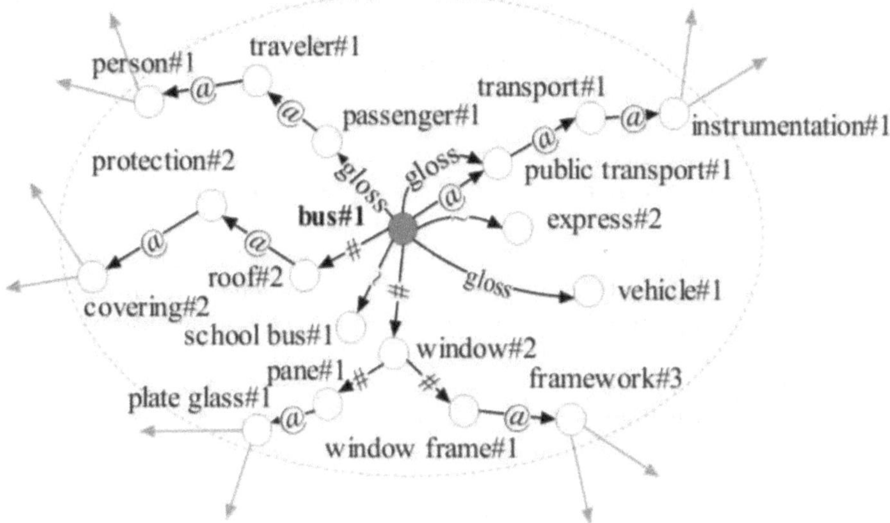

Fig. 2. An example of semantic net created by the Navigli&Velardi (2003) method

Navigli&Velardi (Graphs) [5] is based on the creation of semantic nets (see Figure 2) of candidate concepts starting from a given ontology, to chose in a second step the terms according to how much such nets overlap. Although this approach is extremely interesting, the results have not been encouraging until now;

HO [13] considers the sequences of links through which the terms are bounded to the query terms, taking especially care of terms whose links satisfy given patterns;

Moreover, there exist hybrid systems that exploit ontologies during one or more expansion phases:

Andreou 2005 [14] has a first step in which an expansion based on probabilistic methods is done, and a second one during which an ontology is used to accomplish a revision of terms, getting better results;

Xu et al. 2006 [7] exploits the *Latent Semantic Indexing* technique, comparing *local* and *global* expansion methods. Here, ontologies allow weighting the terms;

Alani 2007 [15] doesn't use ontologies for query expansion, since they are the object of the search. The aim is to find ontologies which refer to a specific domain;

Calegari and Pasi 2008 [16] exploits a local fuzzy ontology to keep trace of past searches made by the user, with the aim to expand searches with the terms that are most frequently inserted in the same query;

Durão et al. 2008 [6] uses query expansion for searching within software source codes. The expansion is accomplished by an ontology created to this aim.

Among these methods, it has been decided to implement three of them to compare: **Voorhees**, **ATJ** and **HO**. In addition, the platform includes **Ancestor**, an approach that calculates the distance between the meanings of two concepts in terms of their distance from a common parent in a tree–like structure.

3 Towards a Platform for QE: Implementation

In the second step, we implemented a generic platform for testing query expansion, and also the previously selected methods. We first defined requirements and motivations for the platform, then we discussed all the implementative choices made during the process. One of the most important choice concerned the decision of which type of ontology to use: we decided to use a *thesaurus*, a restriction of ontologies expressly defined for lexical networks of terms. Next, we'll describe the architecture of the platform, that has been implemented in *Java*. In the implementation process we paid attention to fulfill the requirements and provide the platform with the necessary characteristics: generality, expandability, modularity and uniformity. The platform has been exploited to codify the four methods introduced above.

3.1 Type of Ontologies: Thesauri

Thesauri are structured vocabularies where relationships among terms are specified. They can be considered as a subclass of ontologies, since they represent only a part of the knowledge involved and their power of expression is limited [17]. Anyway, their main features are compatible with most of the ontological methods for QE available at the moment.

There exist two *ISO* standards for the definition of thesauri, the *ISO2788* for single–language thesauri and *ISO5968* for multi–language thesauri, which define the possible relationships among the terms of a thesaurus:

USE / UF (Use For) are two relationships for solving synonymy problems; the *USE* relationships specifies that the term considered is linked to another one that would be preferable (i.e. its synonym), while the *UF* relationship is the inverse of USE. These relationships generate networks are *synonyms clusters* which are be referred to as *synsets* within *WordNet*[2], probably the most known English thesaurus;

BT (Broader Term) is a relationship used for specifying more generic terms. Its inverse is the *NT (Narrower Term).* relationship. BT and NT allows the definition of hierarchical structures among the terms and the synonym clusters;

RT (Related Term) is a relationship which describes a generic association between two terms. This association is neither USE / UF nor BT / NT.

[2] http://wordnet.princeton.edu/

Moreover, the two standards define the following property:

SN (Scope Note) is used to bind a term and its textual description, with the aim to make not ambiguous the usage of such term.

The relationships above are sufficient to create a network of terms which can be very useful in the QE process: for this reason, many approaches are based on WordNet. Moreover, thesauri are simpler to implement and maintain than ontologies and they have no significant drawbacks with respect to the QE approaches we have chosen to implement. For this reasons, it has been decided to design the platform for QE on the basis of thesauri, paying attention to an opportune strategy for adapting it to ontologies adoption in the future.

3.2 Definition of the Ontology Representation Language: SKOS

The initial idea to exploit *OWL (Ontology Web language)* [3] for the ontology representation has been modified with the restriction to thesauri. For this reason, it has been chosen to withdraw to *SKOS* (Simple Knowledge Organization System)[4] a formal language that is going to become the standard de facto for thesauri representation.

As the OWL language, SKOS is based on *RDF*: this is very important from the adaptability of our platform to more general ontologies in the future. As *RDF*, *SKOS* is a language for the definition of graphs (in this case graphs of concepts) by means of 3–ples $\langle subject, predicate, object \rangle$. The *predicate* represents the relationship between the *subject* and the *object* .

The elements of the graph are the *concepts* expressed by the terms, which belong to a specific class named *skos:Concept.* The whole graph is described by the *skos:conceptScheme* class.

The kinds of relationships (i.e. the predicates) and properties fulfill very well the standards previously described; in fact:

- *skos:prefLabel* and *skos:altLabel* allow modeling the synonymy: every concept is described by a *prefLabel*, which is the preferred term to express that concept and a sequence of alternative terms named *altLabel*, which can be considered synonyms. In this way, an opportune *synset*is defined;
- *skos:broader* and *skos:narrower* represent the *BT (Broader Term)* and *NT (Narrower Term)* relationships;
- *skos:related* represents the *RT (Related Term)* relationship;
- *skos:scopeNote* is a representation of the Scope Note property.

Moreover, *SKOS* introduces other relationships and properties: for example, the*skos:definition* property is exploited to give a precise definition of a term.

3.3 Platform Architecture

Figure 3 shows a sketch of the platform architecture, which is divided into three main blocks responsible for the different functionalities:

[3] `http://www.w3.org`
[4] `http://www.w3.org/2004/02/skos/`

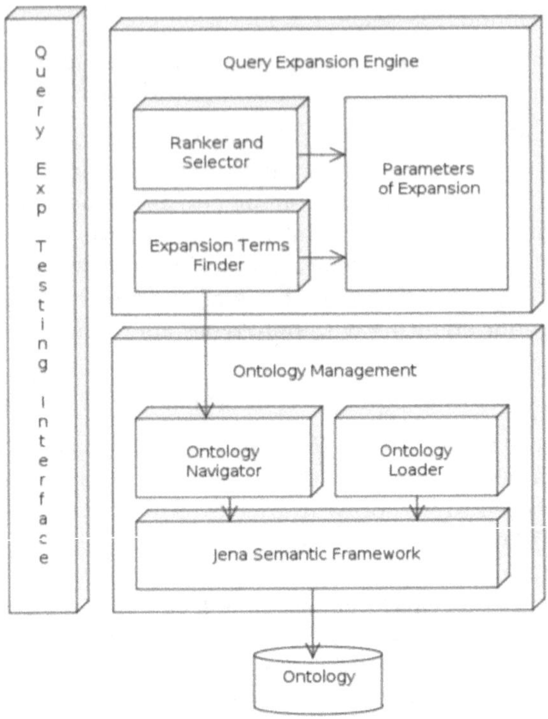

Fig. 3. The architecture of the platform for QE

- the **Query Expansion Testing Interface** implements generic functions to interact with the user;
- the **Query Expansion Engine** implements the engine for the expansion, which is the specific query expansion method chosen among the four approaches introduced above. This component is made of an *Expansion Terms Finder* (ETF) that looks for the closest terms to the subject of QE and a *Ranker and Selector* that focuses on the following expansion phases, weighting the terms returned by ETF and adding the most significant ones to the QE output set;
- the **Ontology Management** allow managing the ontologies by means of Jena, a semantic framework that provides the system with many tools to treat ontologies and RDF language.

  Moreover, also an *Ontology Loader* and an *Ontology Navigator* are supplied: the first component allows the system to load ontologies from files or other sources, the second one allows searching within the ontology the terms to add to the candidate set by means of SPARQL queries.

The platform has been designed to be an effective and flexible tool to test the different methods, to modify their parameters and to combine sets of methods for deriving new approaches to QE.

4 Testing

The last step was the effective testing and comparison of methods. Before that, we had to do some other passages to ensure the consistence of results: we had select properly the underlying thesaurus, the set of query to be tested, the setting of methods and the criteria of evaluation.

About the thesaurus choice, we selected one (see Figure 4) from *http://www.culturaitalia.it*, a site of(*Italian Culture Minister* and we expanded it to make it the most method–neutral possible. We selected the set of query by picking terms with different characteristics and we described the subdivisions made. The parameters of methods were set up in a way that promotes the comparison among them. We defined an approach of evaluation based on scenarios: we thought that is not worth defining an absolute evaluation for methods, but each method or technique could be useful in different situations, so we analyzed the results proposing different scenarios adapted to the results themselves.

Figure 5 shows an example of how the platform works when expanding the search the Italian word *abitazioni*: a complete description of the results is out of the paper scope due to the lack of space; anyway, it is possible to highlight how the four methods are applied and make some considerations for future works. The testing, although not carrying out big and well defined results due to the lack of time for evaluation, has offered a lot of ideas and hints to possible future in-depth examinations and tests. The most general and defined considerations are briefly

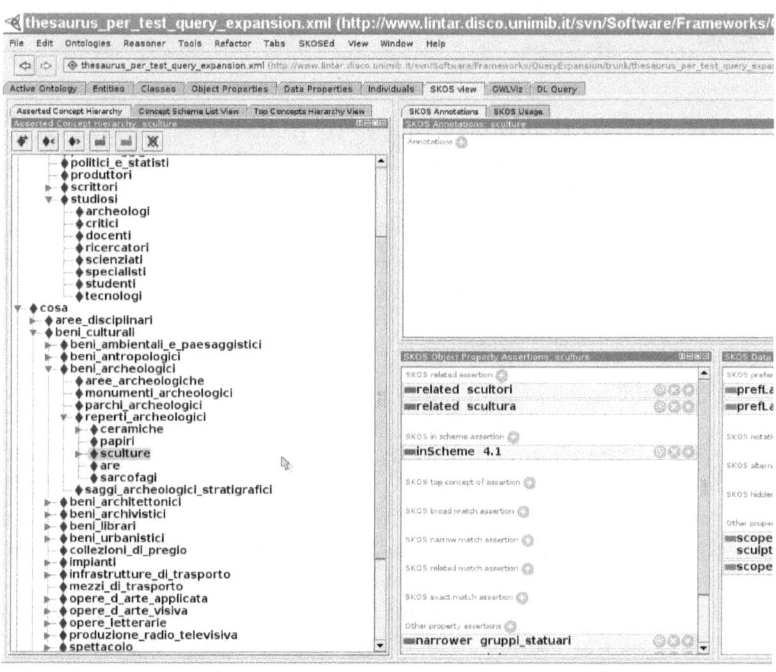

Fig. 4. A screenshot of the target thesaurus used for the platform implementation

```
### QUERY: [abitazioni]

@Metodo: Ancestor
[Appartamenti (0.8), Alberghi (0.8), Ville (0.8), Capanne (0.8), Case rurali
    (0.8), Palazzi (0.8), Grattacieli (0.8), Case (0.5), Alberghi (0.4),
    Ville (0.4), Capanne (0.4), Case rurali (0.4), Palazzi (0.4),
    Grattacieli (0.4)]

@Metodo: Voohrees 1
[Case (0.8)]

@Metodo: Voohrees 2
[Appartamenti (0.3), Alberghi (0.3), Ville (0.3), Capanne (0.3), Case rurali
    (0.3), Palazzi (0.3), Grattacieli (0.3), Case (0.1)]

@Metodo: Voohrees 3
[Beni architettonici (0.5), Appartamenti (0.3), Alberghi (0.3), Ville (0.3),
    Capanne (0.3), Case rurali (0.3), Palazzi (0.3), Grattacieli (0.3),
    Case (0.1)]

@Metodo: Voohrees 4
[Case (0.8), Appartamenti (0.1), Alberghi (0.1), Ville (0.1), Capanne (0.1),
    Case rurali (0.1), Palazzi (0.1), Grattacieli (0.1), Beni
    architettonici (0.1)]

@Metodo: OH
[Beni architettonici (0.8181818), Appartamenti (0.8181818), Alberghi
    (0.8181818), Ville (0.8181818), Capanne (0.8181818), Case rurali
    (0.8181818), Palazzi (0.8181818), Grattacieli (0.8181818), Beni
    culturali (0.72727275), cosa (0.6363636), Architetti (0.54545456),
    Architettura (0.54545456), Nuclei rurali storici (0.54545456),
```

Fig. 5. The application of our platform to the target thesaurus to expand the Italian word *abitazioni*

presented here: it seems that numerosity of relationships is inverse proportional to their importance. That is: if a term has a lot of children each relation is less relevant for expansion than those of a term that has few children. In general, there are some terms that cause the *explosion* of expansion with many irrelevant terms. They seem to be all those terms with a lot of children and brothers, but further tests are required to find a method that properly identifies them.

5 Conclusions

Ontological query expansion is a field in which a lot of improvements are theorically possible. In this work we tried to set up an environment for testing and discovering potentiality of expansion through ontologies: we created a platform that allows to easily implement and analitically compare different methods and we started tests and comparisons to evaluate the usefulness of methods and techniques. The obtained results demonstrated that the platform works and can be useful to the goals for which it has been implemented. Moreover the results showed the possile scenarios in which the four methods implemented could obtain the best performances. We concluded our work by considering possible future developments in both the extension of the platform, the concrete applications and the researches that could be started from the results obtained.

Future work consists in the further development of the platform: to this aim, the first step is the evolution of the analyzed thesaurus of the current version towards a real and complete ontology: the expolitation of more complex relationships and properties should theoretically bring us to better results in QE, but the current development of research in this field is not supported by adequate practical advances.

In this sense, the platform will be soon tested (and, possibly, extended) in the context of the JUMAS project (*Judicial Management by Digital Libraries Semantics*), co–funded by the European Community (grant agreement number FP7-214306) in the context of the 7th Frame Programme. The platform will be used in JUMAS to implement a query expansion interface for users in the legal domain (e.g. judges, witnesses and so on) in order to help them to find significant audio–video transcriptions. A model of knowledge involved is currently going to be defined with the final goal to produce an ontology of legal terms that can be exploited from the platform to guide a next information retrieval phase.

Acknowledgements

The author wishes to thank Andrea Bonomi, Ettore Colombo and Matteo Mondini for their support in preparing this paper and their precious contribution in the QE platform development.

References

1. Tuominen, J., Kauppinen, T., Viljanen, K., Hyvnen, E.: Ontology-based query expansion widget for information retrieval. In: Proceedings of the 5th Workshop on Scripting and Development for the Semantic Web (SFSW 2009), 6th European Semantic Web Conference (ESWC 2009), May 31 - June 4 (2009)
2. Carpineto, C., de Mori, R., Romano, G., Bigi, B.: An information-theoretic approach to automatic query expansion. ACM Trans. Inf. Syst. 19(1), 1–27 (2001)
3. Voorhees, E.M.: Query expansion using lexical-semantic relations. In: SIGIR 1994: Proceedings of the 17th annual international ACM SIGIR conference on Research and development in information retrieval, pp. 61–69. Springer, New York (1994)
4. Ruthven, I., Tombros, A., Jose, J.M.: A study on the use of summaries and summary-based query expansion for a question-answering task (2001)
5. Navigli, R., Velardi, P.: An analysis of ontology-based query expansion strategies. In: Proc. of Workshop on Adaptive Text Extraction and Mining (ATEM 2003), in the 14th European Conference on Machine Learning (ECML 2003), September 22-26, pp. 42–49 (2003)
6. Durão, F.A., Vanderlei, T.A., Almeida, E.S., de L. Meira, S.R.: Applying a semantic layer in a source code search tool. In: SAC 2008: Proceedings of the 2008 ACM symposium on Applied computing, pp. 1151–1157. ACM, New York (2008)
7. Xu, X., Zhu, W., Zhang, X., Hu, X., yeol Song, I.: A comparison of local analysis, global analysis and ontology-based query expansion strategies for bio-medical literature search (2006)

8. Bhogal, J., Macfarlane, A., Smith, P.: A review of ontology based query expansion. Inf. Process. Manage. 43(4), 866–886 (2007)
9. Delort, J.Y.: A user-centered approach for evaluating query expansion methods. In: WI 2005: Proceedings of the 2005 IEEE/WIC/ACM International Conference on Web Intelligence, Washington, DC, USA, pp. 586–589. IEEE Computer Society, Los Alamitos (2005)
10. Manning, C.D., Raghavan, P., Schütze, H.: Introduction to Information Retrieval. Cambridge University Press, Cambridge (2008)
11. Tudhope, D., Alani, H., Jones, C.: Augmenting thesaurus relationships: Possibilities for retrieval. Journal of Digital Information 1, 15 (2001)
12. Alani, H., Jones, C., Tudhope, D.: Associative and spatial relationships in thesaurus-based retrieval. In: Proc. 4th European Conf. on Digital Libraries, pp. 45–58 (2000)
13. Hirst, G., St-Onge, D.: Lexical chains as representations of context for the detection and correction of malapropisms (1997)
14. Andreou, A.: Ontologies and query expansion. Master's thesis, School of Informatics, University of Edinburgh (2005)
15. Alani, H., Noy, N.F., Shah, N., Shadbolt, N., Musen, M.A.: Searching ontologies based on content: experiments in the biomedical domain. In: K-CAP 2007: Proceedings of the 4th international conference on Knowledge capture, pp. 55–62. ACM, New York (2007)
16. Calegari, S., Pasi, G.: Personalized ontology-based query expansion. In: IEEE/WIC/ACM International Conference on Web Intelligence and Intelligent Agent Technology, vol. 3, pp. 256–259 (2008)
17. Wielinga, B.J., Schreiber, A.T., Wielemaker, J., Sandberg, J.A.C.: From thesaurus to ontology. In: K-CAP 2001: Proceedings of the 1st international conference on Knowledge capture, pp. 194–201. ACM, New York (2001)

Exploring Characterizations of Learning Object Repositories Using Data Mining Techniques

Alejandra Segura[1], Christian Vidal[1], Victor Menendez[2], Alfredo Zapata[2], and Manuel Prieto[3]

[1] Univ. del Bio-Bio, Avda. Collao 1202, Concepción, Chile
[2] Univ. Autónoma de Yucatán.Periférico Norte, 13615, 97110 Mérida, Yucatán, México
[3] Univ. de Castilla-La Mancha. ESI. Po. de la Universidad 4, 13071 Ciudad Real, Spain
asegura@ubiobio.cl, cvidal@ubiobio.cl, mdoming@uady.mx, zgonzal@tunku.uady.mx, manuel.prieto@uclm.es

Abstract. Learning object repositories provide a platform for the sharing of Web-based educational resources. As these repositories evolve independently, it is difficult for users to have a clear picture of the kind of contents they give access to. Metadata can be used to automatically extract a characterization of these resources by using machine learning techniques. This paper presents an exploratory study carried out in the contents of four public repositories that uses clustering and association rule mining algorithms to extract characterizations of repository contents. The results of the analysis include potential relationships between different attributes of learning objects that may be useful to gain an understanding of the kind of resources available and eventually develop search mechanisms that consider repository descriptions as a criteria in federated search.

Keywords: Learning objects, metadata, data mining, learning object repositories, association rules, clustering.

1 Introduction

Learning object repositories (LORs) provide a platform for the open sharing of learning resources. In the case of repositories storing only metadata about resources available elsewhere in the Web, they and act as filters of the resources available by providing metadata-based search on learning objects (ADL, 2002). Further, some of these repositories nowadays implement flexible mechanisms for the search and selection of resources based on metadata (Broisin et al., 2005, McLean and Lynch, 2003). However, uses approaching these repositories don't have an *a priori* clear view of the kind of resources stored and to what extent they fit their interests or preferences. This gap might be filled by extracting characterizations of learning objects obtained from the analysis of their metadata. Such characterizations could also be used to enhance search or serve as descriptions of the content bases of repositories, useful both for humans and for software applications.

Machine learning techniques can be used to automatically extract characterizations of learning objects collections. Indeed, the application of *data mining* (DM) techniques to the domain of e-learning have become more frequent in recent years (Romero and

F. Sartori, M.Á. Sicilia, and N. Manouselis (Eds.): MTSR 2009, CCIS 46, pp. 215–225, 2009.

Ventura, 2007, Romero et al., 2008). Data mining in e-learning is mainly oriented to analyze student's behavior, outcomes and interests in their interaction with learning technology and learning resources. Here we focus on analyzing the metadata stored inside learning object repositories (Duval et al., 2001).

For this study we used the metadata base of the AGORA recommender system (Prieto, 2008) and other already existing LORs. Clustering and association rule mining were selected as the two main techniques to approach a first exploration of the characterization of LORs.

Concretely, the study reported here was aimed at providing some preliminary insight on the following questions: (a) which are common characteristics of LO metadata stored in different repositories?, (b) Are there any LO groups which have similar characteristics in their metadata?, (c) from the instructional point of view, which relations between the LO metadata are the most significant?

The structure of the paper is described as follows. Section 2 details methodological issues and describes the technical issues related to data gathering. Then, in the section 3, usefulness of the results for different areas of the LO management and design, especially for the AGORA system, are discussed. Finally, the main conclusions of our research work as well as the future work are highlighted.

2 Methodology

The process of knowledge discovery essentially consists in pre-processing, application of data mining techniques and post-processing stages. Here the methods described in (Romero et al., 2008) were use as the guiding framework.

2.1 Criteria for LOR Selection

A significant number of LORs have been developed in the last years (UNESCO, 2009) and there are some studies that analyze their characteristics (Neven and Duval, 2002). The following criteria that affect the homogeneity of metadata and query interfaces were used to define the scope of our study:

1. Repositories providing query interfaces based on SQI.
2. Repositories conforming to the IEEE-LOM standard (IEEE-LOM, 2002).

As a consequence, our testing group was reduced to the LORs listed in (ARIADNE, 2009) from which three were selected: ARIADNE (Duval et al., 2001), MACE (Stefaner et al., 2007) and LACLO-FLOR (LACLO, 2009). Their metadata was extracted and contrasted also with metadata in the AGORA repository (Prieto, 2008).

The previous learning object repositories use the IEEE-LOM standard (IEEE-LOM, 2002) as a standardized metadata format. IEEE-LOM standard defines 80 fields that are organized in 9 hierarchical categories (Al-Khalifa and Davis, 2006). The study focuses primarily in the educational metadata category. A current trend in the repositories is to publish services for access to their resources. This option promotes interoperability between different applications, LMS systems and other LORs. The SQI (Simon et al., 2005) protocol is a proposed specification for that purpose. The SQI - Simple Query Interface (CEN/ISSS, 2005, Ternier et al., 2008) is defined as a set of methods related to a universal interoperability layer for educational networks.

2.2 Data Collection

The open source tool, SQITester, was used for gathering the required metadata. This allowed us to consume web services provided by the repositories according to SQI (Ortiz Baíllo et al., 2008). The SQITester tool provides access to some of the most important LO repositories, and it allows querying according to a set of specifications for each repository (for example, query language, session ID, result format, etc.) and retrieve LO's metadata that match these queries. The results are presented in a XML compatible format. Queries in ARIADNE, LACLO and MACE repositories were done covering four areas: computing, mathematics, literature and biology, both in Spanish and English language. A known problem in LORs is that most of them store incomplete metadata records (Sicilia et al., 2005). To obtain the greatest number and diversity of LO in terms of their pedagogical metadata, specific queries were held in the repository MACE. MACE is the only repository that can query specific metadata fields with language PLQL 1 (Ternier et al., 2008).

2.3 Preprocessing

The preprocessing consisted in following the activities:

a) *Error removal.* Errors resulting of the interchange between formats were manually removed.
b) *Metadata field standardization* provided by each repository.
c) *Value field standardization* according to IEEE-LOM vocabularies.
d) *Elimination of duplicates.*
e) *Transformation* from a comma-delimited format to the ARFF format (Attribute-Relation File Format) that is directly used by the WEKA data mining tool (Witten and Frank, 2005).

Five data sets were obtained as result of preprocessing[1]. One for each studied repository and also an additional group that integrated all of them. 763 registers were processed. This set are composed of 200 registers from AGORA, 246 from MACE, 179 from ARIADNE and finally 138 registers from the LACLO repository.

2.4 Application of Data Mining Techniques

Clustering techniques and Predictive Association were applied to the just described data set. Concretely, the K-means clustering algorithm (MacQueen, 1967) was applied to each repository separately and then to the data coming from all the repositories. The Apriori Association Algorithm (Agrawal et al., 1993) was applied to each repository separately. Finally the same algorithm was used also with the merged data set.

Since the study focuses on metadata relationships from an educational point of view, processing was done primarily with data from category 5 (*Educational*) in addition to IEEE-LOM elements 1.7 (*structure*), 1.8 (*aggregation level*), 4.1 (*format*) and the repository to which they belong.

---

[1] The data set used in this study is available at http://www.kaambal.com/agora/

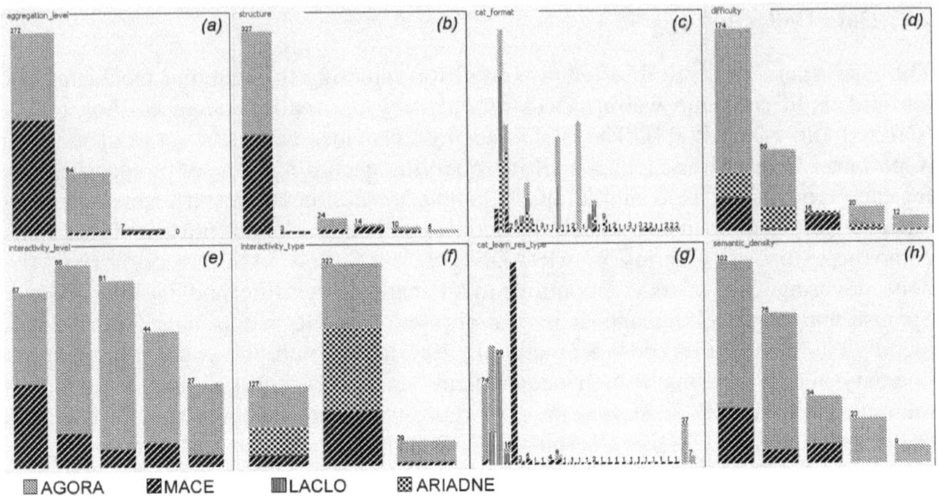

Fig. 1. Data distribution for each analyzed repository versus attributes. The letter of each graph corresponding to attributes described in table 1.

When performing data analysis, it was observed that not all the attributes were filled, with AGORA and MACE being the more complete. Figure 1 shows the data distribution for each repository compared with each attribute used in the processing. The detail of each attribute is shown in Table 1.

Table 1. LOM-Metadata elements used

	Attribute	Description	Values
a)	aggregation_level	The functional granularity of this learning object.	one, two, three, four
b)	structure	Underlying organizational structure of this learning object.	atomic, linear, hierarchical, networked
c)	cat_format	Learning Objects formats shortcuts. Technical datatype(s) of (all the components of) this learning object.	html, img, pdf, rtf, ps, zip, doc, java, swf, xml, xls, mpeg, txt, eps, and their combinations
d)	difficulty	How hard it is to work with or through this learning object for the typical intended target audience.	very easy, easy, medium, difficult, very difficult
e)	interactivity_ level	Interactivity in this context refers to the degree to which the learner can influence the aspect or behavior of the learning object.	very low, low, medium, high, very high
f)	interactivity_ type	Predominant mode of learning supported by this learning object.	expositive, active
g)	cat_learn_res_type	Learning Objects types shortcuts. Specific kind of learning object. The most dominant kind shall be first.	exe, sld, lec, fig, narr, tab, self, rea and their combinations
h)	semantic_ density	The degree of conciseness of a learning object.	very low, low, medium, high, very high

2.4.1 Clustering

The Simple K-means Algorithm for clustering was applied to data set coming from all the repositories. Four clusters were used and the results are shown in Table 2. The evaluation of obtained clusters using the attribute repository showed a 37.0904% error, the best result over other testing algorithms.

Table 2. Clustering Results for MACE, AGORA, ARIADNE and LACLO

	Cluster# + Clustered Instances				
	Full Data	0	1	2	3
Attribute	763-100%	227- 30%	368 48%	97-13%	71 -9%
aggregation_level	one	one	one	one	one
structure	atomic	atomic	atomic	atomic	atomic
cat_format	pdf	ppt	pdf	flash	html
difficulty	medium	medium	medium	medium	medium
interactivity_level	very_low	very_low	very_low	high	very_low
interactivity_type	expositive	expositive	expositive	active	expositive
cat_learn_res_type	narr	lec	narr	exe	narr
semantic_density	medium	medium	medium	high	medium

Correctly classified instances are related to the repository as follows: Cluster 0 = LACLO, Cluster 1 = MACE, Cluster 2 = AGORA y Cluster 3 = ARIADNE.

For each cluster, the grouping and incorrectly classified instances are represented by squares in Figure 2.

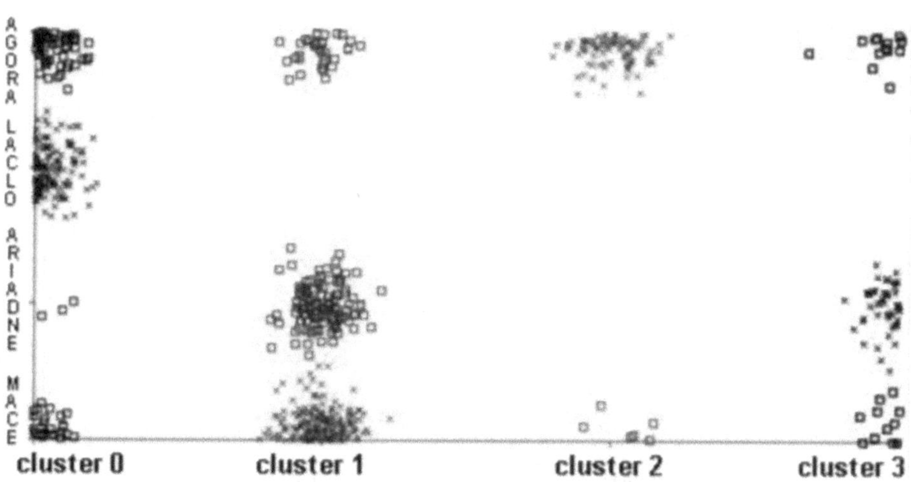

Fig. 2. Visualization of the grouping (cross) and incorrectly classified (square) instances

Due to the higher completeness of metadata records in the repository AGORA, it was important consider to present results of applying clustering. By applying Simple K-means Algorithm 3 clusters were obtained as shown in Table 3.

Table 3. Results of clustering in AGORA

Attribute	Full Data	Cluster# - Clustered Instances 0	1	2
	200 -100%	99 -50%	58- 29%	43- 22%
aggregation_level	one	one	two	one
structure	atomic	atomic	atomic	atomic
cat_format	flash	flash	ppt	pdf
cat_context	high	high	high	high
difficulty	medium	medium	medium	easy
interactivity_level	very_low	high	low	very_low
interactivity_type	expositive	active	expositive	expositive
cat_learn_res_type	sld	exe	sld	rea
semantic_density	medium	high	medium	medium

Table 4. Examples rules for Repository

Repository	Nº	rel	Antecedent	Consequence
MACE	24	0,937	cat_learn_res_type=exe_sim_que_dia_fig_gra_ind_sld_tab_narr	difficulty = medium ; interactivity_type = expositive
ARIADNE	5	0,959	cat_format = html	interactivity_type = expositive
LACLO	4	0,982	cat_learn_res_type = fig	cat_format = jpg_gif_pjpeg_bmp
AGORA	12	0,993	cat_format = ppt ; cat_learn_res_type = sld ; semantic_density = medium	interactivity_type = expositive
AGORA	19	0,993	cat_format = ppt ; difficulty = medium ; cat_learn_res_type = sld	interactivity_type = expositive
AGORA	70	0,986	cat_format = ppt ; cat_learn_res_type = sld	interactivity_type = expositive
MACE	27	0,937	cat_format = html ; cat_learn_res_type = narr	interactivity_type = expositive
MACE	76	0,868	cat_format = html_img ; cat_learn_res_type = narr	interactivity_type = expositive
ARIADNE	2.	0,988	cat_format = html ; difficulty = easy	interactivity_type = expositive
ARIADNE	33	0,322	cat_format = html ; interactivity_type = expositive	difficulty = easy
ARIADNE	16	0,797	difficulty = easy ; interactivity_type = expositive	cat_format = html
MACE	17	0,953	cat_learn_res_type = que	interactivity_type = active
MACE	77	0,868	cat_format = html_img ; cat_learn_res_type = exe_dia	interactivity_type = active
AGORA	64	0,988	cat_format = pdf ; interactivity_level = low	structure = atomic ; interactivity_type = expositive
AGORA	32	0,993	cat_format = ppt ; interactivity_level = low ; interactivity_type = expositive	cat_learn_res_type = sld

Clusters can be described in terms of object grouped as follows:

- Cluster 0: Objects more active and highly interactive for the learner. These are mainly resources of type exercise. They have a high semantic density and high complexity level also.
- Cluster 1: Objects with low interactivity level and expositive. These resources are mainly slides with a medium level of both complexity and semantic density.
- Cluster 2: Expositive objects with very low interactivity. They are mainly resources of type "reading" that are easy to use with medium semantic density.

2.4.2 Associations

The Predictive A Priori Algorithm was applied separately to each repository and then with the full data set also. This generated a set of 100 rules, but the analysis that follows was restricted to those with a reliability (*rel*) greater than 90%. The rules generated for each repository were analyzed considering common rules, redundant rules, rules that reinforce the existing knowledge of the relations among metadata, unexpected rules, interesting rules, and finally questionable rules (from an educational point of view). Table 4 provides some examples of rules extracted. The repository name, identification number and reliability for each rule, are included.

Some examples of analyzing these rules are provided as follows:

- The utility of rule 24 (MACE) is difficult to evaluate since an obvious relation does not exist between resource type with its interactivity and difficulty. Learning resources were grouped by the learning type; this group is wide, from exercises up to tables or narratives.
- Rule 5 (ARIADNE) is an example of simple rule (with a single antecedent and consequent) with a high level of reliability but it is not useful, since it appears with major frequency in repositories with many empty fields.
- Rule 4 (LACLO) appears to be a useful rule, but it shows an obvious or known relation between metadata values.
- There are redundant rules and they can be reduced. Rule 70 is a generalization of rules 12 and 19 (AGORA), with some information loss. In the analysis of predictive attributes it was confirmed that interactivity type is predicted by attributes "resource types" (cat_learn_res_type) and "format" (cat_format).
- Rule 27 is a generalization example of rule 76 in MACE.
- Rules 2, 33 and 16 from ARIADNE are similar. They have equal antecendent and consequent and they can be interchangeable also.
- Finally rules 17 and 77 (MACE) seem useful. They confirm some evidence on metadata relations. For example, the fact resource type that exercise or questionnaire has interactivity type active.

Other examples are rules obtained from AGORA repository. Rule 64 shows that resources with expositive interactivity type and with atomic structure, are mainly related to PDF format and low interactivity level. Rule 32 shows that a slide resource type is related to PPT format, low interactivity level and expositive interactivity type.

2.4.3 Applying Association in Integrated Repositories

The association rules obtained from the analysis of all repositories were more interesting as they have increased explanatory scope (see Table 5).

Table 5. Examples rules in integrated repositories

Nº	rel	Antecedent	Consequence
51.	0.972	structure = atomic ; cat_format = flash ; interactivity_type = active ; cat_learn_res_type = exe	aggregation_level = uno
73.	0.962	structure = atomic ; difficulty = easy ; interactivity_level = medium ; cat_learn_res_type = exe	aggregation_level = uno
42.	0.981	structure = linear ; cat_learn_res_type = sld	aggregation_level = dos ; cat_format = flash
11.	0.990	cat_format = ppt ; cat_learn_res_type = sld ; semantic_density = medium	interactivity_type = expositive
14.	0.989	cat_format = ppt ; difficulty = medium ; cat_learn_res_type = sld	interactivity_type = expositive
33.	0.985	cat_format = flash ; interactivity_level = high ; cat_learn_res_type = sld	semantic_density = high

- Most of the obtained rules were validated using the selection of predictive attributes. For example, the attributes “structure” and “learning resource type” are predictive attributes of “aggregation level” (See rules 42, 51 and 73). These rules are consistent with the principles raised in LOM-ES (LOM-ES, 2008) that establish that: it must exist a relation between resource type and aggregation level.
- In turn, category “format” and category “learning resource type” are predictive attributes of “interactivity type” (See rules 11 and 14).
- The attributes “format”, “interactivity level” and “learning resource type” are predictive attributes of “semantic density”. Rule 33 shows this relation.

One interesting aspect of this study is to analyze if the attribute “repository” is a relevant element to possible classification. Association Rules with high reliability level, relate attributes as “structure”, “format” and “learning resource type”. Examples of these are the rules 68, 2, 1 and 84 in Table 6.

Table 6. Rules example with attribute repository

Nº	Reliability	Antecedent	Consequence
68.	0.985	structure = atomic ; difficulty = medium ; semantic_density = very_high	repository = AGORA
2.	0.994	cat_learn_res_type = lec	repository = LACLO
1.	0.994	cat_learn_res_type = narr	repository = MACE
84.	0.980	cat_format = html_img ; interactivity_level = medium	difficulty = medium ; repository = MACE

It is important to mention that about rule 70 have the attribute “repository” in the consequent. More than half of these rules are related to AGORA repository. This might be attributed to higher completeness of metadata in this repository compared with metadata from LACLO and ARIADNE.

3 Potential Applications

The results of studies like the one presented here can be applied for several purposes in educational technology, including the following:

- *Learning object search based on instructional criteria.* The results of applying clustering to the AGORA repository produced groups of object based in some characteristics of the learning resource type. Instructional design methods (Reigeluth, 1999) requiring some given type can be matched with these groups. It is also possible to build classifiers with the results of clustering, so that new learning objects can be classified automatically in the relevant groups identified.
- *Meta-search strategies in learning object repositories.* Meta-search typically broadcasts a query to several repositories, without considering which of them is more appropriate for the user. Characterizing repositories based mined models ca ne used to direct searches to the "more relevant" repositories for each user (or alternatively, the results of some repositories can be assigned an increased weight), for some given preferences.
- *Automatic metadata generation.* Metadata generation requires some predictive models that help in automatically filling some fields based on the inspection of the available information of the resources. Figure 3 shows the generated predictive relationships in repositories.

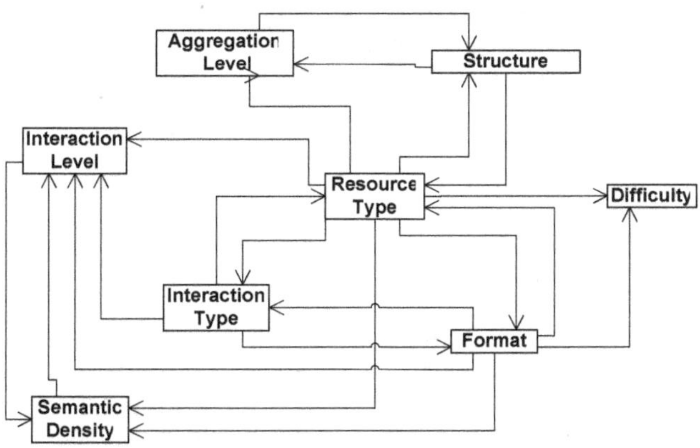

Fig. 3. Associations between metadata fields, as extracted from the study

Metadata generation could be used to suggest values to users editing metadata records, or as a "best effort" approach for incomplete metadata bases. For example, attributes "format" and "resource type" are predictive attributes of "difficulty".

4 Conclusions

This paper has described an exploratory study on four learning object repositories that used data mining techniques to extract characterizations of the repositories from the processing of metadata records. Incomplete metadata records and deviations from the vocabularies in the IEEE LOM standard are two main limitations of this approach, but they could be addressed by providing some simple metadata quality filters inside the repositories.

Two main techniques were applied, clustering and association rule mining. The application of clustering analysis resulted in the identification of three relevant groups of learning objects, which can be roughly characterized by their interactivity level. These characterizations can be applied to filtering search results to subsets of learning resources given some metadata preferences. Association rule extraction resulted in several relationships between metadata elements that have the potential to be useful as characterizations of learning resource bases. These relationships are candidates for automated metadata generation algorithms.

Future work will expand the study reported here to cover a larger number of repositories and a more heterogeneous learning object base. This will eventually allow us to contrast or improve the rules learned and expand the features that define the groups generated. The results of this study are planned to be integrated in the AGORA project and repository, serving as a basis for automated metadata generation and meta-search.

Acknowledgments

This work is partially supported by MECESUP UBB 0305 project, Chile; A/016625/08 AECID project, Spain; "Metodologías para la producción colaborativa de objetos de aprendizaje" project, SINED-ANUIES, México; YUC 2006-C05-65811 project, FOMIX, CONACYT, México.

References

ADL, Emerging and Enabling Technologies for the design of Learning Object Repositories Report (2002), http://xml.coverpages.org/ADLRepositoryTIR.pdf (accessed April 2009)

Agrawal, R., Imieli, T., Swami, A.: Mining association rules between sets of items in large databases. In: Proceedings of the 1993 ACM SIGMOD international conference on Management of data, Washington, DC, United States. ACM, New York (1993)

Al-Khalifa, H.S., Davis, H.C.: The evolution of metadata from standards to semantics in E-learning applications. In: Proceedings of the seventeenth conference on Hypertext and hypermedia, Odense, Denmark. ACM, New York (2006)

ARIADNE, SQI Implementation Registry (2009), http://ariadne.cs.kuleuven.be/SqiInterop/free/SQIImplementationsRegistry.jsp (accessed (15 Enero 2009)

Broisin, J., Philippe, V., Meire, M., Duval, E.: Bridging the Gap between Learning Management Systems and Learning Object Repositories: Exploiting Learning Context Information. In: Advanced Industrial Conference on Telecommunications/Service Assurance with Partial and Intermittent Resources Conference/E-Learning on Telecommunications Workshop (2005)

CEN/ISSS: A Simple Query Interface Specification for Learning Repositories (CEN WorkshopAgreement#15454). Brussels, Belgium (2005)

Duval, E., Forte, E., Cardinaels, K., Verhoeven, B., Durm, R.V., Hendrikx, K., Forte, M.W., Ebel, N., Macowicz, M., Warkentyne, K., Haenni, F.: The Ariadne knowledge pool system. Commun. ACM 44, 72–78 (2001)

IEEE-LOM. Draft Standard for Learning Object Metadata. IEEE P1484.12.1 (2002)

LACLO, Comunidad Latinoamericana de Objetos de Aprendizaje (2009), http://www.laclo.org/ (accessed April 2009)

LOM-ES. Perfil de Aplicación LOM-ES V.1.0 G. G.-S. 36/AENOR Disponible en (2008), http://www.educa.madrid.org/cms_tools/files/ac98a893-c209-497a-a4f1-93791fb0a643/lom-es_v1.pdf

Macqueen, J.: Some methods for classification and analysis of multivariate observations. In: U.O.C. Press (ed.) Proceedings of the fifth berkeley symposium on mathematical statistics and probability, California (1967); Le Cam, L.M., Neyman, J.

Mclean, N., Lynch, C.: Interoperability between Information and Learning Environments: Bringing the Gaps (2003), http://www.imsglobal.org/DLims_white_paper_publicdraft_1.pdf (accessed April 2009)

Neven, F., Duval, E.: Reusable learning objects: a survey of LOM-based repositories. In: Proceedings of the tenth ACM international conference on Multimedia, Juan-les-Pins, France. ACM, New York (2002)

Ortiz Baíllo, A., Tortosa, S.O., Martínez Herráiz, J.J., Hilera González, J.R., Barchino Plata, R.: Estandarización de los Sistemas de Búsqueda Federada: SQI como Interfaz de Búsqueda. In: X Simposio Internacional de Informática Educativa (SIIE), Salamanca, Esapaña (2008)

Prieto, M., Menéndez, V., Segura, A., Vidal, C.: A Recommender System Architecture for Instructional Engineering. In: Lytras, M.D., Carroll, J.M., Damiani, E., Tennyson, R.D. (eds.) WSKS 2008. LNCS (LNAI), vol. 5288, pp. 314–321. Springer, Heidelberg (2008)

Reigeluth, C.M.: Instructional-Design Theories and Models: A New Paradigm of Instructional Theory. Lawrence Erlbaum Assoc., Mahwah (1999)

Romero, C., Ventura, S.: Educational data mining: A survey from 1995 to 2005. Expert Systems with Applications 33, 135–146 (2007)

Romero, C., Ventura, S., García, E.: Data mining in course management systems: Moodle case study and tutorial. Computers and Education 51, 368–384 (2008)

Sicilia, M.A., Garcia, E., Pages, C., Martinez, J.J., Gutierrez, J.M.: Complete metadata records in learning object repositories& #58; some evidence and requirements. International Journal of Learning Technology 1, 411–424 (2005)

Simon, B., Massart, D., Van Assche, F., Ternier, S., Duval, E., Brantner, S., Olmedilla, D., Miklos, Z.: A Simple Query Interface for Interoperable Learning Repositories. In: Saito, N., Simon, B., Olmedilla, D. (eds.) Proceedings of the 1st Workshop on Interoperability of Web-based Educational Systems, Chiba, Japan, CEUR (2005)

Stefaner, M., Vecchia, E.D., Condotta, M., Wolpers, M., Specht, M., Apelt, S., Duval, E.: MACE - Enriching architectural learning objects for experience multiplication. LNCS(LNAI & LNBI). Springer, Heidelberg (2007)

Ternier, S., Massart, D., Campi, A., Guinea, S., Ceri, S., Duval, E.: Interoperability for searching learning object repositories: The proLearn query language. D-Lib Magazine 14 (2008)

UNESCO, O.E.R., Open Educational Resources, useful resources/repositories (2009), http://oerwiki.iiep-unesco.org/index.php?title=OER_useful_resources/Repositories (accessed April 2009)

Witten, I.H., Frank, E.: Data Mining: Practical Machine Learning Tools and Techniques. Morgan Kaufman Publishers, New Zealand (2005)

Assuring the Quality of Agricultural Learning Repositories: Issues for the Learning Object Metadata Creation Process of the CGIAR

Thomas Zschocke[1] and Jan Beniest[2]

[1] United Nations University Institute for Environment and Human Security (UNU-EHS), UN Campus, Hermann-Ehlers-Str. 10, 53113 Bonn, Germany
zschocke@ehs.unu.edu
[2] World Agroforestry Centre (ICRAF), POBox 30677, Nairobi, Kenya
j.beniest@cgiar.org

Abstract. The Consultative Group on International Agricultural Research (CGIAR) has established a digital repository to share its teaching and learning resources along with descriptive educational information based on the IEEE Learning Object Metadata (LOM) standard. As a critical component of any digital repository, quality metadata are critical not only to enable users to find more easily the resources they require, but also for the operation and interoperability of the repository itself. Studies show that repositories have difficulties in obtaining good quality metadata from their contributors, especially when this process involves many different stakeholders as is the case with the CGIAR as an international organization. To address this issue the CGIAR began investigating the Open ECBCheck as well as the ISO/IEC 19796-1 standard to establish quality protocols for its training. The paper highlights the implications and challenges posed by strengthening the metadata creation workflow for disseminating learning objects of the CGIAR.

Keywords: Metadata, learning objects, agricultural learning repository, quality assurance.

1 Introduction

The importance of quality assurance in higher agricultural education has been noted by various experts in the agricultural sector, particularly in international development [1]. eLearning is considered by many to provide innovative ways to help enhance teaching and learning effectiveness and improve the quality of education in agriculture [2], [3]. Research about agricultural learning repositories [4], [5], [6] shows the impact that the sharing and reusing of learning resources within a global infrastructure of learning repositories on agricultural and rural development topics can have in improving the development and welfare of agricultural and rural populations. The implementation of relevant international standards and specifications such as the IEEE Learning Object Metadata (LOM) [7] ensures that these systems are fully interoperable and its learning resources are

F. Sartori, M.Á. Sicilia, and N. Manouselis (Eds.): MTSR 2009, CCIS 46, pp. 226–238, 2009.

truly reusable [8], making progress towards a future "learning object economy" [9]. The use of these standards helps to strengthen technology-enhanced learning [10], [11], [12] by promoting a "large-scale learning object infrastructure" [13], which would also extend into the agricultural sector in international development. However, providing quality metadata about learning resources, which is needed to maintain the interoperability of digital learning object repositories in this infrastructure, remains a challenge [14].

The Consultative Group on International Agricultural Research (CGIAR) recognized the significance of the need to ensure the quality of its learning content and to apply current learning technologies. A recent impact study about the education and training offered by the CGIAR recommended that all of its Centers should develop appropriate quality assurance protocols to be applied at all stages in both formal and informal training [15]. The study also suggested to make more strategic use of eLearning tools and methods to increase the efficiency of CGIAR education and training. In response to these suggestions, the CGIAR training community established a digital repository of its learning resources, integrated with a Web-based learning management system [16]. In order to enhance the search and retrieval facility of the repository the group developed an application profile based on the IEEE Learning Object Metadata (LOM) standard, called the CG LOM Core [17]. The system was established in close collaboration with the ARIADNE Foundation for the European Knowledge Pool, using their knowledge pool system to share its learning resources along with the corresponding educational metadata [18]. In order to ensure the quality of its agricultural learning resources and their metadata, the CGIAR is currently engaged in establishing a quality assurance system by participating in the Open ECBCheck Initiative [19] and adapting the ISO/IEC 19796-1 standard for quality management in education [20].

This paper reports about discussions and research on quality metadata incorporating from information science and computer science to inform quality assurance policies and procedures for LOM in the context of agricultural education in international development. Section 2 of this paper summarizes the discussion about the notion of quality in education. Section 3 analyzes the different dimensions and characteristics of quality metadata. Section 4 introduces the Open ECBCheck initiative and proposes a quality assurance protocol for LOM based on ISO/IEC 19796-1. Section 5 concludes the paper.

2 Issues in Approaching the Nature of Quality in Education

Quality is an equally critical issue in education; it is also no longer a novelty. It matters to all the stakeholders involved in the educational process, including learners, instructors, administrators, employers, funding agencies, to name a few. However, as Harvey & Green [21] have demonstrated there is not a common understanding of what quality actually entails. The term is relative to its user and the context in which it is applied; it means different things to different people.

At the same time the term quality is viewed in relation to some benchmark. This can be seen in terms of absolute thresholds that need to be exceeded to obtain a quality rating, but also relative to the processes that result in the desired outcomes. The different notions of quality in education can be summarized as follows [22]: (1) a traditional notion of exceeding certain high standards of excellence (*quality as excellence*); (2) meeting certain standards of managerial efficiency and institutions effectiveness (*quality as fitness for purpose*); (3) meeting some requirements of certain customer specifications or certain threshold objectives of an institution (*quality as fitness of purpose*); and (4) the effort to permanently improve the performance of an educational institution (*quality as enhancement or improvement*). This shows that there is no coherent, abstract definition of quality in education, but rather a discrete integration of the different elements and functions. Following this assumption, institutions and programs are asked to: (i) meet standards and benchmarks; (ii) set and achieve objectives with given input and context variables; (iii) satisfy the demands and expectations of consumers and stakeholders; and (iv) drive towards excellence [23].

3 Addressing Quality Metadata

Metadata constitute the core elements of any cataloging system, descriptions which provide the most useful information contained in a bibliographic entity. Svenonius [24] has described a core set of seven descriptive cataloging principles to guide such descriptions and controlled forms for accessing the organization of information, including (1) *user convenience*, (2) *common usage*, (3) *representation*, (4) *accuracy*, (5) *sufficiency and necessity*, (6) *standardization*, and (7) *integration*. Bruce et al. [25] suggest to think more of the metadata element as the unit to ensure quality and utility. This shift of attention from metadata records to metadata element is similar to the efforts of moving away from specific metadata schemas towards the use of application profiles [26]. These are schemas consisting of data elements that are drawn and combined together from one or more namespaces, and optimized for specific needs of a local application [27]. An application profile as a documented consensus of a community of practice is designed to help ensure the interoperability requirements between systems by maintaining the conformance with a base standard or specification, and defining any additional requirements as needed [28], [29].

But even with such a general framework there is still a need for certain criteria against which metadata quality can be measured. For instance, Bruce & Hillmann [25] suggest a set of seven general characteristics of metadata quality, including (1) *completeness*, (2) *accuracy*, (3) *provenance*, (4) *conformance to expectations*, (5) *logical consistency and coherence*, (6) *timeliness*, and (7) *accessibility*. These characteristics for quality data are similar to those applied in the context of quality data for official statistics [30], for instance, the Quality Assurance Framework (QAF) for statistical data of Statistics Canada [31]. Very similar quality characteristics are used in quality frameworks of international organizations, such as the Data Quality Assessment Framework (DQAF) of the International Monetary

Fund (IMF);[1] the Quality Framework for European Statistics (ESS) for Eurostat of the European Commission;[2] the Quality Framework for Statistical Activities of the Organization for Economic Co-operation and Development (OECD);[3] or the Data Quality Framework of FAOSTAT of the Food and Agriculture Organization of the United Nations (FAO).[4] Table 1 provides a comparison of these quality dimensions within the conceptual framework of data quality introduced by Wang & Strong [34].

Table 1. Comparison of quality dimensions of Bruce & Hillmann and selected international organizations grouped according to Wang & Strong's quality categories

Quality categories	Quality dimensions	Bruce & Hillmann	Eurostat	FAO	IMF	OECD
Intrinsic	Accuracy	X	X	X	X	X
	Reliability				X	
	Provenance	X				
	Credibility					X
	Conformance	X				
	Integrity				X	
	Soundness				X	
Contextual	Relevance		X	X		X
	Completeness	X		X		X
	Periodicity				X	
	Timeliness	X	X	X	X	
	Punctuality		X			
	Comparability			X		
	Interpretability					X
Representational	Consistency	X			X	
	Coherence	X	X	X		X
Accessibility	Accessibility	X	X	X	X	X
	Clarity		X	X		

According to Wang & Strong [32], information quality is defined as "fitness for use". According to their categorization schema, the intrinsic quality category indicates that data have quality in their own right. Contextual dimensions emphasize the requirements that must be considered within the context of the imminent task based on the context of the user and personal preferences. The representational dimensions relate to the way data is presented. The accessibility category addresses issues of getting access to the information. These quality

[1] http://dsbb.imf.org/Applications/web/dqrs/dqrsdqaf/

[2] http://epp.eurostat.ec.europa.eu/portal/page/portal/quality/code_of_practice/quality_framework_for_european_statistics

[3] http://www.oecd.org/statistics/qualityframework/

[4] http://faostat.fao.org/site/365/default.aspx

dimensions show the most common elements that are accepted across different domains. They serve as a framework to ensure quality in collection-specific schemas and implementations, rather than as a checklist for quality assessments [25]. In their research on measuring the quality of metadata, Ochoa & Duval [33] have applied these quality dimensions to establish quality metrics for LOM. While the quality framework of Bruce & Hillmann [25] was devised for human reviewers, Ochoa & Duval's [33] approach intends to generate automated metrics to assess each of the parameters in the framework.[5] In a related study, Ochoa & Duval [34] have focused on the issue of measuring the relevance of learning objects, where relevance is understood as a multidimensional and dynamic concept. They propose ranking metrics to estimate the relevance of learning objects based on the relevance dimensions in the subjective, context-dependent class according to Borlund [35] and Saracevic [36], that is, topical relevance, personal relevance, and situational relevance. Table 2 maps these relevance dimensions with Duval's [37] "quality in context" or relevance characteristics of learning objects, that is, learning goal, learning motivation, learning setting, time, space, culture, language, educational level, and accessibility and compares it with Wang & Strong's [32] contextual quality characteristics of relevance.

Table 2. Mapping of different contextual quality characteristics of relevance

Wang's contextual category	**Borlund's relevance dimensions** Class	Type	**Duval's "quality in context" characteristics**
Completeness Appropriate amount of data	Subjective or human (user)-based (*context-dependent*)	Topical or subject relevance (*aboutness*)	Learning goal
Relevancy		Cognitive / personal relevance or pertinence	Learning motivation
Value-added			Culture & Language Educational level Accessibility
Timeliness		Situational relevance or utility	Learning setting Learning time Learning space

In correspondence with the notion that the contextual data quality category considers the task at hand, Ochoa & Duval [36] developed metrics to automatically estimate the use of learning objects as well as the context in which they are used along the dimensions of relevance. The authors conclude that the relevance characteristics about "what" a learner wants are mapped to topical relevance. Those relevance characteristics that are intrinsic to the learner are

[5] `http://ariadne.cti.espol.edu.ec/M4M/`

mapped to cognitive relevance or pertinence. And, those relevance characteristics that address physical conditions and limitations as well as time and location are mapped to the situational relevance dimension. This discussion of the notion of data quality and different means for measuring, analyzing, and improving data quality provide a framework for the CGIAR to establish corresponding procedures to help strengthening the provision of quality agricultural learning resources and their metadata. The following section presents an outline of these efforts.

4 Quality Assurance for LOM of the CGIAR

When the CGIAR implemented its digital repository for agricultural learning content, it wanted to ensure that its resources could be accessed as easily and shared as widely as possible. The notion of a multi-dimensional, multi-level, and dynamic concept, covering the different aspects of academic life, equally applies to the provision of quality learning content of the CGIAR and the corresponding educational metadata. This implies not only the descriptive parts of the curricular content provided in the digital repository of the CGIAR, but also the policies and procedures that govern the creation and dissemination of the LOM. As an international provider of agricultural content knowledge for development, the CGIAR needs to ensure information quality of its learning content and the corresponding metadata as they determine the quality of decisions and actions of the different stakeholders of the CGIAR in using this information [38]. As an international organization of 15 international agricultural research institutes working globally with many different national partners in developing countries, the CGIAR needs some mechanism to ensure consistency in the description of its learning resources, hence producing quality metadata, and issue of assuring information quality similar to related open, collaborative information quality work as described by Stvilia et al. in the case of Wikipedia [39]. The CGIAR repository uses a mix of human metadata generation and an automated machine-generated mechanism for metadata generation, which was developed by the ARIADNE Foundation [40]. Following the discussions about the important role of application profiles [27], [28] and compliancy with international standards [10], [11], [12] for quality assurance, the CGIAR developed an application profile, called the CG LOM Core [17], based on IEEE LOM. But, even with these different mechanisms in place, the CGIAR is still confronted with the general problem of lack of appropriate quality metadata. Using IEEE LOM compliant metadata records does not necessarily ensure quality because many of the elements are optional. At the same time, to assess the appropriateness or correctness of metadata as it relates to the content of the learning object requires the intervention of an evaluator or human expert in the appropriate domain. Recent studies [41], [42] on the actual use of LOM and the completeness of metadata records show that only a small amount of the available elements is actually used. However, achieving completeness of metadata as one element of quality is needed to ensure the appropriate (automated) processing of the information. Completeness in this context–as one of the quality dimensions discussed earlier–is a prerequisite of having appropriate metadata records, that is, providing the required metadata values [41].

To address the broader issue of quality in training and education, including eLearning and learning resources, the CGIAR began exploring the implementation of quality assurance systems, initially focusing on the Open ECBCheck (eLearning for Capacity Building) [19] and the ISO/IEC 19796-1 standard [20]. Starting as a project in April 2008, the Open ECBCheck is a new accreditation and quality improvement scheme for eLearning that supports capacity building organizations to measure the success of their eLearning programs and courses. It allows for continuous improvement through peer collaboration and benchmarking. By providing international benchmarks, it will enhance the efficiency and effectiveness of capacity building processes, which are using partly or fully technology-enhanced learning. As an open standard it allows participating organizations to pilot the accreditation scheme with their own programs and validate the developments on an international scale. In 2008-2009, an Open ECBCheck Criteria Framework and Toolkit were developed through community interaction, and participating organizations were requested to apply this to their institution and/or eLearning programs. The Open ECBCheck Toolkit for Programs and Courses allows to measure an organization's eLearning offerings against a set of common areas and corresponding criteria.

Open ECBCheck as a community-oriented approach provides human-generated information, which complements the needs of the CGIAR to adapt a more comprehensive approach that would also provide means to include quality protocols for its digital repository and LOM. To address this need, the CGIAR began exploring the implementation of a quality assurance system based on ISO/IEC 19796-1. This standard provides a comprehensive reference framework (RFDQ) for process-oriented quality management covering all aspects and requirements of educational organizations in providing quality goods and services [45]. The framework allows organizations to develop quality profiles as a generic standard tailored to the needs and requirements of an organization, rather than prescriptive requirements, to guide the process of quality development.

Because of its close collaboration with the FAO and similar national partners in developing countries, it seemed helpful for the CGIAR to adapt the FAO Data Quality Framework as a generic structure for creating quality LOM profile. Noticing the flow of metadata information as shown in Figure 1, the CGIAR would need to incorporate the different approaches for quality measurements and metrics into a comprehensive data quality evaluation and monitoring structure. Similar to the FAO framework, the CGIAR would monitor the data quality at three key points in the metadata creation process, that is, as the metadata is provided by the CGIAR Centers, their regional offices and national partners; as the data is received by a central CGIAR capacity strengthening unit; and as the data leaves the CGIAR. At each stage of the process value-added as well as performance would be assessed. Feedback loops would also be incorporated to provide a mechanism for improving the LOM quality.

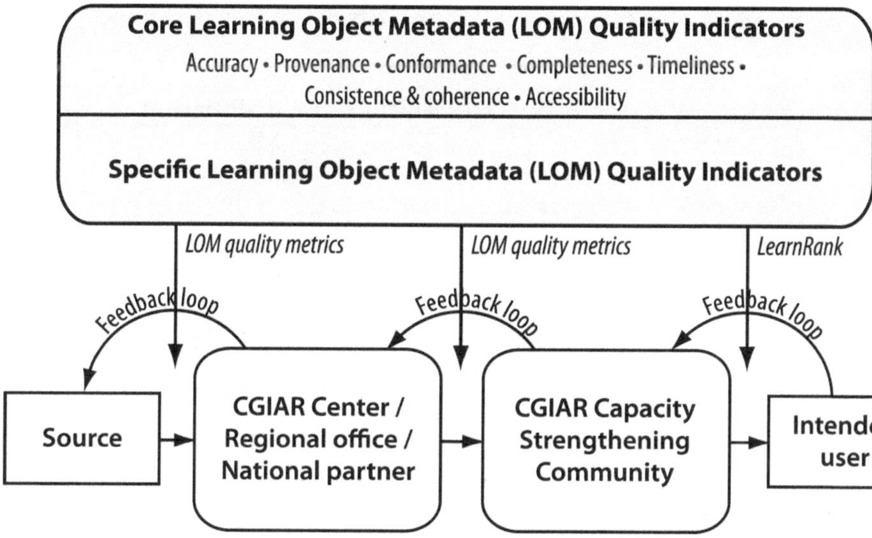

Fig. 1. Proposed LOM Quality Framework for the CGIAR (adapted from the FAO Data Quality Framework)

The CGIAR generates knowledge content on agricultural research by involving many different stakeholders at various locations around the globe. These partnerships have not only enriched this research and development approach of the CGIAR, but also its production of learning content. For instance, one of the research centers of the CGIAR, the International Potato Center (CIP) in Lima, Peru, has developed a methodology called the Participatory Market Chain Approach (PMCA) with national partners in the Andes, and applied and disseminated it in other parts of the world. The methodology was initially described in a user guide in English [43] and later published in Spanish [44]. This example shows that throughout a particular R&D process many different partners in various developing countries are usually involved generating a series of related information outputs. In order to ensure that quality of information is maintained the CGIAR needs a consistent process of capturing and generating this information to describe these information outputs. As illustrated in Figure 1, partners and CGIAR Center staff in regional offices and headquarters would collect and describe the information following the proposed CG LOM Core and its descriptors. The CGIAR capacity strengthening community, which maintains the learning object repository, would set up some mechanism to review the descriptive metadata to ensure that the learning resources following the shared quality protocols when they are initially entered using, for instance, some peer review mechanism. Duval & Ochoa's ranking algorithms can be applied to generate some metrics about this information using an automated mechanism.

Following the recommendations of the Quality Adaptation Model [46] for implementing the ISO/IEC 19796-1 standard, the CGIAR is now in the process of completing individual components of the proposed quality assurance system

Table 3. Proposed profile description for creating quality LOM (based on ISO/IEC 19796-1)

ID	Category	Process	Description	Relation
CD.2	Conception / Design	Concept for contents	Concept for learning and teaching contents	
Sub-processes / Sub-aspects		Metadata creation		
Objective		To establish a quality framework for learning object metadata to guide the metadata creation process		
Method		Monitoring of data quality based on, e.g., Stvilia et al.'s Framework for Information Quality Assessment and Duval & Ochoa's context-dependent ranking algorithm (LearnRank)		
Result		Ranking of LOM records		
Actors		Metadata creator, metadata aggregator, metadata user		
Metrics / Criteria		Core data quality indicators: • Relevance & completeness • Accuracy • Transparency & metadata • Comparability • Standard classification • Coherence • Timeliness & punctuality • Accessibility & clarity Specific data quality indicators (e.g., Duval & Ochoa's "quality-in-context" or relevance characteristics		
Standards		IEEE Standard 1484.12.1-2002, CG LOM Core, FAO Data Quality Framework		
Annotation / Example				

for its education and training, including the creation of quality LOM. Using the generic template structure of ISO/IEC 19796-1, an initial profile has been developed to capture the necessary information on creating quality LOM as shown in Table 3. The suggested profile incorporates the different notions of quality metadata and corresponding measurement criteria as described above.[6]

5 Conclusion

The different quality categories and their underlying dimensions as discussed in this paper provide the constructs to guide the LOM creation process of the CGIAR and their measurement. The proposed profile would permit the CGIAR to anchor its quality assurance efforts within the different notions of quality

[6] This and other completed profiles will be made available on the corresponding website of the CGIAR, `http://learning.cgiar.org/`

of education, that is, excellence of content, fitness for purpose in terms value-added, fitness of purpose in terms of user satisfaction, and enhancement of internal processes. There is a need, though, to develop further applied methods for improving the quality of data as perceived by end users, which would be incorporated into the quality assurance system. Feedback loops would help to assess perceived metadata quality of that kind; the Open ECBCheck would be another mechanism for this type of assessment. The automated estimation of quality metrics such as LearnRank is also critical to be incorporated. It is hoped that the proposed quality profile for LOM is useful as a mechanism for the capacity strengthening community of the CGIAR and its partners to create quality LOM, but also as a checklist during metadata requirements analysis in order, for instance, to train metadata compilers or end users of the repository. The CGIAR would need to explore further how it can adapt its proposed quality assurance protocols to Stvilia et al.'s [45] framework for information quality assessment. Further research is needed to assess how successful the proposed quality assurance measures have been applied in the context of the CGIAR integrate it with the metrics for ranking of learning objects and information quality assessment from the literature.

References

[1] Schmidt, P., Rojas-Briales, E., Pelkonen, P., Villa, A. (eds.): Proceedings of the SILVA Conference 2006: Quality assurance and curriculum development in forestry and agriculture related sciences. SILVA Publications 4. University Press University of Joensuu, Joensuu (2007), `http://www.silva-network.uni-freiburg.de/content/publications/2006_Proceedings_Valencia_Silva_Publications_4_2007.doc.pdf`

[2] Murphrey, T.P., Lindner, J.R., Elbert, C., Malagamba, P., Piña Jr., M.: Assessment of Readiness to Utilize E-Learning at the International Potato Center, Lima, Peru. In: Lindner, J.R., Wingenbach, G.J., Christiansen, J.E. (eds.) Proceedings of the 18th Annual Conference of the Association for International Agricultural and Extension Education (AIAEE 2002), Durban, South Africa, May 26-30, pp. 509–514 (2002), `http://www.aiaee.org/2002/murphrey509-514.pdf`

[3] Zschocke, T., Beniest, J.: On-line Learning Resources for International Agriculture and Natural Resources Management. In: ISHS Acta Horticulturae 762: XXVII International Horticultural Congress – IHC2006: International Symposium on Horticultural Plants in Urban and Peri-Urban Life, Seoul, Korea, August 13-18, pp. 393–400. International Society for Horticultural Science (ISHS), Leuven (2007)

[4] Manouselis, N., Salokhe, G., Keizer, J., Rudgard, S.: Towards a harmonization of metadata application profiles for agricultural learning repositories. In: Proceedings of the IAAL AFITA WCCA 2008 World Conference on Agricultural Information and IT, Tokyo, Japan, August 24-27 (2008)

[5] Manouselis, N., Salokhe, G., Keizer, J.: Comparing Different Metadata Application Profiles for Agricultural Learning Repositories. In: Sicilia, M.-A., Lytras, M. (eds.) Metadata and Semantics, pp. 469–479. Springer, New York (2008)

[6] Manouselis, N., Kastrantas, K., Salokhe, G., Najjar, J., Stracke, C.M., Duval, E.: Results from a Study of the Implementation of Metadata Application Profiles in Agricultural Learning Repositories. In: Proceedings of the 4th International Conference on Information and Communication Technologies in Bio and Earth Sciences (HAICTA 2008), Athens, Greece, September 18-20, pp. 549–557 (2008)
[7] IEEE Standard 1484.12.1-2002. IEEE Standard for Learning Object Metadata. Institute of Electrical and Electronics Engineers (IEEE), New York, NY (2002)
[8] Najjar, J., Duval, E., Ternier, S., Neven, F.: Towards Interoperable Learning Object Repositories: The ARIADNE Experience. In: Proceedings of the IADIS International Conference on WWW/Internet 2003, Algarve, Portugal, November 5-8, vol. I, pp. 219–226 (2003)
[9] Downes, S.: Learning Objects: Resources for Distance Education Worldwide. International Review of Research in Open and Distance Learning 2(1), 35 (2001), http://www.irrodl.org/index.php/irrodl/article/view/32
[10] Duval, E., Hodgins, W.: Standardized Uniqueness: Oxymoron or Vision of the Future? IEEE Computer 39(3), 96–98 (2006)
[11] Duval, E., Verbert, K.: On the Role of Technical Standards for Learning Technologies. IEEE Transactions on Learning Technologies 1(4), 229–234 (2008)
[12] Duval, E.: Learning Technology Standardization: Making Sense of It All. ComSIS 1(1), 33–43 (2004)
[13] Duval, E., Hodgins, W.: A Lom Research Agenda. In: Proceedings of the 12th International Conference on the World Wide Web, Budapest, Hungary, May 20-24 (2003), http://www2003.org/cdrom/papers/alternate/P659/p659-duval.html
[14] Barton, J., Currier, S., Hey, J.M.N.: Building Quality Assurance into Metadata Creation: An Analysis Based on the Learning Objects and e-Prints Communities of Practice. In: Proceedings of the International Conference on Dublin Core and Metadata Applications (DC 2003), Seattle, WA, September 28 - October 2, 10 p. (2003), http://dcpapers.dublincore.org/ojs/pubs/article/view/732
[15] CGIAR, Evaluation and Impact of Training in the CGIAR, CGIAR Science Council Secretariat, Rome, IT (2006), http://www.sciencecouncil.cgiar.org/fileadmin/user_upload/sciencecouncil/Reports/Evaluation_and_Impact_of_Training.pdf
[16] Beniest, J., Zschocke, T.: Developing a learning object repository for international agricultural research. In: Kommers, P., Richards, G. (eds.) Proceedings of the ED-MEDIA 2005 World Conference on Educational Multimedia, Hypermedia and Telecommunications, Montreal, Canada, June 27-July 2, pp. 4553–4555. Association for the Advancement of Computing in Education (AACE), Chesapeake (2005)
[17] Zschocke, T., Beniest, J., Paisley, C., Najjar, J., Duval, E.: The LOM Application Profile for Agricultural Learning Resources of the CGIAR. International Journal of Metadata, Semantics and Ontologies 4(1/2), 13–23 (2009)
[18] Duval, E., Forte, E., Cardinaels, K., Verhoeven, B., Van Durm, R., Hendrikx, K., Wentland Forte, M., Ebel, N., Macowicz, M., Warkentyne, K., Haenni, F.: The ARIADNE Knowledge Pool System. Communications of the ACM 44(5), 73–78 (2001)
[19] InWEnt & EFQUEL: Open ECBCheck. Quality Certification for E-Learning in Capacity Building, http://www.gc21.de/ibt/GC21/en/site/gc21/ibt/permanent/text/What-is-ECB-Check.pdf (n.d.)

[20] ISO/IEC 19796-1:2005. Information Technology – Learning, Education and Training – Quality Management, Assurance and Metrics – Part 1: General Approach. International Organization for Standardization (ISO) Geneva, Switzerland (2005)
[21] Harvey, L., Green, D.: Defining quality. Assessment & Evaluation in Higher Education 18(1), 9–35 (1993)
[22] Vlasceanu, L., Grünberg, L., Pârlea, D. (eds.): Quality Assurance and Accreditation: A Glossary of Basic Terms and Definitions, 2nd edn. UNESCO European Centre for Higher Education (CEPES), Bucharest (2007), `http://www.cepes.ro/publications/blurbs/glossary.htm`
[23] Van Damme, D.: Standards and Indicators in Institutional and Programme Accreditation in Higher Education: A Conceptual Framework and a Proposal. In: Vlasceanu, L., Conley Barrows, L. (eds.) Indicators for Institutional and Programme Accreditation in Higher/Tertiary Edcuation, pp. 127–159. UNESCO European Centre for Higher Education (CEPES), Bucharest (2004), `http://www.cepes.ro/publications/pdf/Indicators.pdf`
[24] Svenonius, E.: The Intellectual Foundation of Information Organization, pp. 67–86. MIT Press, Cambridge (2000)
[25] Bruce, T.R., Hillmann, D.I.: The Continuum of Metadata Quality: Defining, Expressing, Exploiting. In: Hillmann, D.I., Westbrooks, E.L. (eds.) Metadata in Practice, pp. 238–256. American Library Association, Chicago (2004)
[26] Hillmann, D., Dushay, N., Phipps, J.: Improving Metadata Quality: Augmentation and Recombination. In: Proceedings of the International Conference on Dublin Core and Metadata Applications (DC 2004), Shanghai, China, October 11-14, 8 p. (2004), `http://dcpapers.dublincore.org/ojs/pubs/article/view/770`
[27] Heery, R., Patel, M.: Application Profiles: Mixing and Matching Metadata Schemas. Ariadne (25) (2000), `http://www.ariadne.ac.uk/issue25/app-profiles/`
[28] Hillmann, D., Phipps, J.: Application Profiles: Exposing and Enforcing Metadata Quality. In: Proceedings of the International Conference on Dublin Core and Metadata Applications (DC 2007), Singapore, August 27-31, pp. 53–62 (2007), `http://dcpapers.dublincore.org/ojs/pubs/article/view/866`
[29] Duval, E., Smith, N., Van Coillie, M.: Application Profiles for Learning. In: Proceedings of the Sixth International Conference on Advanced Learning Technologies (ICALT 2006), Kerkrade, Netherlands, July 5-7, pp. 242–246. IEEE Press, New York (2006)
[30] Elvers, E., Rosén, B.: Quality Concepts for Official Statistics. In: Kotz, S., Balakrishnan, N., Read, C.B., Vidakovic, B. (eds.) Encyclopedia of Statistical Sciences, Update vol. 3, pp. 621–629. John Wiley, New York (1997), `http://dsbb.imf.org/vgn/images/pdfs/Encyc.pdf`
[31] Statistics Canada's Quality Assurance Framework 2002. Minister of Industry, Ottawa, Ontario, Canada (2002), `http://www.statcan.gc.ca/pub/12-586-x/12-586-x2002001-eng.pdf`
[32] Wang, R.Y., Strong, D.M.: Beyond Accuracy: What Data Quality Means to Data Consumers. Journal of Management Information Systems 12(4), 5–34 (1996)
[33] Ochoa, X., Duval, E.: Quality Metrics for Learning Object Metadata. In: Pearson, E., Bohman, P. (eds.) Proceedings of the ED-MEDIA 2006 World Conference on Educational Multimedia, Hypermedia and Telecommunications 2006, Orlando, FL, June 26-30, pp. 1004–1011. Association for the Advancement of Computing in Education (AACE), Chesapeake (2006)
[34] Ochoa, X., Duval, E.: Relevance Ranking Metrics for Learning Objects. IEEE Transactions on Learning Technologies 1(1), 34–48 (2008)

[35] Borlund, P.: The Concept of Relevance in IR. Journal of the American Society for Information Science and Technology 54(10), 913–925 (2003)
[36] Saracevic, T.: Relevance: A Review of the Literature and a Framework for Thinking on the Notion in Information Science. Part II: Nature and Manifestations of Relevance. Journal of the American Society for Information Science and Technology 58(13), 1915–1933 (2007)
[37] Duval, E.: LearnRank: Towards a Real Quality Measure for Learning. In: Ehlers, U.-D., Pawlowski, J.M. (eds.) Handbook on Quality and Standardisation in E-Learning, pp. 458–463. Springer, New York (2006)
[38] Stvilia, B., Gasser, L., Twidale, M.B., Smith, L.C.: A framework for information quality assessment. Journal of the American Society for Information Science and Technology 58(12), 1720–1733 (2007)
[39] Stvilia, B., Twidale, M.B., Smith, L.C., Gasser, L.: Information quality work organization in wikipedia. Journal of the American Society for Information Science and Technology 59(6), 983–1001 (2008)
[40] Cardinaels, K., Meire, M., Duval, E.: Automating Metadata Generation: The Simple Indexing Interface. In: Proceedings of the 14th International World Wide Web Conference (WWW 2005), Chiba, Japan, May 10-14, pp. 548–556 (2005), `http://www2005.org/cdrom/docs/p548.pdf`
[41] Sicilia, M.-A., García, E., Pagés, C., Martínez, J.-J., Gutiérrez, J.-M.: Complete Metadata Records in Learning Object Repositories: Some Evidence and Requirements. International Journal of Learning Technology 1(4), 411–424 (2005)
[42] Najjar, J., Duval, E.: Actual Use of Learning Objects and Metadata. IEEE Technical Committee on Digital Libraries (TCDL) Bulletin 2(2) (2006), `http://www.ieee-tcdl.org/Bulletin/v2n2/najjar/najjar.html`
[43] Bernet, T., Thiele, G., Zschocke, T. (eds.): Participatory Market Chain Approach. User Guide. International Potato Center (CIP), Lima, Peru (2006), `http://www.papandina.org/fileadmin/PMCA/User-Guide.pdf`
[44] Antezana, I., Bernet, T., López, G., Oros, R. (eds.): Enfoque Participativo en Cadenas Productivas (EPCP). Guía para capacitadores. International Potato Center (CIP), Lima, Peru (2008), `http://www.papandina.org/fileadmin/documentpool/Institucional/Libro/Guia_Capacitadores_EPCP.pdf`
[45] Stracke, C.: Process-oriented quality management. In: Ehlers, U.-D., Pawlowski, J.M. (eds.) Handbook on Quality and Standardisation in E-Learning, pp. 79–96. Springer, New York (2006)
[46] Pawlowski, J.M.: The Quality Adaptation Model: Adaptation and Adoption of the Quality Standard ISO/IEC 19796-1 for Learning, Education, and Training. Educational Technology & Society 10(2), 3–16 (2007)

Ontology Design Parameters for Aligning Agri-Informatics with the Semantic Web

C. Maria Keet

Faculty of Computer Science, Free University of Bozen-Bolzano, Italy
keet@inf.unibz.it

Abstract. In recent years there have been many efforts in the development of bio-ontologies, where the applied life sciences can see the benefits reaped from, and hurdles observed with, such early-adopter efforts. With the plethora of resources, where should one start developing one's own domain ontology, what resources are available for reuse to speed up its development, for which purposes can the ontology be developed? We group inputs that determine effectiveness of ontology development and use into four types of parameters: purpose, ontology reuse, ways of ontology learning, and the language and reasoning services. We illustrate this for the agriculture domain by building upon experiences gained in previous and current projects.

1 Introduction

Only six years ago, multiple modelling issues for the applied life sciences were documented [1], which are currently being addressed, such as with the W3C's incubator group on modelling uncertainty in the Semantic Web (SW), or even surpass the required solution up to a point that is has generated new ones. The most notable advances are the mushrooming of freely available bio-ontologies, the notion of ontology design patterns [2] to save oneself of re-inventing the wheel, and the W3C standard for OWL as common ontology language in the SW. However, solving one problem moves the goal-posts. For instance, which ontologies are reusable for one's own ontology, what are the consequences choosing one over the other? The successor of OWL, draft OWL 2 [3], actually has 4 languages tailored for different purposes: which one should be used for what and when? We structure the main ontology design parameters to provide a brief and clear overview of the principal development options. Ontology development, in particular for highly specialised subject domains in the applied biosciences, is a challenging task and any reuse of information in some way can alleviate this bottleneck. One can both reuse ontologies and ontology-like artifacts, and carry out bottom-up development of ontologies through ontology learning. There are, however, interfering design choices due to the purposes of the ontology and the representation language and reasoning services. We illustrate these parameters with examples taken from the agriculture domain, based on prior and current experimentation with bacteriocins for food processing, semi-automated ontology development in ecology, and ontology-based data access in molecular ecology with horizontal gene transfer (e.g., [1,4,5]), and related literature.

F. Sartori, M.Á. Sicilia, and N. Manouselis (Eds.): MTSR 2009, CCIS 46, pp. 239–244, 2009.

2 Design Parameters

2.1 Purposes of the Ontologies

Arguably, one could take into account the possible aims for which the ontology will be developed. For the ontology purist, however, this is anathema, because an ontology is supposed to be implementation independent—even irrespective if an application will be linked to it or have any computational use at all—and as such has the sole purpose of representing reality. In the practice of ontology engineering, it does have an impact and, based on a literature review and survey [5], the different types of purposes can be summarised as follows:

A. Ontology-based data access through linking data to ontologies [6,5];
B. Data(base) integration, most notably the strand of applications initiated by the Gene Ontology Consortium and a successor, the OBO Foundry [7,8];
C. Structured controlled vocabulary to link database records and navigate across databases on the Internet, also known as 'linked data';
D. Using it as part of scientific discourse and advancing research at a faster pace [4,9], including experimental ontologies in a scientific discipline and usage in computing and engineering to build prototype software;
E. As full-fledged discipline "Ontology (Science)" [10], where an ontology is a formal, logic-based, representation of a scientific theory;
F. Coordination and integration of Web Services;
G. Tutorial ontologies to learn modelling in the ontology development environment (e.g., the wine and pizza ontologies).

A real caveat with choosing explicitly for a specific goal is that a few years after initial development of the ontology, it may get its own life and be used for other purposes than the original scope. This, then, can require a re-engineering of the ontology, as is currently being done with the GO and FMA.

2.2 Reusing Ontologies and Ontology-Like Artefacts

With the mushrooming of ontology development, ontology repositories and semantic search systems, such as Swoogle [http://swoogle.umbc.edu/] and the TONES Ontology Repository [http://owl.cs.manchester.ac.uk/repository/] can be helpful. However, not all ontologies are just more of the same. The principal types of ontologies and ontology-like artifacts that can have a good potential for reuse in part or whole are:

1. Foundational ontologies that provide generic top-level categorisations;
2. 'Reference ontologies' that contain the main concepts of a subject domain;
3. Domain ontologies that have a (partial) overlap with the new ontology;
4. Legacy representations of information systems: conceptual data models of database and application software (sometimes called 'application ontologies'), terminologies, and thesauri;
5. For each of items 1-4, resource usage considerations, such as
   (a) The availability of the resource, such as openly available, copyright, and usage restrictions;

(b) If the source is being maintained or an abandoned one-off effort;
(c) The ontology is a result of a community effort, research group, or if it has already some adoption or usage;
(d) If it is subject to standardization policies or has stable releases.

The foundational ontologies can give a head-start by providing a basic structure, such as endurants being disjoint from perdurants, types of processes, attributes (qualities), and a set of basic relations; e.g., GFO, DOLCE, BFO, RO [11,12]. Reference ontologies, on the other hand, are more restricted in scope of the content, but also intended for reuse, such as an ontology of measurements, of time units and 'top-level' ontologies for a domain, such as BioTop [http://www.imbi.uni-freiburg.de/biotop/] and an ontology of biological investigations (OBI, under development). Domain ontologies, in turn, can build upon such foundations and expand on it for the particular subject domain at hand, such as for traits of rice in Gramene extending GO and marine microbial loops reusing DOLCE [13,4]. The applied life sciences domains have many terminologies and thesauri legacy material, of which a few are being adapted for the SW, such as the reengineering of AGROVOC [14] and reconfiguring and linking for the fisheries domain by using OneFish, AGROVOC, ASFA, and FIGIS with a DOLCE foundation [2]. Other candidates are AOS, and thesauri such as CAB International and CAT. In addition, one can 'ontologise' a conceptual data model and extend the contents. An example for bacteriocins, which are non-therapeutical antibiotics used for food preservation and food safety, is shown in Fig.1. The icons hide the OWL 2 DL in the Protégé ontology development tool, such as `Bacteriocin` $\sqsubseteq \exists$ `inhibits.MicroOrganism`, whereas the grey arrows denote a few of the myriad of possible extensions.

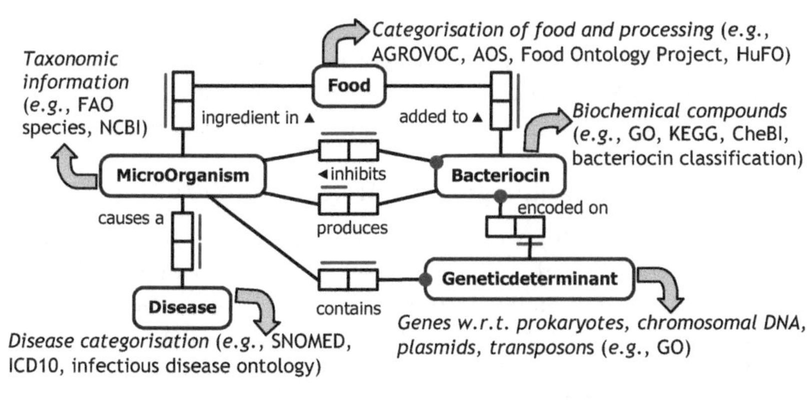

Fig. 1. Section of the conceptual model of the bacteriocins database [1], with reuse of names for relations (e.g., *contains*) and where ontologies, terminologies, and thesauri can be added. This central part about bacteriocins is a candidate for an ontology content design pattern to structure and simplify adding new contents to the ontology. AOS: Agricultural Ontology Service; ChEBI: Chemical Entities of Biological Interest; GO: Gene Ontology; HuFO: Human Food and Nutrition ontology; KEGG: Kyoto Encyclopedia of Genes and Genomes; NCBI: National Center of Biotechnology Information.

2.3 Bottom-Up Development of Ontologies through Ontology Learning

Although one will find something of use in the currently available ontologies, people often will have to develop at least part of the ontology themselves. There are several strategies to speed up this labour-intensive task, which focus on extracting in a semi-automatic way the subject domain semantics present in other legacy sources. The principal techniques are:

I. Extraction of types from data in database and object-oriented software applications, including database reverse engineering, least common subsumer, and clustering;
II. Abstractions from models in textbooks and diagram-based software;
III. Text mining of documents, including scientific articles and other Digital Libraries, to find candidate terms for concepts and relations;
IV. Wisdom of the crowds and usage of those tagging techniques;
V. Other (semi-)structured data, such as excel sheets and company product catalogs.

Reverse engineering is well-known in software development, which is being augmented with a logic-based approach to facilitate the step toward domain and application ontologies [15]. A similar approach in spirit is text mining that seeks to learn the candidate concepts and relations from documents [16]. This is, however, a highly iteratively process [17] that still requires considerable domain expert input (see [16] for a discussion). A different option is to extract knowledge from biological models, such as STELLA models for ecology and environmental sciences made with ISEE software, where, e.g., a STELLA "flow" is a perdurant (the *grazing* process by mesozooplankton) and "stock" corresponds to endurant (e.g., *Plankton*) [4]. One also can try to squeeze out the little semantics available in, say, excel sheets (but see also [9]). If also this fails to extract useful terms and relations, one could resort to the 'wisdom of the crowds'; however agriculture is highly specialised and perhaps not close to the hearts of many online users so that a controlled tagging game with agronomy students may yield better results.

2.4 Representation Languages and Reasoning Services

Depending on the purpose(s)—and, in practice, available resources, such as time, money, domain experts, and available baseline material—one tends to end up with either (a) a large but simple ontology, i.e., mostly just a taxonomy without, or very few, properties (relations) linked to the concepts, where 'large' is, roughly, > 10000 concepts, so that a simple representation language suffices; (b) a large and elaborate ontology, which includes rich usage of properties, defined concepts, and, roughly, requiring OWL-DL; or (c) a small and very complex ontology, where 'small' is, roughly, < 250 concepts, and requiring at least OWL 2 DL. That is, a separate dimension that interferes with the previous parameters, is the choice for a representation language. Moreover, certain choices for reusing ontologies or legacy material, or goal, may lock one into the language that will be used to represent the ontology.

Different from OWL that divided itself between two Description Logics-based versions, OWL-DL and OWL-Lite, and an more liberal RDFS version, the final W3C draft of its successor, OWL 2, has one 'DL' version and three 'lighter-DL' versions [3]. The main motivation for including four DL languages in the standard is to allow tailoring the choice of ontology language to fit best with the usage scope in the context of a *scalable* and *multi-purpose* SW. At the time of writing, no applications exist yet that lets one seamlessly and transparently change one ontology language for another for a given OWL 2-formalised ontology. OWL 2 DL is most expressive and based on the DL language $\mathcal{SROIQ}$ [18], whereas OWL 2 EL and OWL 2 QL are smaller, 'computationally well-behaved', fragments to achieve better performance with larger ontologies and ontologies linked to large amounts of data in secondary storage (databases), respectively; OWL 2 RL has special features to handle rules. Differences between expressiveness of the ontology languages and their trade-offs are discussed in [19]. For instance, OWL 2 DL has the following features that OWL 2 QL does not have: role concatenation, qualified number restrictions, enumerated classes, covering constraint over concepts, and reflexivity, irreflexivity, and transitivity on simple roles. On the other hand, with the leaner OWL 2 QL one can obtain similar performance as with relational databases, whereas for OWL 2 DL one never can achieve that. In addition, not all reasoning services are possible with all languages, either due to theoretical or practical limitations. The current main reasoning services fall into three categories:

i. The 'standard' reasoning services for ontology usage: satisfiability and consistency checking, taxonomic classification, instance classification, and querying functionalities including epistemic and (unions of) conjunctive queries;
ii. Additional 'non-standard' reasoning services to facilitate ontology development: explanation/justification, glass-box reasoning, pin-pointing errors;
iii. Further requirements for reasoning services identified by users (e.g. [20]), such as hypothesis testing, reasoning over role hierarchies, and discovering type-level relations from ABox instance data.

Then, in a software-supported selection procedure, one should be able to select the desired purpose and reasoning services to find the appropriate language, or decide on purpose of usage of the ontology and one's language, and obtain which reasoning services are available. For instance, purpose A or B goes well together with OWL 2 QL and query functionalities, whereas for purposes D and E, OWL 2 DL and the non-standard reasoning services will be more useful.

3 Conclusions

To enhance the efficiency and effectiveness of the recent commencement of developing agri-ontologies, we described the four influential factors. These are (i) seven types of purpose(s) of the ontology, (ii) what and how to reuse existing ontologies and ontology-like artefacts, (iii) five different types of approaches for bottom-up ontology development from other legacy sources, and (iv) the interaction with choice of representation language and reasoning services. Future works pertain to setting up a software-mediated guidance system that can make suggestions how to proceed with ontology development given particular requirements; hence, to structure and make accessible more easily the 'soft' knowledge about ontology development, which then could feed into design methodologies such as methontology.

References

1. Keet, C.M.: Biological data and conceptual modelling methods. J. of Conceptual Modeling 29 (October 2003), `http://www.inconcept.com/jcm`
2. Gangemi, A.: Applying ontology design patterns to practical expertise: roles, tasks and techniques in the agricultural domain. In: CSBio Reader. Free University of Bozen-Bolzano, December 2005, vol. 1, pp. 47–57 (2005)
3. OWL 2. Working draft, W3C (December 2008), `http://www.w3.org/TR/owl2-syntax/`
4. Keet, C.M.: Factors affecting ontology development in ecology. In: Ludäscher, B., Raschid, L. (eds.) DILS 2005. LNCS (LNBI), vol. 3615, pp. 46–62. Springer, Heidelberg (2005)
5. Alberts, R., Calvanese, D., DeGiacomo, G., et al.: Analysis of test results on usage scenarios. Deliverable TONES-D27 v1.0, Tones Project (October 10, 2008)
6. Poggi, A., Lembo, D., Calvanese, D., De Giacomo, G., Lenzerini, M., Rosati, R.: Linking data to ontologies. Journal on Data Semantics X, 133–173 (2008)
7. Gene Ontology Consortium: The Gene Ontology GO database and informatics resource. Nucleic Acids Research 32(1), D258–D261 (2004)
8. Smith, B., et al.: The OBO Foundry: Coordinated evolution of ontologies to support biomedical data integration. Nature Biotech. 25(11), 1251–1255 (2007)
9. Madin, J.S., Bowers, S., Schildhauer, M.P., Jones, M.B.: Advancing ecological research with ontologies. Trends in Ecology & Evolution 23(3), 159–168 (2008)
10. Smith, B.: Ontology (science). In: Proc. of FOIS 2008. IOS Press, Amsterdam (2008)
11. Masolo, C., Borgo, S., Gangemi, A., et al.: Ontology library. WonderWeb Deliverable D18, v1.0, 31-12-2003 (2003), `http://wonderweb.semanticweb.org`
12. Smith, B., et al.: Relations in biomedical ontologies. Genome Biol. 6, R46 (2005)
13. Jaiswal, P., et al.: Gramene: development and integration of trait and gene ontologies for rice. Comparative and Functional Genomics 3, 132–136 (2002)
14. Soergel, D., Lauser, B., Liang, A., et al.: Reengineering thesauri for new applications: the AGROVOC example. J. of Digital Information 4(4) (2004)
15. Lubyte, L., Tessaris, S.: Extracting ontologies from relational databases. In: Proc. of the 20th Int'l Workshop on Description Logics (DL 2007), pp. 387–395 (2007)
16. Gliozzo, A.M., et al.: Results from experiments in ontology learning including evaluation and recommendation. Deliverable 7.3.1, NeoN Project (December 15, 2007)
17. Alexopoulou, D., et al.: Terminologies for text-mining; an experiment in the lipoprotein metabolism domain. BMC Bioinformatics 9(suppl. 4), S2 (2008)
18. Horrocks, I., Kutz, O., Sattler, U.: The even more irresistible $\mathcal{SROIQ}$. In: Proceedings of KR 2006, pp. 452–457 (2006)
19. Keet, C.M., Rodríguez, M.: Toward using biomedical ontologies: trade-offs between ontology languages. In: Proc. of Semantic eScience. AAAI, vol. WS-07-11, pp. 65–68 (2007)
20. Keet, C.M., Roos, M., Marshall, M.S.: A survey of requirements for automated reasoning services for bio-ontologies in OWL. In: Proc. of OWLED 2007. CEUR-WS, vol. 258 (2007)

Developing an Ontology for Improving Question Answering in the Agricultural Domain

Katia Vila[1] and Antonio Ferrández[2]

[1] University of Matanzas, Department of Informatics
Varadero Road, 40100 Matanzas, Cuba
[2] University of Alicante, Department of Software and Computing Systems
San Vicente del Raspeig Road, 03690 Alicante, Spain
{kvila,antonio}@dlsi.ua.es

Abstract. Numerous resources have been developed to have a better access to scientific information in the agricultural domain. However, they are rather concerned with providing general metadata of bibliographic references, which prevents users from accessing precise agricultural information in a transparent and simple manner. To overcome this drawback, in this paper, we propose to use domain-specific resources to improve the results in the answers obtained by an Open-Domain Question Answering (QA) system, obtaining a QA system for the agricultural domain. Specifically, it has been made by (i) creating an ontology that covers concepts and relationships from journal publications of the agricultural domain, (ii) enriching this ontology with some public data sources (e.g the Agrovoc thesaurus and the WordNet lexical database) in order to be precisely used in an agricultural domain, and (iii) aligning this enriched ontology with articles from our case-study journal, i.e. the Cuban Journal of Agricultural Science. Finally, we have developed a set of experiments in order to show the usefulness of our approach.

Keywords: Ontology development, ontology alignment, agricultural ontology, AGROVOC, question answering.

1 Introduction

Nowadays, the number and complexity of websites and the amount of textual information they offer is rapidly growing. Due to this fact, processing and accessing required information is inherent difficult. The scientific information and more specifically the agricultural information is not exempt of this problem [7]. Although in the last few years numerous agricultural repositories or databases have been developed to ameliorate this situation, as the AGRIS repository[1] or the FAO's bibliographical database[2], they have heterogeneous and different structures and interfaces, and different ways of showing their data. Interestingly, accessing these resources could be a prone-to-fail and time-consuming task,

[1] http://www.fao.org/agris/
[2] http://www4.fao.org/faobib/

F. Sartori, M.Á. Sicilia, and N. Manouselis (Eds.): MTSR 2009, CCIS 46, pp. 245–256, 2009.

since users are mainly agricultural researchers with low information technology skills [9]. Furthermore, it is worth pointing out that current approaches for Information Retrieval (IR) within these bibliographical databases and repositories for the agricultural domain often lack mechanisms to provide a precise answer to satisfy the user's needs, since they are rather concerned with providing general metadata of bibliographic references (title, author, abstract, publisher, etc.) and, if available, a link to the full-text document [16].

To overcome these current drawbacks, users need approaches that allow them to access precise information from a set of different data sources in a transparent and simple manner. Therefore, we argue that the most promising solution consists of implementing a Question Answering (QA) system for the agricultural domain instead of considering an Information Retrieval (IR) system. Our motivation is that a QA system provides concise answers to questions stated by the user directly in natural language. However, Open-Domain QA systems should be optimally adapted to a specific knowledge area (in this case, agricultural domain) to improve results [5], thus obtaining a Restricted-Domain QA (RD-QA) system.

In this paper, we have adapted our previously-developed Open-Domain QA system, called AliQAn [12,4,13] to be used for one of the most important agricultural journals in South America, namely The Cuban Journal of Agricultural Science (Revista Cubana de Ciencia Agrícola – RCCA). This journal, as others do, lacks resources for information retrieval in an efficient and simple way for researchers, thus presenting the aforementioned problems. Since it is a representative example for research journals in the agricultural domain, then it will be our case study throughout the paper[3].

One may wonder whether AliQAn may be sufficient in the environment of the RCCA Journal. However, very low precision in the answers has been obtained, making it inapplicable in the agricultural domain. This is a common situation when an Open-Domain QA is used in restricted domains [5], mainly due to the following characteristics of the restricted domains: (a) implicit complexity of the domain terminology, since the more technical the domain is, the more complexity; (b) heterogeneous formats and styles of the user's questions as consequence of their application in real-world situations; and (c) fixed and reduced size of the collection of documents, so the correct answer is not in many documents.

For example, if a QA system is pretended to obtain the right answer, then the right type of the question must be previously determined. Then in the treatment of the question: "Which are the main metabolites that come from the digestive tract?(¿Cuáles son los metabolitos principales que vienen del tracto digestivo?)", a problem may arise due to the incorrect assignment of the question type (entity-profession instead of entity-substance), since the AliQAn system is not able to recognize that "metabolites" are "biochemical compounds". Therefore, the correct answer (that should be "chemical compounds" like "fatty acids (amino

[3] For further details about The Cuban Journal of Agricultural Science, we refer reader to Sect. 3.1.

acids, propionic acid, acetic acid, organic acids)") is never found. This error emerges as a consequence of the characteristics (a) and (b) previously-mentioned.

Bearing these considerations in mind, the aim of our work is to develop the necessary resources to be able to adapt an Open-Domain QA system to be precisely used in the domain of agricultural journals. Specifically, we have defined an ontology for research publications in the agricultural domain. This ontology has been enriched with some concepts from the Agrovoc thesaurus and WordNet and it has been then aligned with the domain terminology of one agricultural research journal (the RCCA journal) in order to improve the QA process for the agricultural domain.

The remainder of this paper is organized as follows: Section 2 presents some related work about using ontologies in several kind of applications. Afterwards, the design of our ontology and the approach used for the alignment between the ontology and the domain terminology are described in section 3. Section 4 shows our experiments and discusses some results. Finally, Sect. 5 details our conclusions and future work.

2 Related Work

In the last years, ontologies have been widely used for different purposes: as part of semantic search applications, for query disambiguation or reformulation, as knowledge sources for RD-QA, for information retrieval on structured data, etc. The reason is that ontologies are models of organized knowledge in a restricted domain (e.g. fisheries, nutrition, or medicine), thus being useful for describing resources in an efficient and consistent manner when knowledge management and information retrieval applications are developed. Also, ontologies help these applications to work together by providing a common vocabulary that describes all important domain concepts without being tied to particular applications in the domains.

In this section, we focus on briefly describing how ontologies are currently used for information retrieval and QA systems.

One issue is how to tailor ontologies to the cross-language information retrieval task on structured data. The domain-specific track at CLEF[4] uses collections from the social science domain to test monolingual and cross-language retrieval in structured bibliographic databases [15,11]. Special attention is given to the existence of controlled vocabularies for content description and their potential usefulness in retrieval. Until now, the success in this area is related to the use of ontologies in structured data as controlled vocabularies for query expansion or as bilingual dictionaries for query translation.

Also, ontologies are used as knowledge sources for restricted-domain QA. In [2], an ontology is implemented as a resource to store the knowledge that is extracted from the web pages of a tourist domain. The ontology is used to apply technics of relaxation of the user's question and of inference in order to obtain a cooperative system. On the other hand, the system presented in [3] is

[4] CLEF: Cross-Language Evaluation Forum, http://clef-campaign.org/

a domestic robot to predict the weather. It uses a domain ontology to recognize and to label the relevant entities (climatic events, weather and places). Other work [6], presents a system of QA on structured bases of knowledge (specific domain ontologies or databases). However, the possibilities that ontologies offer are not completely exploited in this area. So far, they are only used to reflect constraints of the domain, to recognize relevant entities or to carry out question answering directly on the ontology.

Bearing these works in mind, we can conclude by stating that ontologies are not successfully used in information retrieval and QA systems because their concepts are abstract notions not necessarily linked to any domain term, also they are only used to reflect constraints of the domain or to recognize relevant entities in the QA area. The ontology alignment with the indexed documents can be a solution to take completely advantage of the possibilities that ontologies offer [14]. Our work is related to this viewpoint, since we have defined an ontology for research publications and then it has been aligned with the domain terminology of one agricultural research journal in order to improve the QA process.

3 Improving Question Answering for the Agricultural Domain

Our approach is based on the development of an ontology to be used in an Open-Domain QA system (AliQAn) in order to improve the results in the answers obtained for an agricultural domain. The way of proceeding in our approach is as follows:

1. The first step is to analyze the documents collection from the agricultural domain in depth in order to detect their characteristics and their structure, thus designing an ontology. In this case, we have used articles from the RCCA journal as our documents collection. Also, the experience and expertise of specialists in the bibliographical administration support us to define the elements and concepts that our ontology should contain. Section 3.2 highlights the importance of this phase. Our ontology may be enlarged with knowledge of other sources, provided the documents collection has a structure similar to the scientific publications of the RCCA journal.
2. The second step of our approach is the enrichment of our ontology with some public data sources. To do this, the terms of the domain (i.e. the RCCA journal) are mapped with the AGROVOC thesaurus and the Wordnet lexical database.
3. The third and final step is the alignment of our ontology with the domain terminology: instances of the classes in the ontology are aligned with the general vocabulary of the domain by means of a documents collection (i.e. articles from the RCCA journal). These two last steps will be detailed in Sect. 3.4.

3.1 The RCCA Journal as Agricultural Domain

In 1966, the Cuban Journal of Agricultural Science (RCCA: Revista Cubana de Ciencia Agrícola)[5] was created in both English and Spanish. At this moment the RCCA Journal has published 43 volumes, each one with an average of three or four numbers, that makes a total of 140 numbers and almost 2000 articles. In editions about Agricultural Science it publishes original articles comprising the following topics: Applied Mathematics, Genetics, Animal Science, Pastures and Forages, Rural Development, Environment, Sustainability, Agricultural Economy, Production System, Knowledge Transaction, Technological Transfer and Innovation, Technological Extension.

This international journal is included in 48 international indexes of relevance. Its Editorial Board has been integrated by specialists of high qualification of the Institute of Animal Science and other related centers, thus contributing that it has been uninterruptedly published as a way of spreading the results of this center, of other institutions of the country and from abroad. However, although the RCCA Journal has a great importance and it is visible in the Web, it does not still have resources to carry out an information retrieval in an efficient and simple way. For that reason, the RCCA Journal will be the agricultural domain of application of our work.

3.2 Ontology Construction

Ontologies are conceptualizations of some domain. They are composed of "concepts", "attributes", "relations" and "instances". Concepts correspond to objects to be organized (e.g. article, subject, author, etc.); attributes are the features of those objects (e.g. title, serial, abstract, pages, etc.); relations connect two objects or an object and a property to each other; and the instances represent certain elements of a class. There are several standards that provide structures for sharing common descriptions, definitions and relations within the same domain of the knowledge, such as the Web Ontology Language[6].

Our ontology has been constructed using Protégé[7] ontology editor (Fig. 1 shows a portion of our ontology). Protégé is an open-source platform that provides a suite of tools to construct ontologies. It implements a rich set of knowledge-modeling structures and actions that support the creation, visualization, and manipulation of ontologies in various representation formats (RDF, OWL, XML, etc.). Also, it offers a Java API to programmatically access the stored knowledge structure.

Initially, the ontology consists of those elements or concepts from the AGRIS Application Profile [10] which are needed to describe a bibliographic record. Also, in this step, specialists in the bibliographical administration of the RCCA Journal were interviewed in order to refine our ontology. The AGRIS Application Profile provides a standard data model for bibliographic description of

[5] http://www.ica.inf.cu/productos/rcca/
[6] http://www.w3.org/TR/owl-features/
[7] http://protege.stanford.edu/

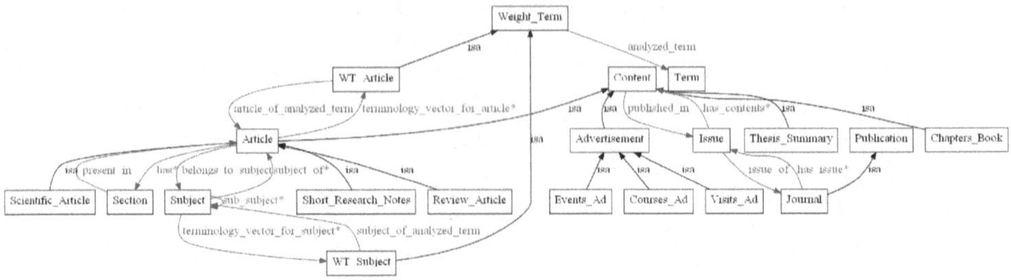

Fig. 1. Excerpt of our ontology visualized with the Ontoviz Plugin for Protégé

resources in the domain of agriculture, covering publications in different areas of the domain (e.g. the RCCA Journal). Next, other classes (concepts) and slots (attributes or relationships) are added to assure the alignment between the domain vocabulary and our ontology (see Fig. 1). For example, the slots (*term_name, synset_wordnet, term_code_agrovoc*) are added in the *Term* class, the *terminology_vector_for_subject* slot is added in the *Subject* class and the *terminology_vector_for_article* slot is added in the *Article* class. The instances of these slots are created by using the results of the alignment process described in Sect. 3.4, while the rest of the instances are created from the current repository of publications of the RCCA Journal.

3.3 The AGROVOC Thesaurus and WordNet

In the last few years numerous public data sources, such as knowledge organization systems (KOS), have been developed, being available for researchers in the agricultural field. KOS are knowledge structures, dictionaries, controlled lists, taxonomies, glossaries, ontologies, thesauri, etc. Samples of KOS are the AGROVOC thesaurus[8] (it represents agricultural knowledge) or WordNet[9] (it represents general knowledge).

The AGROVOC thesaurus covers the terminology of subject fields related to agriculture. Researchers addressing this topic widely use this thesaurus as a unifying vocabulary in order to avoid individual terminology which may hinders further progress in the area. Shortly, AGROVOC is a structured, controlled vocabulary used for indexing and retrieving data in agricultural information systems. To be exact, AGROVOC is a multilingual structured thesaurus covering not only the terminology of agriculture, but also terms in forestry, fisheries, food and other related domains (e.g. environment). These terms are used to unambiguously identify resources. Indeed, the knowledge contained in the vocabulary allows standardizing indexing processes, making searching simpler and more efficient. As in other thesauri, terms are related in AGROVOC, although the kind of relationships supported is generally very limited. The following relationships

[8] http://www.fao.org/agrovoc/
[9] http://wordnet.princeton.edu/

are among the most important in AGROVOC: *Broader Term*–BT (linking a general term to more specific ones), *Narrower Term*–NT (representing the opposite of BT), *Related Term*–RT (used for non-hierarchical relationships between two concepts).

On the other hand, WordNet is a large lexical database of English (also other versions exist in other languages like Spanish), developed at Princeton University. Nouns, verbs, adjectives and adverbs are grouped into sets of cognitive synonyms (synsets), each expressing a distinct concept. Synsets are interlinked by means of conceptual-semantic and lexical relations. The synsets are interconnected with different relational links, such as hypernymy, meronymy, antonymy and others. In this paper, we have only used the noun-taxonomy of WordNet since we are interested in the semantic enrichment of the nouns of our ontology.

Hence, we take advantage of the AGROVOC and WordNet relationships to semantically enrich our ontology for research journals in the agricultural domain. Our ontology will be then better aligned with the agricultural domain.

3.4 Alignment between Ontology and Domain Terminology

In this section, we define our process to align the designed ontology with the domain terminology of the RCCA journal. This alignment is based on different aspects related to the content of the documents used. On one hand, relevance of the terms for a document is considered. It is calculated by applying statistical techniques. The frequency of the terms in each document and the frequency of the terms in the whole document collection are both counted. Then, terms that co-occur more frequently within a document and that do not co-occur frequently within a document collection, at the same time, are hypothesized to be more relevant for a document. On the other hand, if a term appears in most of the documents of the collection then its relevance will be almost null for any document and it is not considered a discriminatory term for the information retrieval. However, we claim that if a term frequently appears in many documents of the collection, but they all belong to the same subject, then it is required to compare with the remaining documents of the group in order to determine its relevance. Keeping in mind this detail is important when weights of the terms are being computed. On the other hand, linguistic techniques are also used, as stemming, which is used to collapse certain terms into a single form.

The ontology alignment is constructed on the basis of a document collection of the agricultural domain (i.e. articles from the RCCA Journal). The overall process consists of three steps (see Fig. 2), which are next described.

The *first step* is to preprocess the documents of the collection during the index phase of the informational retrieval system. In this process, all stop words are removed, and the terms are lightly stemmed by using the IR-n system presented in [8]. For example, the term "metabolites" is transformed in "metabolite" and it is stored as instance of *term_name* slot in the *Term* class. Next, two separate indexes are built for the documents: one is the frequency of appearance for each term in the document and the other is the frequency of each term in all the documents with the same subject where the term appears. The values of

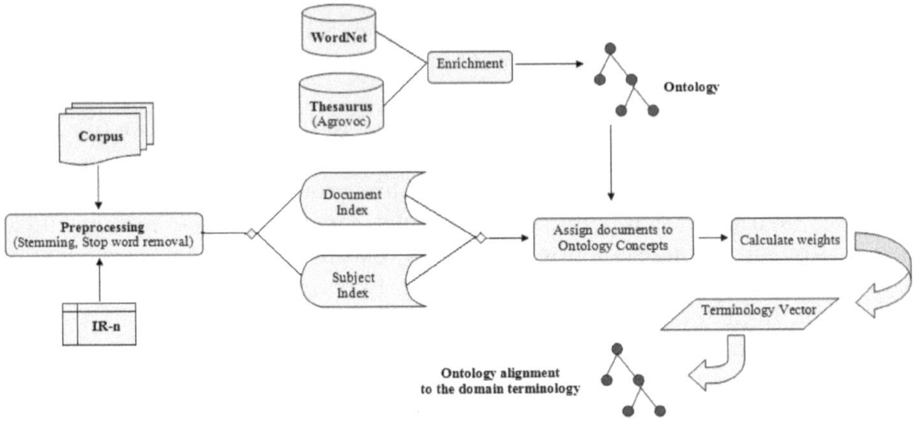

Fig. 2. Process of aligning ontology and domain terminology

the term frequency are used later in order to calculate weights for each term. Finally, in this step, the documents are assigned to ontology concepts (e.g. the properties of one specific article are stored as instances in *Article* class: the article identifier "RCCA1967_T01N01A001" as instance of *record_id* slot, the subject of this article "Animal Science" as instance of *belongs_to_subject* slot, etc.).

Once the documents are preprocessed, the ontology is semantically enriched in the *second step* of our approach. For each term added to the ontology in the previous step the corresponding WordNet synset and the code of the AGROVOC descriptor are obtained and linked to it. For example, the WordNet synset of "metabolite" is "10842652, noun_substance" and it is stored as *synset_wordnet* slot instance in the *Term* class, while the AGROVOC code is "4770" and it is stored as *term_code_agrovoc* slot instance in the same class. Thus, the added term and the AGROVOC descriptor can be matched in order to access other semantically-related terms, for example the concept "metabolites" is a broader term related to more general terms such as "biochemical compounds". It is worth noting that WordNet synsets are also matched. This process is carried out in a semiautomatic manner.

Finally, in the *third step*, the alignment between the ontology and the domain terminology is achieved by associating the instances of the *Article* and *Subject* classes in the ontology with terms from the general vocabulary of the domain. This is achieved by defining the *terminology vector*. For us the terminology vectors are instances of the *terminology_vector_for_article* slot for the *Article* class and instances of the *terminology_vector_for_subject* slot for the *Subject* class. The *terminology_vector_for_article* and *terminology_vector_for_subject* slots are multiple instances of *WT_article* and *WT_Subject* classes, respectively. Also, these classes are characterized by several slots describing the relevance for each term with respect to the article or subject in which it is presented. Both classes contain the *weight* property which is calculated with equation 2 for the *WT_article* and equation 1 for the *WT_Subject*.

Terms in the terminology vectors are weighted in the range [0, 1] where 0 means that the term is not relevant for the instances, while 1 designates the term describing a high relevance. Following, we focus on describing the way of calculating the corresponding weights.

Calculating Weights. The final step in the construction of the ontological alignment is to calculate the weights for the terms assigned to the terminology vectors (previously defined in Sect. 3.4). The basis for the weight calculation is the frequency for each term found in a document (i.e. article) and the frequency for each term found in sets of documents with the same subject (i.e. subject of the article).

Equation 1 shows how to compute the weight for each term with regard to the subject, where $w_{ts_{i,j}}$, is the widely-known $tf * idf$ score [1] for term i in the subject j, $ft_{s(i)}$ is the term frequency for term i in the subject j, $max(ft_{s(l)}, j)$ is the frequency of the most frequent occurring term l in subject j, N_s is the number of subjects and $n_{s(i)}$ is the number of subjects containing term i.

$$w_{ts_{i,j}} = \frac{ft_{s(i)}}{max(ft_{s(l)}, j)} * \log(\frac{N_s}{n_{s(i)}}) \tag{1}$$

The *idf* factor gives more importance to terms that are found in few documents across the document collection, and it is used in an analogue way here for the subject index. The only difference is that we now use groups of documents with the same subject as our documents, meaning that the *idf* factor gives higher weight to terms that are found in few subjects throughout all different subjects presented in the collection (see Equation 1).

Equation 2 shows how to calculate the weight for each term with regard to the document, where $w_{td_{i,j}}$, is the weight for term i in the document j, $tf_{idf i,j}$ is the $tf{*}_{idf}$ score [1] for term i in the document j, $w_{ts_{i,j}}$ is the weight for term i in the subject of the document j and α, and β are constants used for adjusting the relative importance of the two weighting schemes. The best values for α and β proved to be 0.6 and 0.4 respectively for the our collection.

$$w_{td_{i,j}} = \alpha * tf_{idf i,j} + \beta * w_{ts_{i,j}} \tag{2}$$

The alignment between the ontology and the domain terminology is now completed. It is important to highlight that, as consequence of this alignment, the information retrieval process could be done not only by using the frequency of appearance of the question terms in the documents but also by considering the weight of each term with regard to the subject that is interrogated. Further details are shown in Sect. 4.

4 Experiments and Results

A set of two preliminary experiments have been done in order to evaluate the feasibility of our ontological approach for adapting an Open-Domain QA system to the agricultural domain. In this section, these experiments are described and main results are discussed.

The *first experiment* evaluates the usefulness of the alignment between our ontology and the domain terminology in the information retrieval process when the QA system is applied. Firstly, this experiment tries to detect the question subject during the query interpretation phase. Next, it is explained how the question subject is detected. In fact, this means that we map the user query onto a set of one or more subjects in the ontology. In this mapping process the alignment between the ontology and the domain terminology is used as a measure of the coherence between the subjects and the users' query terms.

The query interpretation technique followed produces a ranked list of subjects from which the question subject is chosen. Also, this technique assumes that there is a relation among query terms, which should be recognized by attempting to map all user query terms onto a single subject which suitably represents each of them. To this aim, the candidate subject that contains all query terms is required. The subject which maximizes the score given in Equation 3 is chosen as the question subject for the subsequent question-answering process.

Equation 3 shows how to calculate the score for each subject with regard to the query, where $score_s$ is the score for subject s, $w_{i,s}$ is the weight for query term i in subject s, for more detail of $w_{i,s}$ see Equation 1. The values of these weights are stored in the terminology vector of the subject s (i.e. the *terminology_vector_for_subject* slot instances for the "Subject" class, see Sect. 3.4).

$$score_s = w_{0,s} + w_{1,s} + ... + w_{n-1,s} \tag{3}$$

After the question subject is obtained, we apply the information retrieval process by using the equation 2 as weight of the term in each document. The values of these weights are stored in the terminology vector of the article s (i.e. the *terminology_vector_for_article* slot instances for the "Article" class, see Sect. 3.4). In this way, the information retrieval process could be done not only by using the frequency of appearance of the question terms in the documents, but also by considering the weight of each term with regard to the subject that is interrogated.

This experiment was applied to eight questions that presented the following problem: when the baseline QA system was applied, documents were retrieved where the question terms were more frequent but less important for the subject that was being interrogated. When our approach was applied, six of the questions were improved, because the correct document was returned in first place. The other two questions returned the document with the correct answer in one of the first three positions. For example, the first part of this experiment was applied to the question "Which are the main metabolites that come from the digestive tract?" and it was detected that the question subject was "Animal Science". Thus, we check that the first 10 retrieved documents did not treat the subject of the question and the document with the answer was not among the first 25 retrieved documents when applying the baseline QA system. When our approach was applied, the weight of the question terms in the document with the answer grew (e.g. the weight of term "metabolite" in the article with identifier "RCCA1967_T01N01A001" grew from 0.3 to 0.5 and this article contained the answer). After applying our approach the document with the answer was returned in the third position of the retrieved documents.

The *second experiment* evaluates the usefulness of our approach in the correct assignment of the question type when the QA system is applied. Firstly, the corresponding WordNet synset and the AGROVOC descriptor code are recovered from our ontology for each noun in the query. Afterwards, the associate hypernyms are searched in WordNet, while the associate broader term is found in AGROVOC. Finally, our QA system uses those terms in the detection process of the question type. When we apply our baseline QA system to the group of eight questions used in the first experiment, we obtain an incorrect assignment in six of them. For example, the question type patterns initially assigned to the questions "Which are the main metabolites that come from the digestive tract?" and "Which are the main tissues involved in the lipogenesis?" are *entity-profession* and *entity-event* respectively, which is an incorrect classification. When our approach is applied in this second experiment, then the correct patterns are obtained: *entity-substance* and *entity-part*, because we recover "substance" as one hypernym associated to "metabolite" and "body part" as one hypernym associated to "tissue". The assignment of the question type of the six problematic questions was adequately determined by using our ontology in the pattern classification process, thus returning the correct answer in each case.

Therefore, this evaluation shows that our approach improved the performance of the information retrieval process of our baseline QA system (i.e. AliQAn system) due to the right assignment of the question type.

5 Conclusions and Future Work

Accessing precise information from a set of different data sources in a transparent and simple manner is mandatory for agricultural researchers with low information technology skills [9]. In this paper, we have presented our proposal for using domain-specific resources to improve the results of QA system for the agricultural domain. First, an ontology that covers concepts and relationships from a journal publication domain is designed. Then, this ontology is enriched with agricultural domain concepts by means of the AGROVOC thesaurus and the WordNet lexical database. Finally, this enriched ontology is aligned with articles from the RCCA journal.

The evaluation of our approach has been carried out by a set of experiments. They have shown promising results which support us for keeping on researching on considering ontologies for improving QA applications to be used in the agricultural domain. In the future, we will make a comparison between our approach and those proposed in the literature to deal with answering queries over views and mapping ontologies.

References

1. Baeza-Yates, R., Ribeiro-Neto, B.: Modern information retrieval. ACM Press, New York (1999)
2. Benamara, F.: Cooperative question answering in restricted domains: the WEB-COOP Experiment. Technical report, Institut de Recherches en Informatique de Toulouse, IRIT (2004)

3. Chung, H.Y.-I.S., Han, K.-S., Yoon, D.-S., Lee, J.-Y., Kim, S.-H., Rim, H.-C.: A practical QA system in restricted domains. Technical report, Dept. of Comp. Science and Engineering, Korea University and Dept. of Comp. Software Engineering, Sangmyung University, Korea (2004)
4. Ferrández, S., López-Moreno, P., Roger, S., Ferrández, A., Peral, J., Alvarado, X., Noguera, E., Llopis, F.: Monolingual and Cross–Lingual QA Using AliQAn and BRILI Systems for CLEF 2006. In: Peters, C., Clough, P., Gey, F.C., Karlgren, J., Magnini, B., Oard, D.W., de Rijke, M., Stempfhuber, M. (eds.) CLEF 2006. LNCS, vol. 4730, pp. 450–453. Springer, Heidelberg (2007)
5. Ferrés, D.H.R.: Experiments adapting an open-domain question answering system to the geographical domain using scope-based resources. In: Multilingual Question Answering Workshop of the EACL, pp. 69–76 (2006)
6. Frank, A.H.-U.K., Xu, F., Uszkoreit, H., Crysmann, B., Jörg, B., Schäfer, U.: Querying Structured Knowledge Sources. In: American Association for Artificial Intelligence, German Research Center for Artificial Intelligence, DFKI (2005)
7. Liang, A., Salokhe, G., Sini, M., Keizer, J.: Towards an infrastructure for semantic applications: Methodologies for semantic integration of heterogeneous resources. Cataloging and Classification Quarterly 43(3/4) (2007)
8. Llopis, F., Vicedo, J.L.: IR-n: A passage retrieval system at CLEF-2001. In: Peters, C., Braschler, M., Gonzalo, J., Kluck, M. (eds.) CLEF 2001. LNCS, vol. 2406, p. 244. Springer, Heidelberg (2002)
9. Maliappis, M.T.: Technological Aspects of Using Agricultural Ontologies. In: Proceedings of the Conference on Metadata and Semantics Research (MTSR 2007), Ionian Academy, Corfu, Greece. CCIS Series (2007)
10. Natlacen, M., et al.: The AGRIS Application Profile for the International Information System on Agricultural Sciences and Technology: Guidelines on Best Practices for Information Object Description. Text Version 2.0, Food and Agriculture Organization of the United Nations, Library and Documentation Systems Division, Rome (Italy), 2005–2007
11. Petras, V., Baerisch, S., Stempfhuber, M.: The Domain-Specific Track at CLEF 2007. In: Peters, C., Jijkoun, V., Mandl, T., Müller, H., Oard, D.W., Peñas, A., Petras, V., Santos, D. (eds.) CLEF 2007. LNCS, vol. 5152, pp. 160–173. Springer, Heidelberg (2008)
12. Roger, S., Ferrández, S., Ferrández, A., Peral, J., Llopis, F., Aguilar, A., Tomás, D.: AliQAn, Spanish QA System at CLEF-2005. In: Peters, C., Gey, F.C., Gonzalo, J., Müller, H., Jones, G.J.F., Kluck, M., Magnini, B., de Rijke, M., Giampiccolo, D. (eds.) CLEF 2005. LNCS, vol. 4022, pp. 457–466. Springer, Heidelberg (2006)
13. Roger, S., Vila, K., Ferrández, A., Pardiño, M., Gómez, J.M., Puchol-Blasco, M., Peral, J.: AliQAn, Spanish QA System at CLEF 2008. In: CLEF (2008)
14. Solskinnsbakk, G., Gulla, J.A.: Ontological profiles as semantic domain representations. In: Kapetanios, E., Sugumaran, V., Spiliopoulou, M. (eds.) NLDB 2008. LNCS, vol. 5039, pp. 67–78. Springer, Heidelberg (2008)
15. Stemfhüber, M., Baerisch, S.: Domain-Specific Track CLEF 2006: Overview of Results and Approaches, Remarks on the Assessment Analysis. In: Working Notes for the CLEF 2006 Workshop, Alicante, Spain, September 20–22 (2006)
16. Wildemann, T., Salokhe, G., Keizer, J.: Applying new trends to the management of bibliographic information on agriculture. Zeitschrift fur Agrarinformatik 12(1), 9–13 (2004)

A Service Architecture for Facilitated Metadata Annotation and Ressource Linkage Using agroXML and ReSTful Web Services

Daniel Martini, Mario Schmitz, Jürgen Frisch, and Martin Kunisch

Association for Technology and Structures in Agriculture,
Bartningstrae 49, 64289 Darmstadt
d.martini@ktbl.de

Abstract. ReSTful web services are built by distributing state and functionalities of services across resources. In contrast to RPC services, where a single network object with a (often) large number of method invocations exists, in ReSTful services a large number of network objects, all with the same restricted set of method invocations are available. This allows for scalable and extensible services easily accessible using simple, standardized technology. As semantic web technologies like RDF rely on similar concepts - it is e. g. also possible to use URLs for identification - adding further layers to a service to annotate its content with metadata or to specify relationship between data becomes easy.

1 Introduction

While in the younger days of the World Wide Web most interaction was human-machine, during the last decade web services have increasingly been used to implement machine-machine data exchange between systems. Technologies like the Simple Object Access Protocol (SOAP) and other web services frameworks like XML-RPC have been developed to fulfill special requirements occuring in this context. Traditionally, these web services have been based on the paradigms of message-orientation and remote procedure calls (RPC) known from systems for data exchange common in the late 1980s and 1990s. While in cases where there is a need for short-lived messages this works quite well, this method of building services has severe drawbacks when it comes to scalability and extensibility especially with regard to metadata annotation and semantically rich linkage between data accessible through a single or several services.

In the following paper, we used the ReSTful method to implement a simple service offering data about batches of pigs to be delivered to the slaughterhouse.

2 Materials and Methods

The basic design of the system is oriented on the architecture described by [1]. It relies on the usage of technology components standardized by the W3C. Key

F. Sartori, M.Á. Sicilia, and N. Manouselis (Eds.): MTSR 2009, CCIS 46, pp. 257–262, 2009.

concepts are globally unique identification by URIs, web linking methods, XML as format for the content and a simple protocol, restricted to a few method invocations – the Hypertext Transfer Protocol. The choice for this method has mainly been motivated by the fact, that it suits requirements in a number of agricultural data exchange scenarios very well (see section 3).

The content and document types delivered are provided by agroXML. It offers the necessary elements and datatypes to be able to represent agricultural issues in XML documents. agroXML is defined by a set of XML Schemas and content lists. They are available at http://www.agroxml.de/schema/ and http://www.agroxml.de/content/ respectively.

Unique identification of resources is provided by URIs. We have used the subset commonly known as URLs [2], as the mechanism to dereference them is simple and standardized. To link documents in a service, currently, XLink [3] is used. There are attributes, which allow to further describe arcs and its attached nodes in document graphs or to give hints for applications on how to evaluate the link. The different types of links and the attributes can already be used to introduce certain semantics in document relationships. The Ressource Description Framework (RDF, [4]) could be used to gain further expressivity in describing these interconnections between ressources. However, in the prototype implemented, only the simple link type of XLink and no further attributes apart from href and title have been used.

All together, these technologies provide a ReSTful web service. The term ReST is an acronym for Representational State Transfer and has been introduced by Roy Thomas Fielding [5]. It is based on the assumption that with a few simple operations to read and write data and a system changing its state depending on the operations issued, any use cases in communication can be represented. Variations of this concept are a basic thread present in information technology history up to now. The Turing machine [6] already relied on this simple principle. The SQL language common in database systems with its INSERT, SELECT, UPDATE and DELETE operations is built on this pattern. One of the principles in early UNIX system development was "everything is a file", thus allowing for manipulation of devices and compute resources using simple file operations like open, close, create, read and write. Later on, Kilov coined the term Create-Read-Update-Delete-pattern (CRUD, [7]). The currently widespread Hypertext Transfer Protocol (HTTP, [8]) is built around this assumption as well. So, although the increasing popularity of ReSTful web services is suggesting, that it is a very new technology, it actually has its foundations in methods proven since at least 40 years.

Developing a ReSTful web service involves distributing state and functionalities of a service across a set of network objects. Objects are manipulated using only the small set of basic operations provided by HTTP. HTTP GET and POST-requests are commonly known as these are the calls mostly used in day-to-day work with the protocol e. g. while surfing the web. A number of useful services can be built using only these two requests. A full implementation of a ReSTful web service may however also use the PUT and DELETE requests to allow creation and deletion of resources, respectively. This is in contrast to services using an RPC-paradigm, where a single network object offers a large number of method invocations.

2.1 Prototype

For demonstration purposes, a prototype around the following use case has been developed:

- For a group of animals the fattening process is over.
- The farmer puts together a batch for transport to the slaughterhouse.
- Data about the animals is readily available.
- Data about the location, where the batch is built, is readily available.
- The task is to summarize the information about this batch and present it in a machine-readable form on the network.

agroXML has been used as the format for the content of the service. It offers a collection of datatypes and elements useful for creating XML documents on agricultural issues. The modules available in agroXML allow for building small, self-contained documents able to represent single objects in resources on the web. An XML instance of the farm object containing basic address information (shortened and simplified: the XML prologue, XLink-type und namespace declarations und schema locations are left out) looks like this:

```
<Farm id="FARM_01">
  <Address>
    <StreetAddressOne>At the Field 23</StreetAddressOne>
    <City>Exampletown</City>
    <!-- Further address data might go here -->
  </Address>
  <FarmManager>
    <!-- Farm Manager Data, like his personal address, phone
    number etc. might go here -->
  </FarmManager>
  <!-- Further farm data are possible, like e.g. certifications,
  farming system etc. -->
</Farm>
```

For pigs, data concerning sex, eartag and events related to the animal like weighing or feeding can for example be laid out in XML instances. An (also simplified and shortened) example looks like this:

```
<Pig id="Pig_2348">
  <Events>
    <Birth>
      <DateOrTimestamp>2008-04-23</DateOrTimestamp>
    </Birth>
```

```
    <Weighing>
      <DateOrTimestamp>2008-09-02</DateOrTimestamp>
      <Weight uom="kg">112.3</Weight>
    </Weighing>
  </Events>
  <Sex>m</Sex>
  <EartagNumber>2348</EartagNumber>
</Pig>
```

In the example, two events and their respective data are given in XML encoding: the birth of the pig and a weighing, taking place on the 2nd of September in 2008. For the prototype, we could rely on unique identification of single animals. Nevertheless, it is also possible to address groups of animals.

An (again shortened and simplified) example of an XML instance of a batch might look like this:

```
<Batch id="BATCH_1">
 <LoadLocation xlink:href="http://example.com/farms/farm_01.xml"/>
 <BatchedObject
 xlink:href="http://example.com/animals/pig_2348.xml"/>
 <BatchedObject
 xlink:href="http://example.com/animals/pig_1975.xml"/>
</Batch>
```

The pig data can be fetched by dereferencing the URLs given in the xlink:href attribute. The element <BatchedObject> might also point to a cow or a sack of wheat thus providing a very generic model. The element <LoadLocation> contains a reference to the location, where the loading of the batch took place. Here, it is also possible to use stables or other places instead of the farm. In total, the following URL-structure is used for the service:

farm data: http://example.com/farms/*
animal data: http://example.com/animals/*
batch data: http://example.com/charges/*

The service delivering the XML documents has been built using standard tools available in the class libraries as distributed by Sun together with Java version 1.6.

3 Discussion

The application shown could also have been built using message-oriented remote procedure calls using e. g. SOAP. For this to work, methods for adding pigs to a batch and for retrieving batch data would have to have been defined. This kind of approach has the advantage, that it is conceptually not as demanding in the initial design phase. However, the main disadvantage of message-oriented or RPC approaches like that one to data exchange in the context of the semantic web is, that messages are short-lived and as there is no standardized way to reference objects it is very difficult to add further layers like e. g. a set of RDF statements to relate resources to each other. In contrast, resources in a ReSTful webservice are persistent from a clients point of view. Objects can be fetched at any time and manipulated using a small set of operations. This can be the basis to build a RDF triple store annotating resources with further metadata or describing relationships between objects. As URLs can be used for identifying objects in RDF as well as in ReSTful web services, the technologies play together quite well.

From a functional point of view, services leveraging a mixture of XML documents and RDF triple stores can provide the added value of allowing for semantically rich linkage of tree-oriented documents in a graph structure. Whereas relationship in XML documents is implicitly encoded in the element hierarchies, RDF could provide for more explicit statements. In the agricultural context, e. g. representing statements like "Pig X has been treated with veterinary drug Y by veterinary Z" becomes possible. In XML documents, the relationship can only be expressed using hierarchical structures, which in general is not very clear, as it provides room for interpretation of the reason of a certain parent-child relationship in the tree.

A further disadvantage of the message-oriented approach using technologies like SOAP is, that it is not scalable on a global level. The method calls offered by the server and their parameters have to be known to the client in advance. While this is no problem as long as there is only a limited, strictly controlled set of services available (like e. g. internal to an organisation), it leads to severe difficulties, if an unlimited set of applications is to be added. Every system somehow has to publish its interface. While technologies of service-oriented architectures like UDDI theoretically could provide the necessary functionality, in networks with a large diversity of system environments dynamic binding of clients during runtime is currently practically infeasible. The service architecture described above also coined as web oriented architecture with its restricted set of methods (verbs) allows a client to navigate through an infinitely large set of resources without requiring detailed knowledge of the complete set.

To implement a ReSTful web service, only an XML parser and an implementation of HTTP is required to setup a service or a client. These tools are available for a variety of hardware platforms from energy efficient mobile devices up to powerful server machines and for almost every programming environment. By now, there also exist a number of abstract APIs and frameworks for ReST, which facilitate the setup of a service. But the implementation is also possible with reasonable effort in lower level programming languages e. g. as a server module to the Apache webserver in C. Thus the architecture offers the necessary simplicity required to work in environments like the farm, where the IT infrastructure is not as sophisticated as in large enterprises. An additional bonus comes from the fact, that ReSTful services are - in contrast to services using RPC coded in SOAP - cacheable, which allows them to be used performantly also across high latency links with proxies and caches in between client server. This is often the case in rural areas or developing countries, where the internet infrastructure is not as readily available and reliable as in industrialized areas.

4 Outlook

A prototype leveraging an RDF triple store to interconnect resources to each other will be built. As described above, this will allow for richer semantics in resource relationships than is possible now using XLink. A service may also use RDF to describe itself facilitating dynamic binding and object lookup for clients. A ReSTful web service can be seen as a distributed dataset. A system drawing logical conclusions from this dataset comparable datamining applications on relational database systems could be built. Using technologies like the web ontology language (OWL, [9]), rulesets could be loaded on-demand by client applications for different purposes. Expert systems providing agricultural extension services might profit from such an approach.

Acknowledgement

We thank the Ministry for Education and Research (BMBF) in Germany for funding the work in the IT FoodTrace project (FKZ 0330761).

References

1. World Wide Web Consortium: Architecture of the World Wide Web, vol. One (December 2004), `http://www.w3.org/TR/webarch/`
2. Berners-Lee, T., Fielding, R.T., Masinter, L.: RFC3986: Uniform Resource Identifier (URI): Generic Syntax (January 2005)
3. World Wide Web Consortium: XML Linking Language (XLink) Version 1.0 (June 2001), `http://www.w3.org/TR/xlink/`
4. World Wide Web Consortium: Resource Description Framework (RDF): Concepts and Abstract Syntax (February 2004), `http://www.w3.org/TR/rdf-concepts/`
5. Fielding, R.T.: Architectural Styles and the Design of Network-based Software Architectures. PhD thesis, University of California, Irvine (2000)
6. Turing, A.M.: On computable numbers, with an application to the entscheidungsproblem. Proceedings of the London Mathematical Society 242 (1936)
7. Kilov, H.: From semantic to object-oriented data modeling. In: Proceedings of the First International Conference on Systems Integration (1990)
8. Fielding, R.T., Gettys, J., Mogul, J., Frystyk, H., Masinter, L., Leach, P., Berners-Lee, T.: RFC2616: Hypertext Transfer Protocol – HTTP 1.1 (June 1999)
9. World Wide Web Consortium: OWL Web Ontology Language: Overview (February 2004), `http://www.w3.org/TR/owl-features/`

A Water Conservation Digital Library Using Ontologies

Lukasz Ziemba, Camilo Cornejo, and Howard Beck

Agricultural and Biological Engineering Department, University of Florida,
PO Box 110570, Gainesville, FL 32611, USA
{uki,ccordav1,hwb}@ufl.edu

Abstract. New technologies are emerging that assist in organizing and retrieving knowledge stored in a variety of forms (books, papers, models, decision support systems, databases), but they can only be evaluated through real world applications. Ontology has been used to manage the Water Conservation Digital Library holding a growing collection of various types of digital resources in the domain of urban water conservation in Florida, USA. The ontology based back-end powers a fully operational web interface, available at http://library.conservefloridawater.org. The system has already demonstrated numerous benefits of the ontology application, including: easier and more precise finding of resources, information sharing and reuse, and proved to effectively facilitate information management.

Keywords: Top-level ontology, domain ontology, digital library, web interface.

1 Introduction

The diminishing availability of water in many areas of the world calls for better water resources management. One of such areas is the state of Florida, USA, where ongoing developments in the field of water resources result in increasing amounts of water related information being generated by the university system, state agencies, water utilities, consulting agencies and other organizations. A plethora of information types on water resource management has been produced, such as technical reports and other types of publications, experimental data and other data sets, decision support systems, simulation models and more. To facilitate making water related information available in an organized fashion, the Water Conservation Digital Library [1] was established. The main objectives of the library are identifying, organizing and making accessible various types of information in the domain of urban water conservation in Florida. For this purpose an information management system was developed that integrates all aspects of information and delivers content to decision makers. The core of the system is an ontology containing all relevant information.

Ontology is defined as a formal representation of a body of knowledge formed by a collection of concepts describing a particular domain [2], [3], in this case

F. Sartori, M.Á. Sicilia, and N. Manouselis (Eds.): MTSR 2009, CCIS 46, pp. 263–269, 2009.

the water conservation domain. Ontologies can be used to support a variety of tasks in diverse research areas such as knowledge representation, natural language processing, information retrieval, databases, knowledge management, online database integration, digital libraries, geographic information systems, and visual retrieval or multi agent systems. Ontologies enable knowledge sharing and reuse so that information resources can be communicated between human or software agents [4]. The Water Conservation Digital Library ontology serves as a cataloging back-end and allows for effective management and presentation of a wide variety of information types coming from various sources.

The Water Conservation Digital Library is currently under active development and therefore final results and conclusions are not yet available. This paper shows: 1) the methodology used to construct the ontology with examples in Web Ontology Language (OWL), 2) the ontology's functionality to organize, manage and assist users in locating information, and 3) preliminary conclusions.

2 Ontology Development Methodology

The ontology that constitutes the back-end for the Water Conservation Digital Library was developed following a methodology used previously by various authors [5], [6]. The main steps and rules on this modeling process are:

- Specification: a definition of the scope of the ontology,
- Conceptualization: an agreed and organized set of terms and concepts,
- Formalization: definitions and constraints for terms and relationships used to implement the ontology,
- Implementation: development of the ontology using an ontology editing tool, and
- Evaluation, maintenance and documentation.

Attempts were also made to show how specific concepts in the domain ontology relate to concepts in upper ontologies, including Suggested Upper Merged Ontology (SUMO) [7] and OpenCyc [8]. An upper ontology is limited to generic, abstract and philosophical concepts that are general enough to address a broad range of domain areas [9]. Due to specific goals and limited scope of the project, these efforts can only be continued with project's future expansion.

2.1 Specification and Conceptualization

Water conservation can be used in multiple contexts. However, in this case it is defined as urban water conservation in Florida, and the information managed by the library could be used for regulatory purposes. These conditions limit the scope of the ontology.

The ontology was developed around a set of terms predefined by water management organizations in Florida. Then, that list of terms was related with terms from: Water Science Glossary of Terms [10], United States National Agricultural Library Thesaurus Glossary [11], WaterWiser Glossary of Common Water Terms [12], Water Words Dictionary [13], plus another 20 relevant sources.

2.2 Formalization and Implementation

Two parts of the ontology were developed for the library: a top level ontology that defines all classes of library resources [16] and a domain specific ontology about urban water conservation in Florida. Since the goal of this project is to develop a library, all individuals fall under the top level classes. For example, the class *publication* has subclasses *book*, *manual*, *report*, etc. For each top level class, property restrictions with corresponding data types are defined. The *book* class has the restrictions: *publicationDate* (date type), *title* (string), etc. All individuals are verified against the restrictions of the top level classes. Below is an example of a *book* individual with various properties in Web Ontology Language (OWL) [17]. The property *isReferenceOf* relates the *book* individual with a term in the domain specific ontology under which it is catalogued. Other relationships would point to the publication's author, source organization, etc.

```
<book rdf:ID="AGuidetoFloridaLandscaping">
  <publicationDate rdf:datatype="http://www.w3.org/
    2001/XMLSchema#date">2006</publicationDate>
  <title rdf:datatype="http://www.w3.org/2001/XMLSchema#string">
    A Guide to Florida Landscaping</title>
  <isReferenceOf rdf:resource="#xeriscape"/>
</book>
<owl:ObjectProperty rdf:ID="isReferenceOf">
  <rdfs:domain rdf:resource="#book"/>
  <rdfs:range rdf:resource="#thing"/>
</owl:ObjectProperty>
<owl:ObjectProperty rdf:ID="reference">
  <owl:inverseOf rdf:resource="#isReferenceOf">
</owl:ObjectProperty>
```

In the above example the property *isReferenceOf* indicates that the book "A Guide to Florida Landscaping" contains information about "xeriscape". The inverse property *reference* would help find that book when looking for information about xeriscaping. The application section further explains how relationships facilitate finding related information in the library.

3 Ontology Application

The core of the Water Conservation Digital Library is the ontology containing all relevant information, including various types of publications, datasets, people, organizations, simulation models, news, events and more. The content can be accessed through a dedicated web-based interface available at `http://library.conservefloridawater.org`, as well as parts of other web pages, standard formats and protocols. The two parts of the ontology (top level and domain specific) have different functions in the application.

3.1 Top Level Ontology

The top level ontology contains the classes of various resources collected in the library, for example: *publication*, *book*, *report*, *person*, *organization*, *event*, etc. The resources are related using taxonomic relationships, for example: *book* is a subclass of *publication*. Restrictions on properties are used, for example: title of *publication* must be a string, or first author of *publication* must be a person. Cardinalities are defined as well, for example: *publication* cannot have more than one first author. The top level ontology is utilized for:

- Creation and verification of individuals – all properties must meet the respective class property restrictions, as shown in Fig. 1.
- Displaying the individuals of a specific class – for example all *publication* (and its subclasses) individuals can be displayed with the ability to narrow or widen the selection according to the taxonomic relationships (for example show *book* individuals only).
- Organizing of the displayed individuals according to their class – for example when displaying one of the terms, all related individuals are grouped by class, as shown in Fig. 2 and Fig. 3.

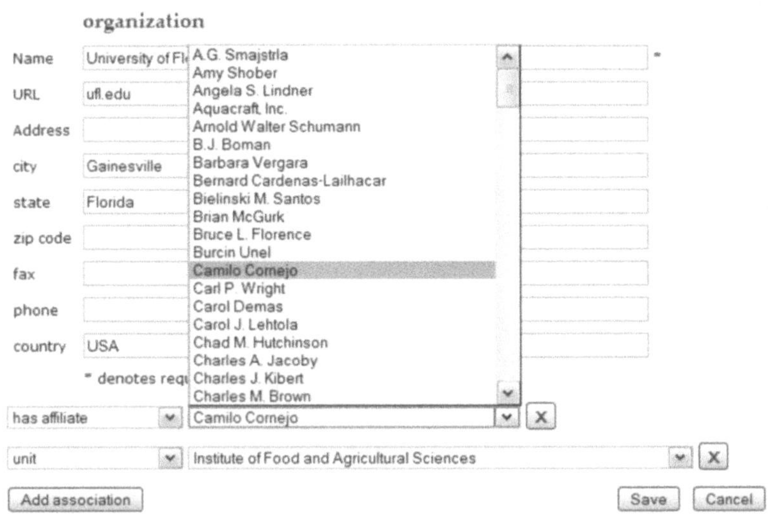

Fig. 1. Class property restrictions used for creating of an individual

3.2 Domain Specific Ontology

The domain specific ontology describes terms, concepts and their relationships within the domain of urban water conservation for Florida. In the web interface the terms are referred to as keywords, since this notion has proved most familiar for the majority of the library users. A definition property is provided for each concept, and taxonomic relationships are used to relate concepts. For example the concept *best management practice* is a subclass of the concept *water conservation measure.* Additionally, the library resources, represented by ontology individuals, are related to relevant concepts, for example *report* "Analysis of Water Conservation Measures for Public Supply" is related to *water conservation measure* concept. At its current stage the Water Conservation Digital Library contains about 1000 terms with definitions and 700 individuals. The domain specific ontology is utilized for:

- Browsing through the concepts in the water conservation domain with the ability to display any of the related concepts, as shown in Fig. 2.
- Showing library resources related to a concept and vice versa, see Fig. 2.
- Finding more relevant resources than a traditional text search.

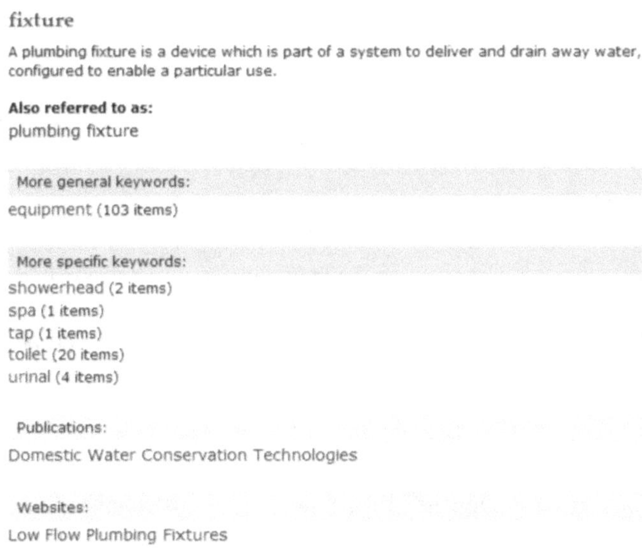

Fig. 2. Domain specific ontology concept in the web interface

3.3 System Architecture

The Water Conservation Digital Library information management system was implemented using the Lyra ontology management system [14], [15], the Java programming platform [18], the XML language [19] and the XSLT transformation [20] technologies. The ObjectEditor tool [14] was used to construct the ontology holding all information content. The ontology is stored within an object database. A Java Servlet queries the database according to the user request passed in the form of HTTP URI and generates XML representation of requested data which is transformed by a XSLT stylesheet into a required output format. This architecture leverages the flexibility and extensibility of the ontology-based back-end by providing the information in many formats and adhering to widely accepted standards as follows:

- most ontology features can be accessed in a dedicated library web site, as shown in Fig. 1 and Fig. 2,
- based on the ontology, many elements of the project main web site are dynamically generated, like current news and events shown in Fig. 3,
- ontology is used to generate RSS feeds according to the specification [21],
- the library exposes its content as Open Archives Initiative Data Provider [22] powered by the ontology.

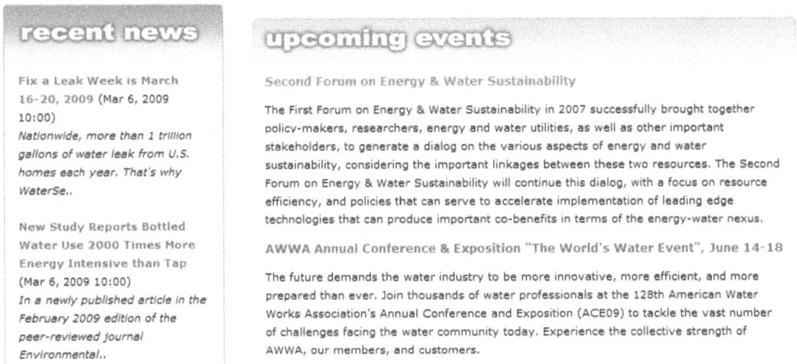

Fig. 3. News and events individuals presented in a web page

4 Preliminary Conclusions

The Water Conservation Digital Library is fully operational and available online. During the development process it demonstrated numerous benefits of the use of ontology. It allows the information to be easily presented in many different ways. The ontology makes finding resources of interest easier and more precise. Furthermore, the flexibility and extensibility of output formats and standards allow for information sharing and reuse. And most importantly, the ontology proved to be an effective aid in organization and management of a growing collection of various types of digital resources. On the other hand, some drawbacks of the approach have been observed. Firstly, the initial workload associated with developing an ontology of adequate complexity is significant and could be discouraging for certain applications. Some of the library users reported a substantial learning period to familiarize with the system. Finally, there are difficulties with keeping the ontology consistent, especially when dealing with a large and dynamic number of concepts. The ongoing research – including the aid of reasoners – is aimed at addressing these issues and further leveraging on the advantages of the ontology application.

References

1. Water Conservation Digital Library, http://library.conservefloridawater.org
2. Gruber, T.R.: A translation approach to portable ontologies. Knowledge Acquisition 5(2), 199–220 (1993)
3. Chandrasekaran, B., Josephson, J.R., Benjamins, V.R.: What are ontologies, and why do we need them? IEEE Intelligent Systems and their Applications 14(1), 20–26 (1999)
4. Kim, S., Beck, H.W.: A Practical Comparison between Thesaurus and Ontology Techniques As a Basis for Search Improvement. J. Agr. Food Info. 7(4), 2–42 (2006)
5. Pinto, H.S., Martins, J.P.: A methodology for ontology integration. In: K-CAP 2001, British Columbia (2001)
6. Uschold, M., King, M.: Towards a methodology for building ontologies. In: Workshop on Basic Ontological Issues in Knowledge Sharing, Montreal (1995)
7. Suggested Upper Merged Ontology (SUMO), http://www.ontologyportal.org
8. OpenCyc, http://www.opencyc.org
9. Standard Upper Ontology Working Group (SUO WG), http://suo.ieee.org
10. United States Geological Survey (USGS). Water Science Glossary of Terms, http://ga.water.usgs.gov/edu/dictionary.html
11. National Agricultural Library, Agricultural Research Service, United States Department of Agriculture. United States National Agricultural Library Thesaurus (NALT), http://agclass.nal.usda.gov/agt/agt.shtml
12. American Water Works Association (AWWA). WaterWiser Glossary of Common Water Terms, http://www.awwa.org/waterwiser/references/glossary.cfm
13. Nevada Division of Water Resources. Water Words Dictionary, http://water.nv.gov/WaterPlanning/dict-1/ww-index.cfm
14. Beck, H.: Evolution of Database Designs for Knowledge Management in Agriculture and Natural Resources. J. Info. Tech. Agr. 3(1) (2008)
15. Beck, H.W.: Integrating ontologies, object databases, and XML for educational content management. In: Proc. of E-Learn 2003, Phoenix, AZ (2003)
16. Beck, H.W., Pinto, H.S.: Overview of Approach, Methodologies, Standards, and Tools for Ontologies. In: The Agricultural Ontology Service. UN FAO, Rome (2002)
17. Web Ontology Language (OWL), http://www.w3.org/TR/owl-ref
18. Sun Microsystems Inc. Java Platform, http://java.sun.com
19. World Wide Web Consortium (WC3). Extensible Markup Language (XML), http://www.w3.org/XML
20. World Wide Web Consortium (WC3). XSL Transformations (XSLT), http://www.w3.org/TR/xslt
21. RSS Advisory Board. RSS 2.0 Specification, http://www.rssboard.org/rss-specification
22. Open Archives Initiative (OAI). The OAI Protocol for Metadata Harvesting, http://www.openarchives.org/OAI/openarchivesprotocol.html

Evaluation of a Metadata Application Profile for Learning Resources on Organic Agriculture

Nikos Palavitsinis[1,2], Nikos Manouselis[1], and Salvador Sanchez Alonso[2]

[1] Greek Research & Technology Network (GRNET S.A.)
56 Messogeion Av., 115 27, Athens, Greece
{palavitsinis,nikosm}@grnet.gr
[2] University of Alcala de Henares (UAH)
Ctra. Barcelona, Km. 33,600, Alcala, Spain
{palavitsinis,salvador.sanchez}@uah.es

Abstract. Metadata specifications and standards serve as the basis for creating metadata application profiles that are particularly adapted to the needs of specific applications. The process of developing such application profiles is usually an iterative one, involving several stakeholders such as technical experts and domain experts. In this process, evaluation should have a pivotal role, by engaging methods and instruments that can ensure that the interests and needs of all stakeholders are reflected in the produced application profile. This paper presents how evaluation is dealt with, in a particular case study of developing a metadata application profile for learning resources. It particularly puts emphasis on the way the domain experts have evaluated the elements of the application profile, on dimensions related to their envisaged usefulness, comprehensibility, and ease to use during content annotation. The methodology followed, the pilot evaluation experiment with the domain experts, and the way the results have been incorporated in the application profile elaboration process, are discussed.

Keywords: Evaluation, metadata, application profile, learning resource, agriculture.

1 Introduction

Metadata has can be usually described as a data record that contains structured information about some resource. The structure of the metadata records aims to facilitate the management, discovery and retrieval of the resources they describe (Al-Khalifa & Davis, 2006). Various metadata standards or specifications can be adapted or "profiled" to meet community context-specific needs (Kraan, 2003). This has lead to the emergence of the metadata application profile concept: application profiles (APs) take one or more base standards or specifications as their starting point. By imposing some restrictions and modifying vocabularies, definitions or elements of the original (base) standard, they tailor the standard to fit the needs of a specific application (Duval et al., 2002).

In the field of technology-enhanced learning, the need for describing resources with information that extends the scope of regular metadata has been early identified

F. Sartori, M.Á. Sicilia, and N. Manouselis (Eds.): MTSR 2009, CCIS 46, pp. 270–281, 2009.

(Recker & Wiley, 2001). Educational metadata schemas have been introduced, adding further fields and elements to metadata records in order to describe information that has particular educational relevance and/or value. The ones that have been more often reported to be implemented and profiled in educational applications are IEEE Learning Object Metadata (IEEE LOM, 2002) and Dublin Core (DC, 2004).

To this end, we present the case of a metadata AP for agricultural learning resources. Although there have been several educational metadata APs developed for agricultural applications (Zschocke et al., 2005; Bio@gro, 2005; Cebeci et al., 2008; Stuempel et al., 2007), the process for involving the domain experts in the evaluation of the AP has not be extensively addressed. In this paper we present how the first version of an educational metadata AP is being evaluated through a pilot evaluation trial from the domain experts that were initially involved in its design. More specifically, the paper introduces the methodology used for the development of this AP, and then presents the results of an expert pilot trial on the conceived usefulness, appropriateness and comprehensiveness of the AP metadata elements by the domain experts. The results lead to a number of revisions to the AP, which are incorporated into its new version.

2 Background

2.1 Generic Process for AP Development

In the related literature various processes of developing an AP have been presented (Friesen et al., 2002; Rivera et al., 2004; Duval et al., 2006; Tambouris et al., 2007). An issue that arises in all of these processes is how metadata APs can be developed in such a way that they will meet the initially stated needs of all involved stakeholders, and especially non-technical ones (such as the domain experts). In the case study that we examine, an educational metadata AP is developed for learning resources related to Organic Agriculture (OA) and Agroecology (AE). This work is taking place in the context of the Organic.Edunet initiative, in order to facilitate access to relevant learning resources that will be stored in a multilingual online federation of learning repositories. The metadata are expected to help domain experts upload and annotate their learning resources, as well as other users (e.g. students) to search, locate and access them. In order to be compliant with major other initiatives developing federations of learning repositories, it has been chosen that the metadata AP of this initiative will be based on the IEEE LOM standard (thus called, the Organic.Edunet IEEE LOM AP).

After studying the various approaches found in the literature on how an AP can be developed, a number of generic steps could be outlined for developing the Organic.Edunet AP:

1. *Definition of own requirements:*
   Depending on the domain of application, the requirements of the AP are documented.
2. *Selection of LOM elements:*
   Taking into account the requirements described in step 1, the LOM elements that will be needed, are selected.
3. *Semantics refinement:*
   The selected LOM elements are refined so as to better fit the needs of the application.

4. *Specification of multiplicity constraints and value spaces of the elements:*
   For each selected element the multiplicity constraints and the possible values are decided. Usually, the LOM values and constraints are kept, but in case the context of a specific application/domain demands, these can be changed.
5. *Specification of relationships and dependencies between elements:*
   The relationships of the elements are decided as well as their dependencies with other elements or sub-elements (i.e. if one element take a specific value, then another one is limited to a small subset of possible values).
6. *Introduction of required extensions to fit specific needs:*
   When all elements are defined according to the needs of the application, then any extra elements can be introduced as extensions to the original metadata schema.
7. *Finalization of AP:*
   The finalization of the AP involves the finalization of the elements, values and constrains as well as the technical representation of the latter.
   a. Conceptual design:
      The design of the whole AP defining how metadata will be obtained, created and stored.
   b. Technical representation:
      The representation of the AP in XML and/or WSML bindings.

It can be noted, that the evaluation of the AP from the stakeholders stating the initial requirements (in our case, the OA and AE domain experts) is not explicitly foreseen in the studied approaches. To this end, we investigate how such an extra step can be introduced in the process.

2.2 Evaluating the Application Profile

When evaluating an AP, problems and opportunities may arise. This information can lead to redesigning certain elements of the AP, adding values in vocabularies, etc. Through a search in the related literature we were able to identify specific studies on the evaluation of metadata APs where users evaluated the APs. Table 1 that follows presents briefly the main surveys that were identified along with their basic parameters.

Table 1. Studies on the evaluation of application profiles

Study	Year	Application Domain	Participants	Tools
Zhang and Li	2008	Metadata schemas for Moving images	100 participants (archivists, educators, librarians, public)	Questionnaires (online & printed)
Krull et al.	2006	Metadata AP for Learning resources	17 experts	Questionnaire (Likert scale)
Howarth	2003	Element names for nine metadata APs	19 experts	Questionnaire
Carey et al.	2002	Metadata AP for Learning Resources	Experts	Questionnaire & interviews
Chang	2001	Web based Portfolio system	35 undergraduate students & 3 experts	Questionnaire (Likert scale) & interviews (with the experts)

Studying all the aforementioned cases, we concluded that, as far as the Organic. Edunet AP is concerned, it would be important to introduce a new step in the process where a group of domain experts will be evaluating the first version of the produced AP. The tools to be used would be a questionnaire with a Likert scale along with Microsoft Excel that will be used to analyze and present the findings.

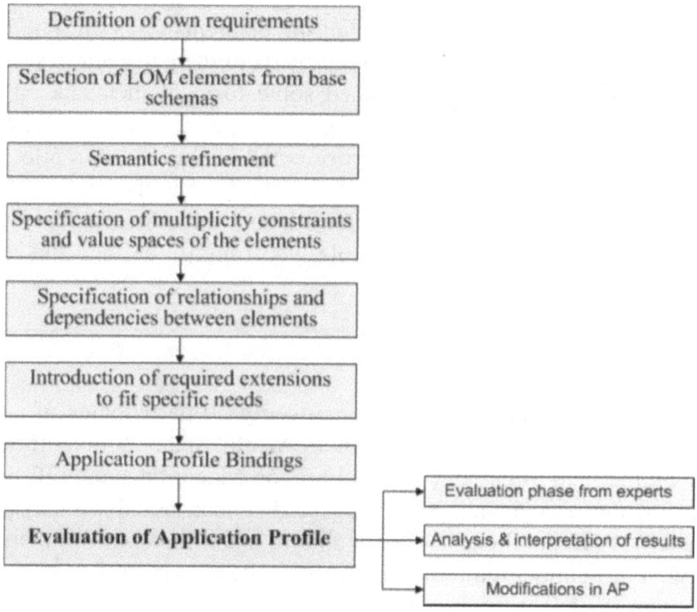

Fig. 1. Proposed stage in the development of an AP

2.3 Revisiting the Process

Based on the discussion on evaluating a metadata AP that has been carried out in the previous sub-section, it has been possible to introduce an expert-oriented evaluation step in the generic process for developing the AP. More specifically, we decided to include at the end of the process (after step 7) an evaluation step that will contain the following activities (Figure 1):

1. Evaluation of the metadata AP by experts
2. Analysis and interpretation of the results
3. Modifications in the AP

In the following section, we describe how the process has been followed for developing and evaluating the Organic.Edunet LOM AP.

3 Development of a LOM-Based AP for Organic Agriculture and Agroecology and Evaluation Study

In the case of Organic.Edunet, we needed an AP to describe the organic agriculture and agroecology domain. This poses particularities in the whole AP development because of the richness and multiplicity of the domain (i.e. issues ranging from cultivation techniques of organic tomatoes to pesticides and agricultural machinery, laws, etc.). The Organic.Edunet AP adopts many of the elements of LOM, but also specializes several of them in order to best match the needs of the particular resources.

Based on relevant studies that introduced some form of metadata AP evaluation (Krull et al., 2006; Carey et al., 2002; Zhang & Li, 2008; Chang, 2001; Moen et al., 1998), we decided to implement the expert-oriented evaluation as a pilot experiment where domain experts will be asked to evaluate one by one the various AP elements and their vocabularies. To this end, feedback about each selected element and its specialization can be collected by the people who initially defined the requirements for the metadata AP.

3.1 Pilot Experiment Settings

The questionnaire was handed out to approximately 20 participants of a content experts' meeting that was held in January 2009 in Alcala, Spain. Among the participants were experts on the field of Organic Agriculture, on education and on ICT. During the specific session, the participants were presented with a detailed description of the Organic.Edunet AP. In parallel, they were able to examine a pilot implementation of the AP though a test annotation tool that has been developed for this purpose. Participants were given sufficient time to carry out some testing annotation of a sample of resources, and then to evaluate the elements.

To collect feedback, a questionnaire was used measuring the users' opinion towards the elements in the AP. For each element, the following questions were asked:

1. Is the element easy for you to understand?
2. Is this element useful for describing Organic.Edunet content resources?
3. Is the selection of the element's possible values clear and appropriate?
4. Should this element be mandatory, recommended or optional?

For questions 1 to 3, the participants were asked to grade each element by using a five point scale with one being the worst (e.g. meaning that this element is difficult to understand in question 1, etc.) and five being the best (e.g. Meaning that this element is very useful for describing resources, etc.). For question 4, the participants would choose between the values "mandatory", "recommended" and "optional" stating each time if they wanted the element to be mandatory in the Organic.Edunet AP (i.e. Anyone that describes a resource using the AP must provide this element), if they wanted the element to be recommended or if they though it should be optional.

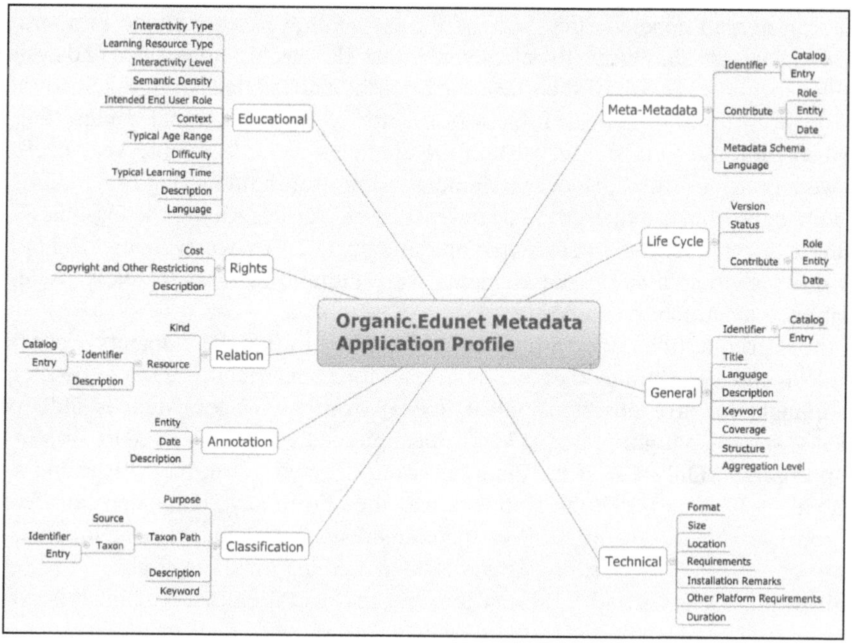

Fig. 2. The Organic.Edunet Application Profile

3.2 Results

The first question of the questionnaire that the participants completed had to do with the comprehension of the proposed elements. More specifically, the participants were asked to evaluate how easy it was to understand the element itself. As we can see in Table 1, 33% of the elements were fully understood by the participants, while 42% of them were understood amounting to a total of 75% that did understand the elements. 21% of the elements were more or less understood and only 4% had a problem with understanding the elements while no element indicated a complete lack in understanding.

Table 2. Overall evaluation of the elements based on aggregated results

	Results				
Question	Totally Disagree	Disagree	Neutral	Agree	Totally Agree
Is the element easy for you to understand?	0%	4%	21%	42%	33%
Is this element useful for describing Organic.Edunet content resources?	0%	12%	33%	41%	14%
Is the selection of the element's possible values clear and appropriate?	0%	4%	37%	50%	9%

We can overall conclude that 96% of the elements were understood in a sufficient level which in turn shows that the elements of the AP were well described and defined.

In the second question of the questionnaire, the participants were asked to rate the elements according to their usefulness in regards to the educational content that they describe (Table 2). In this case, 14% of the elements were deemed as very useful and 41% were pretty useful. 33% of the elements were useful amounting to a total of 88% of the elements that are thought to be at least useful for describing the Organic.Edunet resources. A relatively low percentage of elements (12%) were evaluated as not being that useful whereas none of the elements were thought to be completely useless for describing educational resources.

In the third questions, the participants were asked to rate the elements according to how clear and appropriate their values are. More specifically, 9% of the elements were thought to have absolutely clear and appropriate values whereas 50% of the elements were also highly rated. 37% of the elements were rated as more or less clear and appropriate. Only 4% of the elements were deemed not to have clear and appropriate values while none of the elements was thought to have completely unclear and inappropriate values. Overall, 96% of the elements were evaluated as having clear and appropriate values for describing Organic.Edunet educational resources.

Following, Tables 2 and 3 present the best and worst elements in each one of the three questions that were used in the survey.

Table 3. Best rated elements for each of the three questions

	Best rated			**Rating**
Is the element easy for you to understand?	General. Keyword	Technical. Format	Technical. Size	9.2 out of 10
Is this element useful for describing Organic.Edunet content resources?	General. Identifier	General. Description	Technical. Format	8.8 out of 10
Is the selection of the element's possible values clear and appropriate?	General. Description	Rights. Cost	Format.Size	8.1 out of 10

Table 4. Worst rated elements for each of the three questions

	Worst rated			**Rating**
Is the element easy for you to understand?	Classification. Taxon	Relation. Resource	Educational. Semantic Density	3.1 to 4.8 out of 10
Is this element useful for describing Organic.Edunet content resources?	Classification. Taxon	Annotation. Entity	Annotation.Date	2.3 to 3.1 out of 10
Is the selection of the element's possible values clear and appropriate?	Classification. Taxon	Classification.Purpose	General. Identifier	2.9 to 4 out of 10

Table 5. Overall status of the elements before and after the evaluation process

Results	Mandatory		Recommended		Optional	
Question	Before	After	Before	After	Before	After
Should this element be mandatory, recommended or optional?	19	25	26	21	12	11
Percentile change in overall number of mandatory / recommended or optional elements	+31%		-19%		-8,3%	

Table 6. Changes in the status of the elements based on aggregated results

Results	Mandatory		Recommended		Optional	
Question	Remain	Become	Remain	Become	Remain	Become
Should this element be mandatory, recommended or optional?	90%	40%	69%	33%	75%	27%

Tables 3 and 4 show that most elements of the General category as well as the Technical.Size and Technical.Format were rated as being easy to understand, useful for Organic.Edunet and having clear and appropriate values. On the other hand, as far as the worst rated elements are concerned, experts indicated problems regarding the easiness, usefulness and selection of values especially for the Classification.Taxon element.

As far as the fourth question is concerned, the participants were asked to decide for each element if it should be mandatory, recommended or optional. From the users' choices, some of the elements changed status while some others remained as they were.

By checking the suggested revisions (Table 5), it could be identified that:

- According to the participants, 26% of the total elements should change their status, meaning 15 elements.
- Most mandatory elements kept their status whereas almost 1/3 of the recommended elements changed their status to become mandatory in most cases.
- Generally speaking, a shift towards making more elements mandatory was observed alongside with a decrease in recommended elements. The optional ones remained more or less at the same levels as it is depicted in Table 5.

3.3 AP Revisions

According to the results of the survey, we made some changes to the vocabularies and attributes of certain elements. Tables 7 and 8 present, an overview of the changes made to the Organic.Edunet AP.

Overall, as Table 7 illustrates, changes in the status of 15 elements out of a total of 57 elements that were evaluated were proposed. In 13 cases the changes were implemented to facilitate the use of the AP. Some changes were also made regarding the vocabularies of the AP. These changes came up through the "comments" section of the questionnaire were all the participants gave unstructured comments on the use of the AP. Based on these comments, small adjustments were made to some of the vocabularies. These changes are depicted in Table 8.

Table 7. Changes in the status of the AP elements

Category	Element	Previous Status	Users' Decision	Final Status
General	Structure	Recommended	Optional	Optional
Meta-metadata	Schema	Recommended	Mandatory	Mandatory
	Language	Recommended	Mandatory	Mandatory
Technical	Format	Recommended	Mandatory	Mandatory
	Size	Recommended	Mandatory	Mandatory
	Location	Recommended	Mandatory	Mandatory
	Minimum Version	Mandatory	Optional	Optional
	Maximum Version	Mandatory	Optional	Optional
Educational	Description	Recommended	Optional	Optional
Rights	Cost	Recommended	Mandatory	Mandatory
Annotation	Entity	Optional	Recommended	Recommended
	Date	Optional	Recommended	Recommended
	Description	Optional	Recommended	Recommended

Table 8. Changes in the vocabularies of the Organic.Edunet AP

Category	Element	Changes	Values removed	Values added
Technical	Name	Revised operating systems' list	ms-windows	ms-windows XP, ms-windows Vista, other ms-windows versions, Linux
		Revised browser's list	-	Google chrome, Mozilla Firefox
Educational	Context	Revised environment's list	-	Post-graduate education, Pre-graduate education (*under higher education*)

4 Conclusions

Research on AP development is extended thus providing researchers with many different perspectives on the process of developing an AP. Nevertheless, the issue of non-technical stakeholders' engagement in the process, after the initial requirements are collected, is not extensively covered in literature related to educational AP development. In this context, this paper presented a domain experts' evaluation experiment of a metadata AP that has been developed to describe learning resources related to OA and AE. First of all, the process of developing the AP was presented, and an additional evaluations step has been introduced. Then, the AP itself has been presented, and the initial choice of elements and vocabularies was justified. To continue with, the evaluation experiment that took place and its main results have been introduced, with a particular focus to the changes that have been introduced to the initial version of the AP as an outcome.

This study highlights importance of involving all stakeholders in an iterative manner during the development of a metadata AP. It extends the initial work carried out from researchers working on the development of metadata APs (Howarth, 2003; Carey et al, 2002) by evaluating a number of element attributes (perceived usefulness, clarity and easiness of use), whereas existing studies focused only on one or two dimensions. The added value of the approach has been proven by the nature of revisions that have been finally introduced in the next version of the Organic.Edunet AP. These revisions concerned changing of status for thirteen (13) elements, as well as further elaborating the vocabularies of three (3) elements. Future work will be carried out in order to investigate how other stakeholders could be involved in such experiments. In addition, the questionnaire used in this study could be complemented by additional evaluation methods, such as in-depth interviews with the domain experts, in order to further understand their needs and requirements.

Acknowledgements

The work presented in this paper has been funded with support by the European Commission, and more specifically the project ECP-2006-EDU-410012 "Organic.Edunet: A Multilingual Federation of Learning Repositories with Quality Content for the Awareness and Education of European Youth about Organic Agriculture and Agroecology" of the eContentplus Programme.

References

1. Al-Khalifa, H.S., Davis, H.C.: The evolution of metadata from standards to semantics in elearning applications. In: Proceedings of the Seventeenth Conference on Hypertext and Hypermedia, Odense, Denmark (2006)
2. Bio@gro, Metadata Models for Bio@gro Content Objects (BCOs) Description. Bio@gro Technical Report (2005)
3. Carey, T., Swallow, J., Oldfield, W.: Educational Rationale Metadata for Learning Objects. Canadian Journal of Learning and Technology 28(3) (Fall/Automne 2002)
4. Cebeci, Z., Erdogan, Y., Kara, M.: TrAgLor: A LOM-Based Digital Learning Objects Repository for Agriculture. In: Proc. of the II Convegno "e-LearningPoint 2008", Cassino (FR), Italy (2008)

5. Chang, C.C.: Construction and Evaluation of a Web-Based Learning Portfolio System: An Electronic Assessment Tool. Journal of Innovations in Education and Teaching International, 144–155 (2001)
6. Dublin Core: Dublin Core metadata initiative home page (2004), http://dublincore.org (retrieved 10/3/2009)
7. Duval, E.: Metadata standards: what, who and why. Journal of Universal Computer Science 7, 591–601 (2001)
8. Duval, E., Hodgins, W., Sutton, S., Weibel, S.L.: Metadata Principles and Practicalities. D-Lib Magazine 8 (2002), http://www.dlib.org/dlib/april02/weibel/04weibel.html (accessed January 18, 2008)
9. e-Learning Consortium, Making sense of learning specifications and standards: a decision maker's guide to their adoption (2003), http://www.staffs.ac.uk/COSE/cosenew/s3_guide.pdf (retrieved 01/04/2009)
10. EUN Consortium. The EUN Learning Resource Exchange Metadata Application Profile, version 3.0 (2007), http://insight.eun.org/intern/shared/data/insight/lre/AppProfilev3p0.pdf
11. Friesen, N., Mason, J., Ward, N.: Building educational metadata application profiles. In: Proceedings of the International Conference on Dublin Core and Metadata for e-Communities (DC 2002), Florence, Italy, October 13-17, pp. 63–69. Firenze University Press, Firenze (2002)
12. Haase, K.: Context for semantic metadata. In: Proceedings of the 12th ACM International Conference on Multimedia. ACM, New York (2004)
13. Holden, C.: From Local Challenges to a Global Community: Learning Repositories and the Global Learning Repositories Summit. The Academic ADL Co-Lab Version 1.0 (2003)
14. Howarth, L.C.: Designing a common namespace for searching metadata-enabled knowledge repositories: an international perspective. Cataloging & Classification Quarterly 37(1/2), 173–185 (2003)
15. Hughes, B.: Metadata quality evaluation: Experience from the open language archives community. In: Digital Libraries: International Collaboration and Cross-Fertilization, pp. 320–329 (2004)
16. IEEE LOM. Draft Standard for Learning Object Metadata, IEEE Learning Technology Standards Committee, IEEE 1484.12.1-2002 (July 15, 2002)
17. IMS. IMS Meta-data Best Practice Guide for IEEE 1484.12.1-2002 Standard for Learning Object Metadata v1.3 (2004), http://www.imsglobal.org/metadata/mdv1p3pd/imsmd_bestv1p3pd.html (retrieved 16/10/2004)
18. ISO. Final Report on the International LOM Survey. ISO/IEC JTC1 SC36 (2004), http://jtc1sc36.org/doc/36N0871.pdf (accessed January 18, 2008)
19. Kraan, W.: No one standard will suit all. The Centre for Educational Technology Interoperability Standards (2003), http://www.cetis.ac.uk/content/20030513175232
20. Krull, G.E., Mallinson, B.J., Sewry, D.A.: Describing online learning content to facilitate resource discovery and sharing: the development of the RU LOM Core. Journal of Computer Assisted Learning 22(3), 172–181 (2006)

21. Manouselis, N., Salokhe, G., Keizer, J.: Comparing Different Metadata Application Profiles for Agricultural Learning Repositories. In: Sicilia, M.-A., Lytras, M. (eds.) Metadata and Semantics, Metadata and Semantics, pp. 469–479. Springer, Heidelberg (2009)
22. Moen, W.E., Stewart, E.L., McClure, C.R.: Assessing metadata quality: Findings and methodological considerations from an evaluation of the U.S. Government Information Locator Service (GILS). In: Proceedings of the IEEE International Forum on Research and Technology Advances in Digital Libraries - ADL 1998, pp. 246–255. IEEE Computer Society, Los Alamitos (1998)
23. Rivera, G., Simon, B., Quemada, J., Salvachua, J.: Improving LOM-Based Interoperability of Learning Repositories. In: Proceedings of the OTM 2004 Workshop on Ontologies (2004)
24. SCORM. Sharable Content Object Reference Model, 2nd Edition Overview (2004), http://www.eife-l.org/publications/standards/elearning-standard/scormoverview/english_release/attachment_download/file (retrieved 25/03/2009)
25. Steinacker, A., Ghavam, A., Steinmetz, R.: Metadata Standards for Web-Based Resources. IEEE Multimedia, 70–76 (2001)
26. Stuempel, H., Salokhe, G., Aubert, A., Keizer, J., Nadeau, A., Katz, S., Rudgard, S.: Metadata Application Profile for Agricultural Learning Resources. In: The Proc. of the 2nd International Conference on Metadata and Semantics Research (MTSR 2007), Corfu, Greece, October 11-12 (2007)
27. Tambouris, E., Manouselis, N., Costopoulou, C.: Metadata for Digital Collections of e-Government Resources. The Electronic Library, Special Issue on Metadata and Semantics for Digital Libraries and Information Centres, Emerald 25(2), 176–192 (2007)
28. Tzikopoulos, A., Manouselis, N., Vuorikari, R.: An Overview of Learning Object Repositories. In: Northrup, P. (ed.) Learning Objects for Instruction: Design and Evaluation, pp. 29–55. Idea Group Publishing, Hershey (2007)
29. Zhang, Y., Li, Y.: A user-centered functional metadata evaluation of moving image collections. Journal of the American Society for Information Science and Technology 59(8), 1331–1346 (2008)
30. Zschocke, T., Paisley, C., Duval, E., Beniest, J.: CG Learning Object Metadata (LOM) Core. CGIAR ICT/KM OLR Project (2005)

Ontology for Seamless Integration of Agricultural Data and Models

Ioannis N. Athanasiadis[1], Andrea-Emilio Rizzoli[1], Sander Janssen[2], Erling Andersen[3], and Ferdinando Villa[4]

[1] Dalle Molle Institute for Artificial Intelligence, USI-SUPSI, Lugano, Switzerland
[2] Alterra, Wageningen University and Research Centre, Wageningen, The Netherlands
[3] Danish Centre for Forest, Landscape and Planning, University of Copenhagen, Copenhagen, Denmark
[4] Ecoinformatics Collaboratory, University of Vermont, Burlington, VT, USA

Abstract. This paper presents a set of ontologies developed in order to facilitate the integration of a variety of combinatorial, simulation and optimization models related to agriculture. The developed ontologies have been exploited in the software lifecycle, by using them to specify data communication across the models, and with a relational database. The Seamless ontologies provide with definitions for crops and crop products, agricultural feasibility filters, agricultural management, and economic valuation of crop products, and agricultural and environmental policy, which are in principle the main types of data exchanged by the models. Issues related to translating data structures between model programming languages have been successfully tackled by employing annotations in the ontology.

1 Introduction

The study of agricultural systems requires data spanning across several domains, including ecology, crop science, agronomy, meteorology, economy, policy and demographics. Any modelling framework that aims to integrate crop biophysical models and agro-economic models, at different scales of time and space, obviously needs to offer processes and tools for the seamless and sound management of data. Accessing data is just one side of the problem, as different sources need to be homogenized, documented and properly annotated, before been made available. The other side is persistent storage of simulation results, which again requires rich meta-data to ensure transparency and provide some degree of quality control. We faced such issues in the development of the Seamless-IF framework, where a community of more than one hundred scientists were in need to achieve consensus in their data and model conceptualizations.

This paper presents a remedy to tackle the complexity of agricultural data management issues, by developing and utilizing a set of ontologies for the development of knowledge bases related to agriculture. In the following section we discuss in short the Seamless-IP project and its supporting software infrastructure, *SeamFrame*, from the perspective of data integration and annotation. Next,

F. Sartori, M.Á. Sicilia, and N. Manouselis (Eds.): MTSR 2009, CCIS 46, pp. 282–293, 2009.

we document the empirical process followed in the development of the Seamless ontologies by a community of scientists. In section 4 the main constructs of the developed ontologies are presented along with their use to facilitate the integration of a set of constituent models with a relational database. We also report how the ontology development was integrated in the software lifecycle.

2 Ontologies in Integrated Assessment Studies

2.1 The Seamless Integrated Project

The *Seamless Integrated Project (Seamless-IP)*[1] *develops an integrated framework for assessing and comparing, ex-ante, alternative agricultural and environmental policy options, allowing analysis across different scales, dimensions of sustainability and for a broad range of issues and agents of change* [1].

A large community of more than a hundred scientists from different disciplines was involved in Seamless-IP to study the phenomena involved, develop new (or adopt existing) computer models to quantify them, discover and organize appropriate data required for model calibration and execution, and develop a computer-based integrated framework that is capable to execute the model chain and apply it to various regions of Europe. Certainly, the goals of the project are highly complex, as it is required to bring together an array of heterogeneous models, which are developed following different paradigms (continuous-time simulation models, combinatorial models, market and farm optimization models) accessing data provided by diverse sources. Agricultural, economic, meteorological and landscape data, at different temporal and spatial scales, are fed into the models.

The wide diversity of modeling paradigms and data sources underline the need for cross-disciplinary conceptual integration, by facing the challenge of scientific integration, while providing with practical solutions that can be applied in the software development process. The approach adopted in Seamless-IP was to employ Semantic Web techniques for specifying the domain of agriculture. Specifically, this was achieved by developing a set of domain ontologies in order to:

- build a shared view on the systems modeled, through identifying and resolving ambiguities in terms and data structures;
- facilitate model integration in a sound way, by overcoming scaling problems that are typically remain hidden in low levels (i.e. at the coding phase);
- contribute with added value to the model development, by targeting reusability, interoperability and extensibility of model components.

Mutual understanding across disciplines is often hindered by jargon, language, past experiences and presumptions of what constitutes persuasive argument, and different outlooks across disciplines or experts of what makes knowledge or information salient for policy makers or policy assessments [2].

[1] The Seamless-IP project website is: `http://www.seamless-ip.org`

2.2 A Platform for Agro-Environmental Impact Assessments: SeamFrame

SeamFrame is the software platform used to develop the Seamless integrated modelling framework. Seamframe aims to facilitate model integration through scientific workflows, transparent data access and storage, and end-user interaction. Seamframe's architecture and components are detailed in [3,4,5]. On the basis of SeamFrame, Seamless-IF has been built, following a layered, client-server architecture. The end user interacts with the server by means of a Graphical User Interface (GUI) that is executed as a web client. The server-client architecture of Seamless-IF allows for future applications to be developed and linked to the existing server, in order to cater for specific needs of different user groups.

The Seamless-IF GUI is structured from an end-user perspective in three phases. First comes the pre-modelling phase that involves the interaction with policy experts. At this stage, a Seamless assessment project and narrative experiments are defined and related indicators are selected. Then comes the modeling phase, that consists of the selection of model chains through scales, the detailed specification of experiments, and the execution of model chains. The SeamFrame server provides most of these functionalities, while the SeamFrame client supports remote invocation of model chain execution and retrieving of model results. The last step involves the post-modelling activites, which are supported through the visualization of model results and indicators.

3 Ontology Development for Scientific Workflows

3.1 Model Chains and Scales – Interoperability Issues

In Seamless there are several models to be integrated, which are of different types, follow different modeling paradigms, operate at different scales, and are implemented using different programming languages. More specifically, in Seamless we find [5]:

- combinatorial models, as those required for the generation of agricultural management alternatives;
- biophysical models for crop growth simulation;
- economical models dealing both with farmer income optimization and agricultural product market equilibrium;
- decision making models, including social, economic and environmental indicators;
- databases, providing with reference agro-economic, meteorological and landscape data, at various temporal and spatial scales.

As a result, interoperability issues played a major role in model integration, as different models were originally developed in different programming languages, platforms and operating systems. There is often a distinction between syntactic, structural and semantic interoperability [6]. *Syntactic interoperability* is the ability of two or more systems to exchange and share information by marking up data in a similar fashion, in order to overcome technical heterogeneity (e.g. by using XML). *Structural interoperability* means that the systems share common data models to structure and exchange information. *Semantic interoperability* ensures that the communication between systems is sound, as data models are formally defined and using logical operations the peers can verify the content exchanged [7].

In Seamless, we opted for a solution that lies between the two latter options. For our developments, we adopted a common schema for all data that is exchanged by models. In this respect, model chain composition was based on binding contracts in terms of data structures. In order to enable future extensions to benefit from rich semantics, and enable semantic interoperability, the common data schema was defined in terms of ontologies. In this respect, ontologies were employed to document binding decisions that specify data types exchanged.

3.2 Community Ontology Development in Seamless-IP

A community process for knowledge elicitation and representation was deployed, with the goal to come up with ontologies that specify the data structures exchanged by the models. Following [8], at first researchers were asked to compile a list of concepts they considered relevant for model coupling, e.g. concepts that were shared between the models. These concepts were supplemented with examples and comments, in order to exemplify the meaning. In this first step, we captured an ad hominem response of the researchers about the models to be linked. Then, all list of concepts were merged into one full list of concepts, which served as a sort of lexicon [8]. In this full list of concepts, conflicts between concepts and unclear concepts were indicated through iterative discussions in smaller groups. In these group discussions also the relationships between concepts were discussed. After some iterative rounds of discussion the common ontology was created, which included concepts, properties of concepts and instances of concepts. As experienced by [9], the common ontology can rapidly increase in size through iterative rounds with additional specifications that might make the ontology over-comprehensive. As models were developed in parallel, the synchronisation of the development of the ontology is therefore a difficult task.

4 The Seamless Ontologies

4.1 The Seamless Model Chain

The collaboration of scientists resulted to a shared ontology, covering scales, models, indicators and dimensions relevant to the Seamless project. Instead of making one large ontology, spanning across different sub-domains of the project, we developed eleven small ontologies, each one of which refers to a distinct aspect of the project. In this respect, common concepts and relationships are shared across granular ontologies. In Figure 1, a simplified view of the scientific workflow is presented, along with the ontologies that specify model communications.

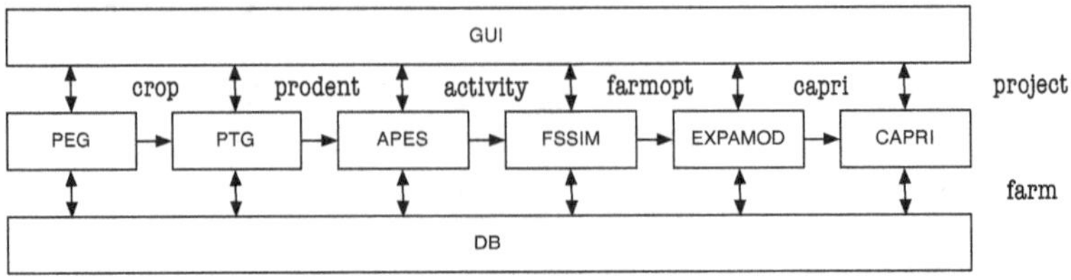

Fig. 1. Seamless model chain and ontologies involved

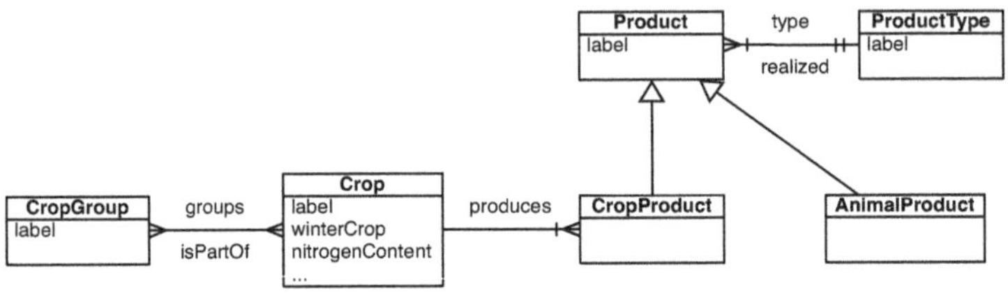

Fig. 2. Partial view of the crop ontology

First come two combinatorial models: Production Enterprise Generator (PEG) and the Production Technique Generator (PTG), which generate alternative arable activities and are coded in Java. Then follows APES[2], which is a biophysical crop growth model, and is written in C#. The workflow concludes with two optimization models (FSSIM and CAPRI[3]), and an extrapolation model (EXPAMOD) written in GAMS.

4.2 Crops and Products Ontology

The crop ontology provides with a conceptualization of crops and crop products and their relationships, as they have been defined in Seamless. The concept of a crop is central for the project, as many components rely on it. We consider industrial crops, which are grown and harvested for producing one or more products, that are of different types. Also, crops form groups according to several criteria. Crop groups are used for abstracting crop production to higher levels in optimization models. Key concepts of the crop ontology are depicted in Fig. 2. As an example consider 'winter soft wheat', 'spring soft wheat', 'winter durum wheat', 'spring durum wheat' as crops grouped in the 'wheat' crop group. 'Winter soft wheat' produces two products: 'winter soft wheat grain' and 'winter soft wheat straw', of type 'Grain' and 'Straw' respectively.

---

[2] For more on the Agricultural Production and Externalities Simulator (APES) see: `http://www.apesimulator.it`

[3] CAPRI stands for Common Agricultural Policy Regionalized Impact Modeling System. See also: `http://www.capri-model.org`

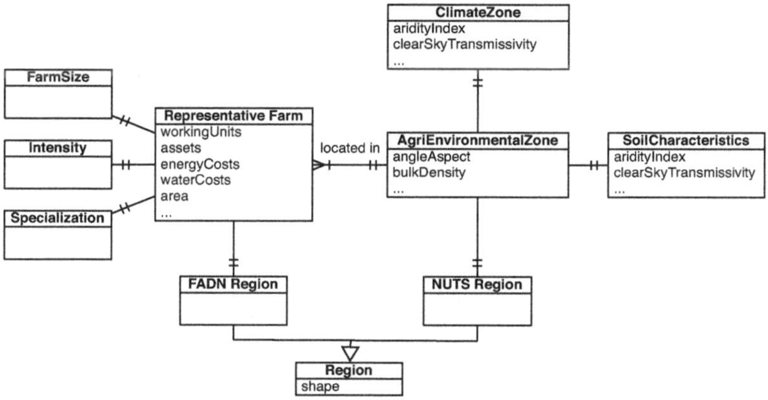

Fig. 3. Partial view of the farm ontology

4.3 Farm Ontology

The farm ontology provides with concepts about farms, geographical regions, in which farms are located, and soil and climate information. A representative farm corresponds to an average farm associated with a FADN region (and agro-economical data)[4] and with a unique specialization, intensity and size classification. As an example consider a large farm specialized in arable crops, of medium intensity production, in the region of Flevoland, The Netherlands. Each representative farm is located in one agri-environmental zone, through which is associated to climate, soil and administrative information. An agri-environmental zone identifies a location within Europe as a unique combination of a soil type, an environmental zone and a NUTS region[5]. Key concepts are depicted in Figure 3. Farm ontology and its use for developing an common database for European agricultural data is further discussed in [10].

4.4 Production Enterprise

The production enterprise ontology specifies concepts related to crop rotations and cultivation choices available to farmers. A rotation is a crop succession scheme. Typically, artificially generated crop rotation schemes start from all possible re-arrangements of the available crops that are subsequently filtered with respect to cyclic equivalence and crop succession suitability requirements. Crop-specific cultivation restrictions are defined in a generic concept `CropRequirements`, which define conditions that have to be met for a crop to be able to grow. Crop requirements

[4] FADN stands for Farm Accountancy Data Network, which is an instrument of the European Commission.

[5] NUTS stands for Nomenclature of Territorial Units for Statistics. It provides with geographical references of the administrative divisions of European countries for statistical purposes.

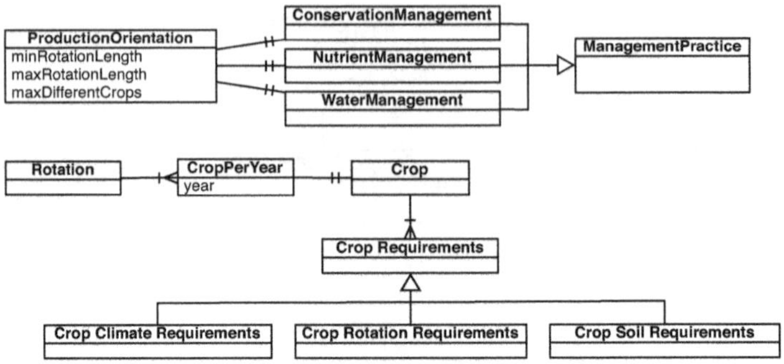

Fig. 4. Key concepts of the production enterprise ontology

can be specified with respect to climate, soil or rotation. Other types of restrictions are defined with respect to production orientations, which quantify stakeholders' preferences on agricultural production. Examples of production orientations include integrated, organic, or conventional farming, which are quantified in terms of management practices available in the farm and restrictions related to the rotation size and structure. Key concepts of the production enterprise ontology are depicted in Figure 4.

4.5 Agricultural Activities

Another ontology is devoted to agricultural activities. An agricultural activity is a coherent set of crops (or animals or grass or trees) with operations and associated inputs, which (when applied on a farm) result in the delivery of a marketable

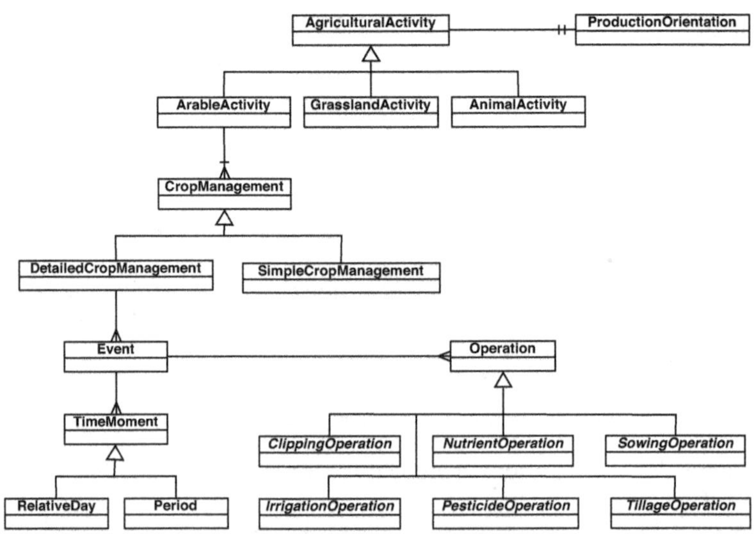

Fig. 5. Key concepts of the activity ontology

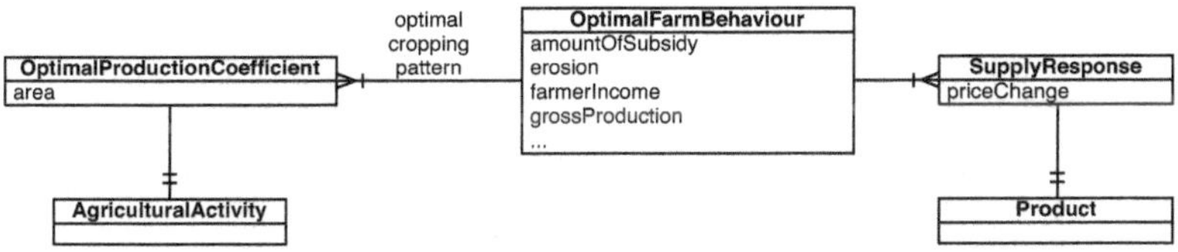

Fig. 6. Key concepts of the farm optimization ontology

product. An arable activity is a type of agricultural activity, and refers to a set of crop management entities. A crop management concept is further specified as a simple crop management entity that realizes a compact, though simplified view of crop management alternatives. A detailed crop management concept provides with a more detailed definition, as management operations and their timing are defined as events are defined as events. Operations can be of different types (e.g. irrigation, clipping, sowing, tillage, etc). Each operation is associated with the necessary inputs, which include among others fertilizers, water, and seeds, and implements which include sowing implements, irrigation methods and tools and fertilizer application methods.

Agricultural activities specifications span across three ontology files in Seamless: `activity.owl` provides with the generic framework, while `agrirule.owl` and `livestock.owl` specify respectively arable and livestock activities in more detail. Key concepts of the activity ontology are depicted in Figure 5.

4.6 Economic Valuation of Agricultural Activities and Optimal Farmer Behaviour

In the Seamless scientific workflow, agricultural activities are used as inputs to a biophysical simulation model, namely APES, that results to yields and environmental effects, which are associated to agricultural activities, constructing the production coefficient concept. The production coefficients are in turn the input of a farm optimization model that takes under account economic, environmental and policy constraints for allocating the optimal farm area to each activity. The Optimal Farm Behavior concept aggregates optimal farm behaviors and provides with an economic and environmental valuation of the optimal production pattern. It also links to a set of supply-response values that quantify changes in production levels due to price changes. Supply-response values are associated with products of the crops and products ontology.

4.7 Agricultural Policy Assessment

Agricultural policy assessment has been modeled in Seamless using the CAPRI model. CAPRI is a spatial economic model that makes use of non linear mathematical programming tools to maximize regional agricultural income with explicit consideration of the Common Agricultural Policy instruments of support

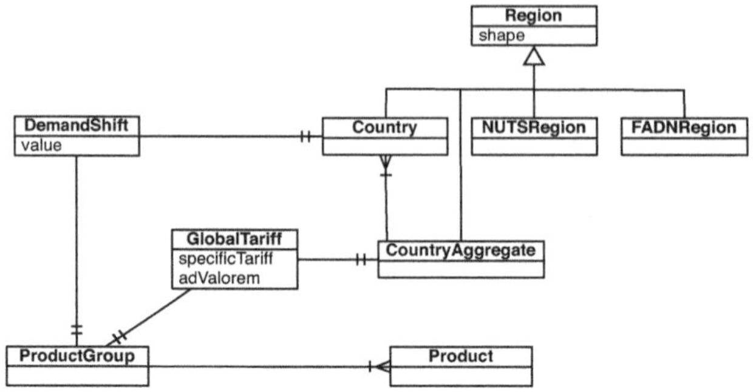

Fig. 7. Key concepts of the CAPRI ontology

in an open economy where price interactions with other regions of the world are taken into account. The corresponding ontology specifies a subset of the CAPRI parameters that are relevant for SEAMLESS assessments, including tariffs, energy prices, price elasticities, basic premiums, inflation rate and exchange rates. Each one of those has a geographical reference to a country or a country-aggregate. As an example in Figure 7 'demand shift' and 'global tariff' are illustrated along their geographical references. Note that CAPRI follows a coarser definition of crops and products therefore 'product groups' have been introduced as an aggregation of 'products'.

4.8 Assessment Project Ontology

An integrated assessment project refers to the process of assessing policy or technological innovations impact on the sustainability of agricultural systems as was adopted in Seamless (see discussion in [11]). The project concept encompasses several textual information fields documenting the process, and most importantly, it is associated with the problem at hand. The scientific problem defines its spatial and temporal scale, a set of associated models that can be used for its solution, a set of experiments to be evaluated and a set of indicators that are appropriate to measure the phenomena involved. Temporal and spatial scale define problem boundaries in terms of extent and resolution in time and space. An experiment is one of the alternative configurations of the scientific workflow to be evaluated, and it is composed of two configurations: one for biophysical simulation and one for policy assessment. The biophysical configuration is composed of a single context and a single outlook. The context specifies the boundaries of the biophysical simulation in terms of products of interest and agricultural management options available, including production orientations. The outlook is defines foreseen changes to the system that are not modeled endogenously. This includes climate, economic or societal trends (i.e. atmospheric CO_2 concentration, energy prices). The policy assessment concept gathers together the attributes required for the ex-ante the impacts of a policy on agricultural

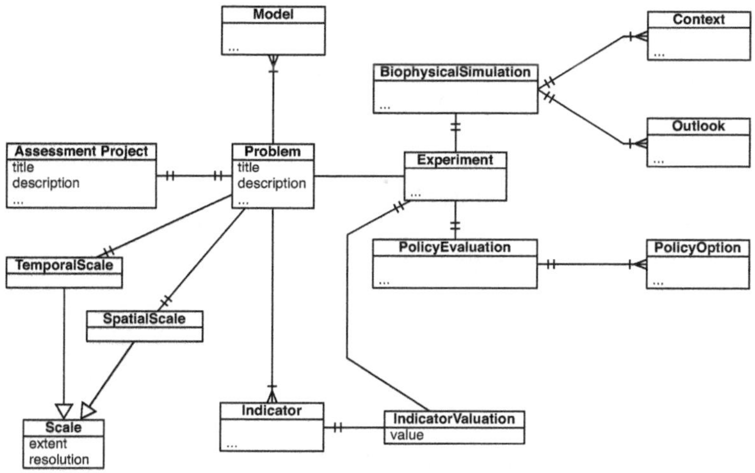

Fig. 8. Partial view of the project ontology

sustainability. It consists of policy parameters within a given timeframe, that include quotas, tariffs, set-aside regulations, subsidies and premiums.

5 Contribution of Ontology Development in Software Engineering Tasks

5.1 Cross-Programming Language Transliterations

One of the technical challenges that we addressed with the adoption of ontologies was to enable data sharing across different simulation and optimization models, each one of which is implemented in a different programming language. This diversity in programming paradigms is typical in environmental applications for various reasons, as prior developments and legacy, expertise of the developers, availability of supporting libraries and tools, and most importantly performance issues. The '*right tools*' are needed to be used for solving each particular problem, thus we ended up having pleiad of different programming paradigms. In the Seamless project, the scientific workflow included models implemented in three programming languages: GAMS, C# and Java, plus the data that were stored in a PostgreSQL relational database. Through the community process followed, we achieved a consensus on the common data structures that specify the data exchanged by the model components of the workflow. The agreement was specified as a set of ontologies described above (in section 4). The challenge that remained open was to develop wrappers for the workflow components to enable data exchange across models, ensure the bi-directional communication with the database, and facilitate the execution process of the scientific workflow. This process is described in detail in [5].

In the Seamless ontologies, we annotated shared concepts with their names in different programming languages. For example, the Crop concept is annotated with the corresponding variable name (`Crop`) in C# for APES model and set name (`C`) in GAMS used by FSSIM, as:

```
<owl:Class rdf:ID="Crop">
    <rdfs:label xml:lang="en">Crop</rdfs:label>
    <rdfs:label xml:lang="gms">C</rdfs:label>
    <rdfs:label xml:lang="aps">Crop</rdfs:label>
</owl:Class>
```

These annotations have been further exploited by tools generating code for accessing data in different programming languages, as discussed in [12].

5.2 Persistent Storage of Simulation Results

Data schema definitions in Seamless ontologies have been exploited for the persistent storage of model results. Using the SeRIDA framework [13] we have transliterated OWL ontologies into relational schemas and corresponding data access classes. A software layer for transparent reading and writing data in the database was derived from the ontology definitions, and upon which the enterprise software was built. The technical solution is based on JavaBeans for structuring data and Hibernate for object-relational mapping, which was favored for the development of model wrappers and the GUI.

5.3 Documentation of Source Code and Database Schemas

As data access classes and database schema was generated from the ontology specifications, the comments of the ontology have been included into both JavaBeans and Hibernate mappings. In this respect, the ontology served as a single entry point not only for specifying data structures, but also for their documentation, which is propagated in the object oriented and the relational design.

6 Discussion

In this paper we presented a set of ontologies related to agricultural models of different types and discussed their use for model integration and environmental software development. Ontologies proved a powerful medium for specifying the data structures involved in model integration. Ontologies were used as an abstract framework for conceptual modeling of data exchanged by models, and assisted in software and database development. As conceptual models with ontologies are more rich compared to object oriented and relational models, we used ontologies to generate code across different programming languages (namely C#, GAMS and Java) and a common underlying relational schema for data storage.

Acknowledgement. We thank the scientists involved in the Seamless-IP project, who contributed to the development of the shared ontology. This work has been carried out as part of the Seamless Integrated Project, EU 6th Framework Programme, Contract No. 010036-2.

References

1. van Ittersum, M.K., Ewert, F., Heckelei, T., Wery, J., Olsson, J.A., Andersen, E., Bezlepkina, I., Brouwer, F., Donatelli, M., Flichman, G., Olsson, L., Rizzoli, A.E., van der Wal, T., Wien, J.E., Wolf, J.: Integrated assessment of agricultural systems - A component-based framework for the European Union (SEAMLESS). Agricultural Systems 96(1-3), 150–165 (2008)
2. Cash, D., Clark, W., Alcock, F., Dickson, N., Eckley, N., Guston, D., Jäger, J., Mitchell, R.: Knowledge systems for sustainable development. Proceedings of the National Academy of Sciences 100(14), 8086–8091 (2003)
3. Rizzoli, A.E., Donatelli, M., Athanasiadis, I.N., Villa, F., Huber, D.: Semantic links in integrated modelling frameworks. Mathematics and Computers in Simulation 78(2-3), 412–423 (2008)
4. Wien, J.J.F., Rizzoli, A.E., Knapen, M., Athanasiadis, I.N., Janssen, S.J.C., Ruinelli, L., Villa, F., Svensson, M., Wallman, P., Jonsson, B.: A web-based software system for model integration in agro-environmental impact assessments. In: Environmental and agricultural modelling: integrated approaches for policy impact assessment. Springer, Heidelberg (2009)
5. Athanasiadis, I.N., Janssen, S.: Semantic mediation for environmental model components integration. Information Technologies in Environmental Engineering 1, 3–11 (2008)
6. Ouksel, A.M., Sheth, A.: Semantic interoperability in global information systems. SIGMOD Rec. 28(1), 5–12 (1999)
7. Heiler, S.: Semantic interoperability. ACM Comput. Surv. 27(2), 271–273 (1995)
8. Musen, M.A.: Dimensions of knowledge sharing and reuse. Comput. Biomed. Res. 25(5), 435–467 (1992)
9. Holsapple, C.W., Joshi, K.D.: A collaborative approach to ontology design. Communications of ACM 45(2), 42–47 (2002)
10. Janssen, S., Andersen, E., Athanasiadis, I.N., van Ittersum, M.K.: A database for integrated assessment of European agricultural systems. Environmental Science and Policy (2009); article in press
11. Janssen, S., Wien, J., Li, H., Athanasiadis, I.N., Ewert, F., Knapen, M., Huber, D., Thérond, O., Rizzoli, A., Belhouchette, H., Svensson, M., van Ittersum, M.: Defining projects and scenarios for integrated assessment modelling using ontology. In: MODSIM 2007 Int'l Congress on Modelling and Simulation, pp. 2055–2061 (2007)
12. Athanasiadis, I.N., Rizzoli, A.E., Donatelli, M., Carlini, L.: Enriching environmental software model interfaces through ontology-based tools. Int. J. of Applied Systemic Studies (accepted for publication)
13. Athanasiadis, I.N., Villa, F., Rizzoli, A.E.: Ontologies, JavaBeans and Relational Databases for enabling semantic programming. In: Proc. of the 31st IEEE Annual International Computer Software and Applications Conference (COMPSAC), Beijing, China, vol. 2, pp. 341–346. IEEE, Los Alamitos (2007)

Assessment of Food and Nutrition Related Descriptors in Agricultural and Biomedical Thesauri

Tomaz Bartol

Agronomy Dept., Biotechnical Fac., Ljubljana Univ., Jamnikarjeva 101,
1001 Ljubljana, Slovenia
tomaz.bartol@bf.uni-lj.si

Abstract. Food- and human nutrition-related subject headings or descriptors of the following thesauri-databases are assessed: NAL Thesaurus/Agricola, Agrovoc/Agris, CAB Thesaurus, FSTA Thesaurus, MeSH/Medline. Food concepts can be represented by thousands of different terms but subject scope of a particular term is sometimes vague. There exist important differences among thesauri regarding same or similar concept. A term that represents narrower or broader concept in one thesaurus can in another stand for a related concept or be non-existent. Sometimes there is no clear implication of differences between scientific (Latin) and common (English) names. Too many related terms can confuse end-users. Thesauri were initially employed mostly by information professionals but can now be used directly by users who may be unaware of differences. Thesauri are assuming new roles in classification of information as metadata. Further development towards ontologies must pay constant attention to taxonomic problems of representation of knowledge.

Keywords: Descriptors, metadata, thesauri, ontologies, taxonomy, classification, semantics, food, nutrition, agriculture, biomedicine.

1 Introduction

Food and nutrition-related topics can be found in a variety of information systems and databases, ranging from agriculture and medical sciences to social sciences and economics. These topics are covered especially well by all three global agricultural databases Agricola (National Agricultural Library/NAL), Agris (Food and Nutrition Organization of the United Nations/FAO), and CAB Abstracts (CAB International/CABI). The food-specific database FSTA (Food Science and technology Abstracts) is produced by IFIS (International Food Information Service). There exists a certain degree of cooperation between the IFIS and CABI, with CABI being one of the sponsors of IFIS. Medline (National Library of Medicine/NLM), which also covers extensively this subject area, puts more emphasis on food in human nutrition.

All of the above databases employ complex thesauri, i.e. descriptor systems or subject headings organized in complex hierarchical tree structures. Similar coverage notwithstanding, there exist important topical differences and even discrepancies among different subject headings in representing same or similar concepts. Our assessment will focus on some more noticeable variations among thesauri with regard to

F. Sartori, M.Á. Sicilia, and N. Manouselis (Eds.): MTSR 2009, CCIS 46, pp. 294–305, 2009.

the lexical representation of food-and-nutrition-related subjects. The purpose of the study is to analyze and interpret these differences which can be rather significant, and can impact subsequent searching and retrieval. These inter-thesauri differences have gradually emerged during the sustained ongoing compilation of tens of thousands of specialized terms and can't be overcome by simple harmonization of terms, such as synonyms because important inconsistencies occur on different hierarchical levels.

The advent of agricultural databases in the 1970's was followed by creation of two general agricultural thesauri, CAB (CABI) and Agrovoc (FAO). NAL decided to employ CAB Thesaurus which enabled searches in Agricola [1] but NAL also worked with FAO/Agris to develop tables for converting CAB terms to Agrovoc [2]. There existed differences between CAB Thesaurus and Agrovoc. Variations in terminology were noted on the examples of food-related dairy subjects [3] or alternative agriculture [4]. There even existed efforts towards a creation of a unified agricultural thesaurus [5]. But such a tool never materialized. Moreover, NAL even began to construct its own thesaurus, partly because it was not able to change input of the proprietary CAB Thesaurus [6]. In addition to these large thesauri there also exist other food classification systems and categorizations which, however, also exhibit certain conflicting features [7]. In trans-disciplinary areas such as agromedicine, biomedical database Medline is also considered an important information resource besides the above databases such as FSTA [8]. Medline employs Medical Subject Headings (MeSH). This have also been investigated in comparison with agricultural thesauri [9]. But MeSH strategy can not be freely shifted to an agricultural database [10]. Also, some MeSH deficiencies were identified on occasions. Some commonly used medical concepts were found not to be sufficiently associated with specific MeSH headings [11]. That ontologies and thesauri are not always very efficient tools to express end-user information needs was also shown on the case of nutrition-related concepts in both General Finnish Ontology and Finnish version of MeSH [12]. In order to bridge some differences among the existing classification schemas there were some experimental attempts to create a glossary with a more specialized scope such as functional foods [13]. Every so often the needs for unification of concepts regarding food, nutrition and human health are expressed [14]. To these end, new attempts are being made in investigating semantic coherence and possibilities of mapping terms between different vocabularies such as Agrovoc and NAL Thesaurus [15]. Differences among thesauri left aside, the advent of the Internet and metadata brought about extensive new opportunities in the area of vocabularies and taxonomies. New subject indexing practices can be implemented with new metadata standards [16] so large agricultural thesauri can now play a role of veritable knowledge organization systems for classification of information and are evolving towards agricultural ontologies [17]. Development of the semantic Web offers new possibilities for the uses of multilingual agricultural ontologies such as Agrovoc [18].

2 Materials and Methods

The principal aim of thesauri is indexing and subsequent retrieval of database records or documents. We therefore address the utility of thesauri in their primary function. In order to abbreviate the phrasing in the text, thesauri under investigation will be referred to with an acronym which we employ for a respective thesaurus-related database.

2.1 Thesauri and Databases under Study

(AC) Agricola database - *NAL Thesaurus* (NAL) - 45,000 descriptors and 28, 000 non-preferred terms. http://agclass.nal.usda.gov/agt/agt.shtml
(AS) Agris database - *Agrovoc* (FAO) - 17,000 descriptors and more than 10,000 non-preferred terms. http://www.fao.org/aims/ag_intro.htm
(CB) CAB Abstracts database - *CAB Thesaurus* (CABI) - 48,500 descriptors and 10,500 non-preferred terms. Available only for licensing.
(FS) FSTA database (Food Science and Technology Abstracts) - *FSTA Thesaurus* (IFIS) - 10,000 descriptors and 1,000 non-preferred terms. Av. only for licensing.
(ME) Medline database - *MeSH* (NLM) - Medical Subject Headings - 25,000 descriptors and 160,000 non-preferred terms. http://www.nlm.nih.gov/mesh/

In our survey we used freely available products by NAL (AC), FAO (AS), and NLM (ME). We used Ovid Technologies IP-restricted access to log onto full versions of CAB (CB) and FSTA (FS) which are only available on a subscription basis.

Table 1. Thesaurus and database-specific abbreviations for narrower terms (NT), broader terms (BT), non-descriptors (Used For), related terms (RT) and definition and scope of a descriptor

Thesaurus	NT	BT	Used For	Related	Definition
AS	NT	BT	UF -non descriptors	RT	scope note
AC	NT	BT	UF -non-preferred terms	RT	definition
CB	NT	BT	UF -non-preferred synonyms	RT	scope note
FS	NT	BT	UF -synonyms	RT	scope note
ME	tree	tree	entry term	see also	annot. + scope note

In Table 1 we present the names for descriptors and related concepts as they are referred to in respective thesauri and databases. Descriptors in all five thesauri are organized in similar tree-structures, based on Narrower Terms (NT), Broader Terms (BT), and Related Terms (RT). Non-descriptors or non-preferred terms are linked to descriptors with a Used For (UF) expression. In the structure of MeSH there are some differences and descriptors are invariably called Subject Headings. In order to identify the BTs and NTs it is necessary to open the numerical hierarchical tree. MeSH non-descriptors are referred to as Entry Terms. MeSH employs as many as 160,000 non-descripting entry terms. All thesauri are equipped with a possibility to present a definition or scope note of a descriptor. However, most thesauri, excepting MeSH, only rarely provide such definitions. MeSH scope notes are also enriched with further information on the context of a descriptor, i.e. Annotation.

3 Results

There are thousands of food-related terms so we will only present a few selected items relating both to animal- and plant-based foods used in human nutrition. We identified broader, narrower, and related terms (hereafter referred to as BTs, NTs, and RTs) stemming from a selected primary descriptor-subject heading. We also collected non-preferred terms or synonyms. The NT and RT groups were sometimes very large, containing dozens of terms, especially in CB, so the results present only a few selected terms, furnished with the information for the total number of terms in each particular group. In order to present results clearly and consistently we italicize and capitalize descriptors in the text. However, for reasons of clarity we don't capitalize them in tables, scientific names notwithstanding. Presentation in a form of original tree structures would be more comprehensive, but these trees are too complex and would span several pages. The selected terms attempt to give a general idea regarding relationships in a particular thesaurus and offer an analogy with other thesauri.

3.1 Descriptors Related to Food in General

The first logical term we selected was the concept *FOODS* which is used in most database thesauri (Table 2). It usually stands for non-descriptors *Foodstuffs* and *Food products*. Agricultural databases, including FS, distinguish between separate descriptors *Foods* (pertaining to human nutrition) and *Feeds* (pertaining to animal nutrition). MeSH, however, employs NT2-level *Animal Feed* (at NT1-level *Agricultural Crops*) stemming from the *Food* tree. It is important to highlight this difference: food-related NT *Animal foods* in FS stands for foods *of* animal origins, but food-related NT *Animal feed* in ME denotes food *for* animals.

Table 2. Hierarchical, associative and preferential relations for the term food

	DE	NT	RT	UF
AC	foods (BT agric. products, food & human nutr.)	animal-based foods, appetizers, batters ... weaning foods, whipped foods, wild foods /93/	diet, food industry, food preparation, food prices, food processing ... food transport /7/	food products
AS	foods (BT 0)	bakery products, beverages, confectionery ... simulated f., soups /21/	cereal products, cereals, dried products ... sugar, vegetable products, vegetables /23/	food contamin., f. products, f. quality, foodstuffs, freshness of f. /5/
CB	foods (BT food)	batters, beverages,... tropical foods, unconventional foods, wild foods/37/	codex alimentarius, diets, edible cultivars ... plant products, vegetables /35/	0
FS	foods (BT 0)	animal foods, fresh produce ... plant foods, processed f., sea f. /10/	0	food products, foodstuffs /2/
ME	food (BT food and beverages)	bread , cereals, agricultural crops (animal feed) honey/24/	diet ... nutritional requirements, edible plants /5/	nutrients

Foods in AS and FS has no BTs. It does have BTs in AC and CB. The AC BT term *Agricultural products* exists also in AS but in AS it has no NT *Foods*. *Foods* in AC spreads into as many as 93 different NTs but has only 10 NTs in FS. These, however, are subdivided into narrower logical groups such as *Animal foods* or *Plant foods*, standing for foods of animal or plant origin, respectively. There are no such groups in

AS or CB. *Vegetables*, for example, are a NT only to *Crops* in CB, and in AS the same *Vegetables* are a NT to *Plant products*. But in CB the term *Foods* has a BT at *Food*, which is subdivided into 20 NTs such as *Foods*, and also *Diet planning*, *Food analysis, Food consumption* etc. In ME the *Food* has a BT at *Food and Beverages*, but has no specific narrower plant and animal group. It is immediately subdivided into 24 even more specific groups such as *Cereals, Dairy Products* etc.

AC also uses two group-terms *Animal-based foods* and *Plant-based foods*, but at the same level this thesaurus uses 93 different terms, mostly related to technological concepts which can pertain either to animal or plant-based products. So *Food* has a NT at *Plant-based foods* but these are then related to *Vegetable products* which has a NT at *Vegetables*. But terms *Vegetables* and *Vegetable products* have no BT at *Foods*. In AS *Cereals* are only related to *Foods* but in ME Cereals are a NT stemming from *Food*. In FS *Cereals* are only a NT to *Grain crops*, unrelated to *Food*.

3.2 Descriptors Related to Vegetables

The term *VEGETABLES* is presented in a different form (Table 3). There exists an important number of vegetables so we present only selected items. There exist several different RTs which also include a great number of vegetable-related concepts such as

Table 3. Hierarchical and associative relations for selected vegetable terms

AC - vegetables (BT1: vegetable products, BT2: plant products, BT3: agricultural products)
NT1: artichokes, asparagus spears, avocados, bamboo shoots ... edible fungi (nt2: mushrooms , truffles), eggplants, fennel, onions (nt2: shallots, welsh onions), peppers (nt2: hot peppers, sweet peppers) ... squashes (nt2: winter squashes, zucchini) , sweetcorn, tomatoes /67/
RT: vegetable crops, vegetable gardens, vegetable juices /3/

AS - vegetables (BT1: Plant products)
NT 1: adzuki beans, asparagus beans, asparagus ... onions, palm hearts ... peas, plantains, potatoes ... squashes (nt2: marrows, pumpkins, winter squashes), sweet corn, sweet peppers, ... tomatoes (nt2: cherry tomatoes), turnips, urd, watercress, yams /74/
RT: feeds, foods, fresh prod., legumes ... vegetable crops, veget. legumes, vegetable prod. /8/

CB - vegetables (BT1: crops)
NT1: bulbous vegetables (nt 2: chives, garlic, leeks, onions, shallots, welsh onions)
NT1: fruit vegetables (nt 2: aubergines, cucurbit vegetables, ... vegetable legumes)
NT1: leafy vegetables (nt 2: broccoli, brussels sprouts, cabbages ... watercress)
NT1: root vegetables (nt 2: beetroots, carrots, celeriac ... turnips, yams)
NT1: stem vegetables (nt 2: bamboo shoots, cardoons, kohlrabi, rhubarb) /50 TERMS/
RT: canned veget., cucurbit fruits, field crops, foods, horticultural crops ... vegetable pulps, vegetable stores, vegetable washers, vegetarians /18/

FS - vegetables (BT1: crops-> BT2: plants; BT1: plant foods-> BT2: foods)
NT1: dried vegetables, NT1: wild vegetables, NT1: vegetable specific -> NT2: allium, amaranthus, anu, arracacha ... brassica, broccoli, brussels sprouts, buffalo gourds, burdock, butterbur, cabbages, canna, capsicums, cardoons, carrots ... green vegetables ... kudzu, leaf beet, leafy vegetables ... leeks, legumes ...... onions ... zedoary /130/ *RT: 0*
ME- vegetables (BT1: Plants, Edible-> BT2: Plants)
NT1: allium, brassica , capsicum , chicory , chive ..., cucumis sativus , daucus carota , fabaceae, lettuce , lycopersicon esculentum , mustard plant , rheum , shallots , solanum tuberosum , spinacia oleracea , vegetable proteins /16/ *RT: 0*

Vegetable crops, *Vegetable legumes* or *Horticultural cops*. The concepts are arranged very differently among thesauri. It is frequently difficult to know why a concept is sometimes defined as RT and not NT. We present a vegetable concept *legumes*: in AC the term *Legumes* does exist but is a NT only at a descriptor *Plant products.* It is not mentioned in relation to *Vegetables*. In AS the *Vegetable legumes* are a RT to *Vegetables*, but according to CB the same *Vegetable legumes* are not related to anything and exist as a NT at *Fruit vegetables* and *Legumes*. In AC there is NT vegetable term *Onions* (genus *Allium*), but *Garlic* (also *Allium*), however, is placed among *Herbs*. In AS it is placed among *Spices*, but in CB *Garlic* is very well a vegetable, just like *Onion*. In FS there are two separate NT terms *Allium* and *Onions*. But *garlic (*also *Allium)* is placed in a different group - *Spices*. It is unclear what is the scope difference between *Allium* and *Onions*.

In AC *Pumpkins* and *Squashes* are separate (although being the same species), but in AS *Pumpkins* are a NT at *Squashes*. The *Peppers* are classified as *Vegetables* both in AC and AS. But in FS the *peppers* are presented only with a scientific name *Capsicums* at *Vegetables*. In CB, however, *Capsicum* has only the scientific BT at *Solanaceae*. In FS there is a term such as *Leafy vegetables*, but is not subdivided into narrower concepts like in AC or CB. On the other hand, FS has as many as 130 different vegetable-specific concepts.

We examine some specific plants. Term *BEANS* is an interesting example of thesauri discrepancies between common-names, technological, culinary uses and scientific taxonomy (Table 4). In AC *Beans* has no NTs for *Lentils, Peas, Soybeans,*

Table 4. Broader hierarchical, associative and preferential relations for the terms pertaining to beans and Phaseolus vulgaris

	DE	BT1	BT2	RT	UF
AC	beans	legumes	plant prod.	0	0
	kidney beans	beans	legumes	dry beans, phas. vulgaris /2/	0
AS	0-> grain legumes	legumes	plant prod.	adzuki beans ... kidney beans ... urd /18/	pulse crops, pulses
	kidney beans	vegetables	plant prod.	grain legumes, ph. vulgaris, vegetable legumes /4/	bush beans ... string beans, wax b. /12/
CB	beans	plant products	products	grain legumes	0
	phaseolus vulgaris	phaseolus	papilionoid.	grain legumes, veget. legumes /2/	beans (phaseolus) ... green b., haricot b., kidney beans /8/
FS	beans	legumes	vegetables specific	0	0
	kidney beans	common beans	legumes	0	0
ME	fabaceae	angiosperms	plant families	0	afzelia ... beans
	phaseolus	fabaceae	angiosperms	0	kidney bean ..., phas. vulgaris /8/

even though these descriptors do exist, but it has 16 other different NTs related to this group of plants, for example *Chickpeas, Faba Beans, Kidney Beans*. But the scientific *Phaseolus* term is only related to "Kidney Beans". It is not clear what is the criterion for such a placement. And in AS there is no occurrence for *beans* at all. There is only the descriptor *Grain legumes.* But there is *Grain legumes* RT at *Kidney Beans* which, in turn, has RT at *Phaseolus.* Descriptor *Grain legumes* has in fact no NTs but only RTs such as *Lentils, Peas, Kidney beans, Soybeans.* But these RTs have BTs only at *Vegetables* and not at *Grain legumes*, as would be normally expected. *Phaseolus* has a BT at *Papilionoideae.* In CB, *Kidney Beans* is a non-descriptor of *Phaseoulus*, but *Beans* has no relation to a particular plant, and only has BT at *Plant Product*.

In FS *Beans* shares the same BT (*Legumes*) with *Common beans, Chick peas, Lentils, Red beans* etc. But *Common beans* has 14 further NTs, including *Kidney b.* Some general taxonomic confusion exists with *beans* as this term can stand for different plants belonging to different species and even genera (*Phaseolus, Vigna*). In ME the concept of *Phaseolus* is indexed only in Latin. However, ME does employ English terms *Soybeans* and *Peas* as subject headings. In ME there sometimes seems to be no consistency of English *vs.* Latin. Phaseoulus has only a broader botanical concept. But ME has a non-descriptor *Beans* which is represented only at *Fabaceae.* This botanical family has then BTs at *Vegetables* and *Food.*

We present an example of a fruit - APPLE (Table 5), pertaining to plant MALUS (M. domestica). There exists a difference in indexing of a fruit (e.g. apples) or a plant (e.g. apple tree), so plants are frequently presented as RT at a particular fruit. AC and CB relate *Apples* to a species, AS to a genus, which has then three more narrower species. But unlike vegetables, which are classified as food in ME, this thesaurus does not see apples as food.

Table 5. Hierarchical and associative relations for the terms pertaining to apples

	DE	BT1	BT2	RT
AC	apples	pome fruits	tree fruits	apple cider, apple juice ... malus domestica /5/
AS	apples UF crab apples	pome fruits	fruits	apple juice, malus, temperate fruits
CB	apples	pome fruits	temperate tree fruits	apple chlorotic leaf spot virus ... malus domestica /24/
FS	apples	fruits specific	fruits	apple musts, apple pectins ... cider apples /8/
ME	malus UF apple, malus domestica	rosaceae	angiosperms	0

3.3 Descriptors Related to Milk

Let's now take an example of an animal-based product MILK (Table 6). In AS this is the highest descriptor without further BTs. But it is a RT both to *Animal products* and *Beverages*. In some other thesauri *milk* is considered a *product*, unlike AS where *Milk products* are only a RT to *Milk*. The NTs in AS, CB and FS are based on an animal species, and also on some technological aspects. In ME an important NT concept is *Cultured milk products*. These are called *Cultured milk* in AS, and *Fermented*

Table 6. Hierarchical, associative and preferential relations for the terms pertaining to milk

	DE	BT	NT	RT	UF
AC	milk	dairy prod., maternal m.	low fat milk, skim milk, whole milk /3/	beverages, lactose, milk industry, milk prices, m. quality /5/	cow milk
AS	milk	0	buffalo milk, camel milk, colostrum, human milk, mare milk, milk fat /9/	animal prod., beverag., body fluids ... milk products /11/	raw milk, whole milk /2/
CB	milk	milk products	buffalo milk, camel milk, concentrated milk ... residual milk, skim milk, sterilized milk /23/	beverages, casein, chymosin ... perishable products, rennet, whey /65/	cow milk
FS	milk	dairy products	acidophilus milk, ass milk, buffalo milk, ..., uht milk, whole milk, yak milk /36/	0	cow milk
ME	milk	beverages, dairy prod.	cultured milk products, infant formula, human m. /3/	lactation	0

milk in AC, CB and FS. *Acidophilus milk* is considered a NT to *Cultured milk* in AS but not in FS, where this is a separate descriptor. FS has no *milk* RTs, but CB has as many as 65 RTs, tackling a variety of issues, also related to production economics or animal physiology. In other thesauri assignment of RTs seems to be more restraint.

3.4 Descriptors Related to Seafoods

We present another interesting animal example, SEAFOODS (Figure 1, Table 7). This model exhibits a very low level of intra-thesaurus consistency, especially in

Table 7. Hierarchical, associative and preferential relations for the terms pertaining to seafoods

	DE	BT	NT	RT	UF
AC	seafoods	foods	calamari, edible seaweed, raw seafoods ...	crustacea, fish, fish products, macroalgae, molluscs, sea vegetables, seaweed products, shellfish, sushi	see[1]
AS	seafoods	foods	sea squirts, sea cucumbers, octopuses ...	fish, fish products, krill, shellfish	0
CB	seafoods	products	0	clams (NT2 hard clams), raw fish, shellfish, squids	0
FS	sea foods	aquatic f., foods	see[2]	see[3]	0
ME	seafood	meat	fish products, shellfish		

[1] abalone as food, clam meat, crabmeat, fish as food, fish meat, lobster meat, mussel meat, oyster meat, scallops as food, shellfish as food, shellfish meat, shrimp meat, snails as food

[2] ascidians, echinod., fish, fish prod., ika shiokara, jellyf., jeotgal, laverbread, nukaz., plankton, sea cucumb., s. food prod., s. foods industr., s. squirts, s. urchins, seaweeds, shellfish, squid

[3] amnesic shellfish poisoning, aquaculture products, diarrhoetic shellfish poisoning, mussel poisoning, neurological shellfish poisoning, paralytic shellfish poisoning, psp

relation to the fish-related concepts. *Fish* is NT at *Sea Food* in FS and also in ME (as *Fish Products*). It is RT in AC and AS. CB has no NTs at *Seafoods* but that doesn't mean that there are no such descriptors. They are merely organized in different trees. In CB *Fish* as food stands as an independent descriptor which, however, has no narrower terms. Particular fish species have different taxonomic tree relations.

Another example of discrepancies and even confusion are cephalopod-derived *Seafoods*. AC offers *Cuttlefish, Octopuses* and *Squids*. AS has only *Calamari*, and ME has none. But ME does have *Shellfish* as a separate descriptor, just like CB. *Shellfish* is only a RT in AC and AS. But in fact *Shellfish* is taxonomically broader than *molluscs* (e.g. *Cuttlefish, Calamari*). CB has in addition a separate descriptor *Raw fish*. ME seems to be the most consistent, given that the broad general term *Shellfish* applies to a majority of edible non-vertebrate aquatic animals, including cephalopods, mussels and crustaceans. *Shellfish* in ME has no further NTs. ME also contains more specific aquatic animal terms for crustaceans and molluscs, but these are classified in taxonomic descriptor trees, unrelated to *Seafood*. Interestingly, ME classifies *Seafood* as BT *Meat*. This is rather rare. In other thesauri the concept *Meat* only relates to mammals and birds (poultry).

In order to better illustrate the original architecture of tree-structures in the respective online thesauri we also present some selected descriptors stemming from *seafoods* in original form (Figure 1). Entire indexing trees being large, we only show alphabetically closest terms to *seafoods*. But even the very term *seafoods* presents different variants such as *Seafoods, Sea foods* and *Seafood*. Although seemingly identical, these terms will come about in a different word order in word-lists. A correct

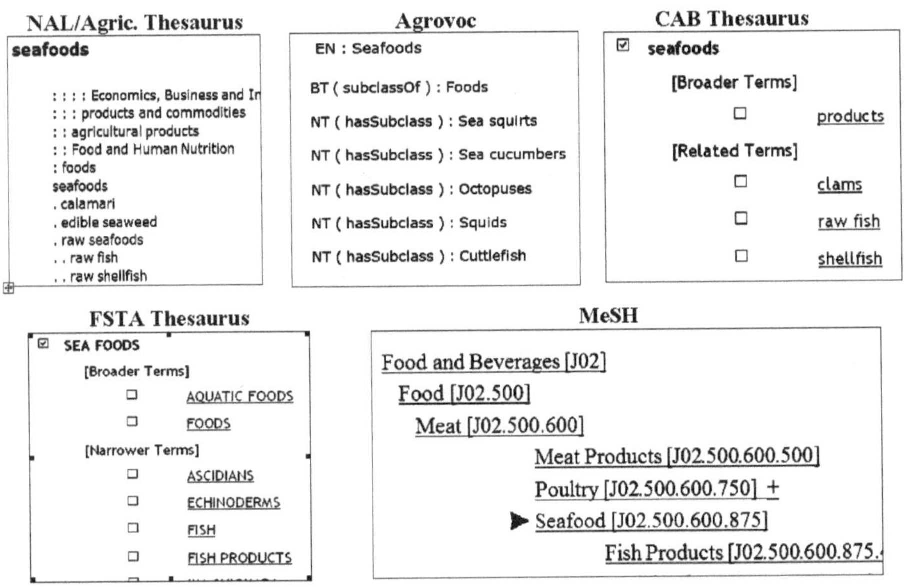

Fig. 1. Details from the original thesauri web pages and selected subject headings with regard to the *Seafoods*-related terms

variant in one system may even produce no hits in another system if an end-user ignores a particular thesaurus-defined correct spelling which should be used in that particular database searches.

It is also worth mentioning the case of a fish *trout (genus Salmo),* one of the most commonly used fresh-water species in human nutrition. In CB there exists a descriptor for genus *Salmo*, with a narrower term *Rainbow trout*. But there also exists a separate descriptor *Trout* as RT at *Salmo*, with no NTs leading to *Rainbow trout*. It is difficult to guess which is indexing/retrieval difference between *Salmo* and *Trout*. In AC there is a hierarchical path: *Trout* -> BT1 *Fish* -> BT2 *Aquatic organisms*, where *Fish* is used in a sense of *food*, according to the UF explanation. But there is yet another, taxonomic tree *Salmo trutta -> BT Salmo*, where *Trout* is not even mentioned as a possibly related concept to *Salmo trutta*. In AS there is *Trout -> BT1 Freshwater fishes*, and another, taxonomic branch *Salmo trutta -> BT1 Salmo*. It is not specifically mentioned that the English term *Trout* apparently stands for culinary issues. End-users are probably not aware of the rule which also implies distinctions between *plants* (e.g. *Fragaria*) and *plant products* (e.g. *Strawberries*).

4 Discussion and Conclusions

The concept of *foods* used in *human nutrition* is represented by thousands of different terms. It is sometimes difficult to know, however, which was a criterion for associative and hierarchical relationships in glossaries, which may contain almost 50.000 descriptors, all connoting agriculture-related subject matter.

Subject scope of many terms in a particular thesaurus can be vague and is frequently not clarified to a sufficient detail. There exist particularities within a particular thesaurus and important differences among different thesauri. Same terms that stand for a narrower or broader concept in one thesaurus will in another thesaurus stand for a related term, and may in yet another thesaurus be referred to as a non-preferred term. Nonexistence of a certain term may merely signify that a subject had been arranged differently. But it is quite possible that a concept in one thesaurus is not at all present in another. But such differences can probably not be statistically evaluated because there are no clear taxonomic criteria.

It is already a challenge to try to identify a logical topical group in one thesaurus, let alone compare such groups among thesauri, what was shown on the example of the so called *seafoods*. The choice of descriptors and related terms seems quite arbitrary and very incomplete. Based on thesauri under study we can't even determine what the seafoods really are. We can only vaguely establish what is supposed by this term in a particular thesaurus, not necessarily reflecting possible definitions elsewhere. Also, the existing tree-structures are frequently inaccurate and do not correctly reflect taxonomical relations within a group of hierarchically related terms. Sometimes there is no clear logic of differences between scientific, common, technical and culinary names. This is perceived quite differently in different thesauri. These differences hamper attempts at mappings between different thesauri and cross domain searching.

It was not our intent to judge particular thesauri because the background and history of each such glossary has its own logic. CAB Thesaurus has many more descriptors than Agrovoc but it is accessible only for a subscription. CAB descriptors are provided with many more related terms. But such an abundance of subjectively defined terms may confuse end-users. On the other hand Agrovoc is a freely accessible product and also serves as a very powerful multilingual tool. NAL Thesaurus, having its origins in CAB, is a more recent product adapted for American usage. FSTA Thesaurus is a tool specialized precisely for food-and-nutrition. MeSH is a vast biomedical indexing tool, also having an important stake in this subject area.

Initially, these thesauri were used mostly by information professionals for indexing purposes, and consequently, for information retrieval. Gradually, end-users have begun to use databases independently of information-services. But are end-users aware of important differences among thesauri and of a thesaurus-dependent subject scope of a particular term?

At this point it seems very difficult to reconcile differences among the existing thesauri as differences can be attributed to several factors, such as gradual particularized database-specific development of thesauri during an extended period, specific needs of a particular discipline (agriculture vs. biomedicine), scope and depth (10,000 vs. 50,000 terms), etc. A very important factor are "historic" errors or inaccuracies. It appears to be very difficult to modify a structure of a thesaurus once the hierarchical trees have been constituted. The differences can't be overcome by simple harmonization of terms, such as synonyms, because the inconsistencies and incompatibility occur on different hierarchical levels. The best solution would be a creation of a single international life-sciences thesaurus. Such a project, however, is at the moment not realistic.

In the age of the Internet and metadata, thesauri have assumed a new role in classification of knowledge and have gradually begun to evolve into complex ontological tools. The existing indexing systems, however, appear to be too inconsistent to be used as a sole criterion for representation of knowledge. But they can still serve as a good reference tool for end-users to compile as many relevant terms as possible to be used in later searches. In order to use these taxonomic tool efficiently, end-users need to gain comprehensive understanding of metadata vocabularies and implications thereof. Further steps in development of ontologies must continue to pay strong attention to general taxonomic differences and discrepancies regarding representation of knowledge.

References

1. Hood, M.: Using the CAB thesaurus to search AGRICOLA. Agricultural Libraries Information Notes 14, 15–16 (1988)
2. Thomas, S.E.: Bibliographic control and agriculture. Library Trends 38, 542–561 (1990)
3. Oide, N., Moriwaki, N.: Comparisons of indexing words used in CAB ABSTRACTS and AGRIS. Bulletin of the Japan Association of Agricultural Librarians and Documentalists 78 (1990)
4. Weintraub, I.: The terminology of alternative agriculture searching AGRICOLA, CAB and AGRIS. Quarterly Bulletin of IAALD 37, 209–213 (1992)
5. Andre, P.: Towards a unified agricultural thesaurus. Quarterly Bulletin of IAALD 37, 224–226 (1992)

6. Agricultural Economics Reference Organization (AERO). Report. Website. In: 13th Workshop. Iowa State University, Iowa (2002), http://are.berkeley.edu/AERO/iowa.html
7. Ireland, J.D., Moller, A.: Review of international food classification and description. Journal of Food Composition & Analysis 13, 529–538 (2000)
8. Kawasaki, J.L.: The interdisciplinary field of agromedicine: searching the literature. Quarterly Bulletin of IAALD 44, 191–195 (1999)
9. Bartol, T., Baricevic, D.: Bibliometric analysis of agricultural and biomedical bibliographic databases with regard to medicinal plants genera Origanum and Lippia in the period 1981-1998. In: Kintzios, S.E. (ed.) Oregano: the genera Origanum and Lippia. Medicinal and aromatic plants-industrial profiles, pp. 245–267. Taylor and Francis, New York (2002)
10. Murphy, S.A.: Applying methodological search filters to CAB Abstracts to identify research for evidence-based veterinary medicine. Journal of the Medical Library Association 90, 406–410 (2002)
11. Portaluppi, F.: Consistency and accuracy of the Medical Subject Headings (R) thesaurus for electronic indexing and retrieval of chronobiologic references. Chronobiology International 24, 1213–1229 (2007)
12. Poikonen, T., Vakkari, P.: Lay persons' and professionals' nutrition-related vocabularies and their matching to a general and a specific thesaurus. Journal of Information Science 35, 232–243 (2009)
13. Juvan, S., Bartol, T., Boh, B.: Data structuring and classification in newly-emerging scientific fields. Online Information Review 29, 483–498 (2005)
14. Lange, M.C., Lemay, D.G., German, J.B.: A multi-ontology framework to guide agriculture and food towards diet and health. Journal of the Science of Food & Agriculture 87, 1427–1434 (2007)
15. Pesce, V., Maru, A., Salokhe, G., Keizer, J.: A Distributed Architecture for Harvesting Metadata Describing Organizations in the Agriculture Sector. Website. In: International Conference on Metadata and Semantics Research, Corfu, Greece (2007)
16. Kaloyanova, S., Onyancha, I., Salokhe, G., Ward, F.L.H.: Information technologies and standards for agricultural information resources management: AGRIS application profile, AGROVOC and LISAGR. Quarterly Bulletin of IAALD 52, 17–21 (2007)
17. Sicilia, M.A.: Linking learning technology with agricultural knowledge organization systems. Website. In: Workshop on Learning Technology Standards for Agriculture and Rural Development, Athens, Greece (2008)
18. Manouselis, N., Kastrantas, K., Sanchez-Alonso, S., Cáceres, J., Ebner, H., Palmer, M., Naeve, A.: Architecture of the Organic. Edunet Web Portal. International Journal of Web Portals 1, 71–91 (2009)

Networked Ontologies from the Fisheries Domain

Caterina Caracciolo, Juan Heguiabehere, Margherita Sini, and Johannes Keizer

Food and Agriculture Organization of the United Nations (FAO),
v.le Terme di Caracalla 1,
00154 Roma, Italy
{Caterina.Caracciolo,Juan.Heguiabehere,Margherita.Sini,
Johannes.Keizer}@fao.org

Abstract. In this paper we report on ongoing work concerning the creation of a network of ontologies based on metadata for time series relative to the domain of fisheries, and hint at the possibility of exploiting the network for web service applications. The results obtained so far show that the reengineering of classification systems stored as relational databases is possible, although some technical problems is still to be addressed.

Keywords: Ontologies, statistics, metadata, fisheries, web services.

1 Introduction

The Food and Agriculture Organization of the United Nations (FAO) is deeply involved in the collection and dissemination of data concerning all areas related to food and agriculture. The data of interest to FAO includes statistical data, and structured, non-structured and semi-structured documents. Since metadata are essential to all these data sets, FAO is dedicating great attention to the opportunities offered by semantically-oriented technologies to model, exploit and exchange metadata. The focus of the work we report on in this paper is on metadata for time series, i.e. the "reference data" used by the Fisheries and Aquatic Information and Statistical Service (FIES) department of FAO.[1]

Reference data is used to identify the dimensions of any piece of statistical data. For example, time series on fish catch are identified by the species caught, the country reporting the catch, the water area where the catch happened, and the year the data refers to. All these piece of information (except the year) are organized into hierarchical tables, called reference tables, stored as in a relational database.

FAO carried out some previous work[2] [1] on converting reference data into ontologies. Now FAO is experimenting with networking the ontologies together [2]. A network of ontologies is a collection of ontologies related to each other via a variety of different relationships such as mapping, modularization, version, and dependency relationships. Elements of this collection are called "networked ontologies" [3].

---

[1] This collection of reference data stays at the core of the use case of the NeOn project (http://www.neon-project.org), dedicated to the development of a "Fish-stock Depletion Assessment System".

[2] Available at: http://www.fao.org/aims/neon.jsp

F. Sartori, M.Á. Sicilia, and N. Manouselis (Eds.): MTSR 2009, CCIS 46, pp. 306–311, 2009.

When the relationship requires the existence of correspondences between ontologies (and thus overlap in their modeled domain), it is called mapping. When the relationship involves two disjoint ontologies, it is called e-connection (however, throughout this document we prefer to use the more general term "link").

The rationale for moving to networked ontologies is that they can allow for more flexibility in the queries that can be asked to the time series collection[3] and exploit the semantics hidden in the available data. Classification systems typically describe single domains, independently of other, related domains. Data about the correspondences between domains, as well as data about the correspondences between different classifications concerning the same domain are usually rather difficult to represent, and may be exploited only by users having deep knowledge of the area. The consequence of this state of affairs is that "constraints" about what queries are meaningful to ask the statistical database are not explicitly modeled. For example, it is possible to compose queries like: "catch of yellowfin tuna (Thunnus Albacares) in the Mediterranean Sea", despite the fact that yellowfin tuna is found in open waters of tropical and subtropical seas worldwide, but not in the Mediterranean Sea. Furthermore, appropriate linking of reference data would allow systems to integrate missing data. For example, a network of ontologies may easily embody the notion of proximity between countries or water areas. This information can be extremely useful, as when data about one area (land or water) is missing, one may look at neighboring areas, or at regions sharing some specific features (e.g. climatic zones, shore on the same sea, contiguous or noncontiguous water areas where a given species can be found).

Other reasons to move to (networked) ontologies are that the NeOn Toolkit[4] provides support for the whole data lifecycle, as it supports editorial workflow, collaboration, ontologically-driven access to relational databases, and different visualizations for domain experts and ontology experts. Moving to networked ontologies would then have the additional advantage that the entire lifecycle could be supported in a coherent way and avoid pipelining different systems addressing different users, roles and activities. Finally, the advantage deriving from using standard languages for both the domain modeling and data encoding is expected to facilitate data reuse and sharing. This would also improve the possibility of using off-the-shelf tools.

In the rest of this paper we describe the process followed to create the first network of fisheries ontologies (Sec. 2), and introduce the main feature of the network (Sec. 3). We also draw some preliminary conclusions and present the direction of future development (Sec. 4).

2 The Process of Reengineering Reference Data for Time Series

The process of reengineering the reference data closely followed the NeOn methodologies for reengineering non-ontological data into ontologies [4, 5], by applying a tight iteration of modeling, population, and evaluation. We extensively used the NeOn Toolkit (version 1.0) and several of its plug-ins (in particular, ODEMapster[5], a tool

[3] See for example the online FAO statistical databases available at: http://www.fao.org/fishery/statistics/en

[4] http://www.neon-toolkit.org

[5] See http://www.neon-toolkit.org/wiki/index.php/ODEMapster

that allows one to access relational data by means of ontological models). Whenever possible, i.e. whenever the development of the tools available allowed for that, the ontology models accommodated the ontology design patterns[6] developed within NeOn [6].

Any work of this type, and the work presented here makes no exception, has to cope with a natural tension between: (a) the "nature" of the domain described, in particular the classification systems used to reference the time series; (b) the available guidelines and best practices for ontology modeling; (c) the support provided by available state-of-the-art tools, to be considered together with the characteristics of the underlying database; (d) the requirements coming from the need to ensure backward compatibility with current applications; and (e) the requirements coming from the applications that are going to exploit the data. While all constraints coming from (a) to (d) have been taken into account in the current version of the network of ontologies, future work will aim at exactly pinning down the requirements coming from (e), and implementing them in the future release of the network.

The data included in the network of ontologies was mostly stored in the relational data base, but the network also includes the Geopolitical Ontology[7], and data extracted from other sources, mainly fact sheets in the domain of fisheries[8]. The data linking the ontologies to one another was extracted either from the database of reference data, or from the fact sheets.

3 The First Network of Fisheries Ontologies

The first network of fisheries ontologies is available at: http://www.fao.org/aims/neon.jsp and includes ontologies about: biological species (taxonomic classification and ISSCAAP classification), commodities (ISSCFC, HS, and ISSCAAP[9] classifications), division of water areas (FAO water areas for statistical reporting, large marine ecosystems, national jurisdiction areas), land (FAO geopolitical ontology), and stocks. Also, three single ontologies (about fishing gear, vessel type and size and an ontological version of the ASFA thesaurus) were released and are going to be included in the next version of the network. Note that the ontologies released should not be considered as replacing the official corresponding classification systems. All ontologies contain comments concerning this and other aspects of their design and content.

The approach followed was to keep apart the OWL ontologies (T-box) from the corresponding data[10] (A-box). The data files import their corresponding ontology. When concepts are defined compositionally from concepts defined in other ontologies, the corresponding T-boxes are imported. A typical example is the concept

[6] See http://ontologydesignpatterns.org/

[7] A FAO resource collecting data about countries and regions, and available at: http://www.fao.org/countryProfiles/geoinfo.asp?lang=en

[8] See http://www.fao.org/fishery/factsheets/en

[9] ISSCAAP classification (with links to taxonomic classification of biological entities and ISSCFC commodities).

[10] The data extracted from the database of reference data is returned by ODEMapster as RDF that was manually formatted as OWL ontologies for reasons of compatibility with the NeOn Toolkit.

"stock", the definition of which would not be possible without referring to the concept of "species" and of "water area". The stock ontology then imports the corresponding ontologies on taxonomic classification and FAO divisions of water areas (and the A-box on stocks imports the corresponding A-boxes on taxonomic classification and water areas). The separate publication of T-box and the A-box allows for the use of the T-box for run-time use with the database, and their reuse within other applications and with the corresponding data. All files are serialized as RDF/XML.

Both A-box and T-box share the same namespace, but have different URIs. All ontological elements (classes, instances, properties) are defined in a dedicated namespace, one per domain[11]. Instances of a given class are then given the URI formed by the namespace where that class is defined, concatenated with the relevant pieces of information (including their ID) coming from the database. The idea behind this is to have classes and instances defined in the same namespace, even though they are actually contained in different ontologies/files, with different URIs. So far, namespaces have been kept the same through the various versions and revisions, so there is no notion of version embodied in the namespaces and in the URIs based on them.

Names of ontology elements (classes, properties) are based on English, while names of instances are formed by unique IDs that do not have explicit meaning for the human user. At the time when the work was carried out, no better solution was possible, because in many cases English names are missing (e.g. water areas according to FAO division for statistical reporting), and because the tools used did not support the rdfs:labels. However, one expected advantage of this choice is to smooth future connection with existing applications (requirement (d) mentioned in Sec. 1). Names in natural languages are rendered as value of datatype properties; the names of those datatype properties also contain two-character ISO codes of the language used (e.g. hasNameEN). This is done to comply with a widely used FAO convention, to facilitate the reading of the ontologies by human users, and to account for the fact that typically there is more than one name for each language (e.g. a "name," a "long name" and a "full name").

In order to comply with requirement (d) mentioned in Sec. 1, all pieces of information of most common use by existing applications using the reference data have been included in the ontologies (including the ID of all items in the database and the meta code used to identify the "type" of reference data at hand[12]). Arguably, the most important existing applications using the reference data are the query systems[13] that allow one to retrieve answers to questions such as: "what is the quantity of frozen crustacean imported by the Russian Federation in 1997?", or "what is the total amount of capture production of Japan in South West Pacific?" The network of ontologies based on this reference data, together with the web service applications that are going to be deployed, are expected to continue to provide this functionality. Currently, a prototype for a demonstrator of such web application (FSDAS) has been delivered [7] and is going to be improved in the coming months.

[11] For example, the concept "stocks" is defined in the namespace http://www.fao.org/aims/fi/stock# and it uses concepts defined in other namespaces, such as the concept "species" defined in http://www.fao.org/aims/fi/taxonomic# and the concept "FAOdivision" defined in http://www.fao.org/aims/fi/water#

[12] Full details are provided in [2].

[13] See for example the online query panels available at: http://www.fao.org/fishery/topic/16140/en

4 Conclusions and Future Work

In this paper we presented an ongoing work carried out at FAO, aimed at devising a network of ontologies based on reference data for time series. To the best of our knowledge, this is the first experiment of this type. We found that the reengineering of relational databases into well-formed ontologies following NeOn methodologies and best practices is possible. However, the tool available (ODEMapster), still does not allow smooth run-time access is it is, as in several occasions we had to pre-process the data, sometimes intensely. In most cases we decided to create views (which imply the use of a copy of the database, unless full rights on the database are granted).

The owl:imports mechanism is used everywhere to access these ontologies. This mechanism may become expensive, depending on the size of the imported ontologies. One possibility to minimize the impact of this point is to push forward the modularization of the network and publish separately the ontologies and the links among them. One advantage of this is also that one could more easily load ontologies in tools that request the imported ontologies to be loaded before the importing ones (as in the case of the NTK).

In the first network of fisheries ontologies, all links between ontologies are at the level of instances. When the data about linking is in the database, we found that the creation of the network should be streamlined by improving the processor (ODEMapster) in charge of the lifting of the data. In particular: it should be possible to have several range values per each object properties;[14] it should be possible to automatically update URIs and add import statements; it should be possible to use rdfs:label. The case in which linking between ontologies can be established from the ontologies directly represent an application of the work commonly known as "ontology alignment". The application to the FAO data of techniques developed in that area is an open and interesting are for future work.

Future work includes a thorough validation process of the network of ontologies by domain experts in fisheries. Feedback will also be gathered from partners and the whole community using the network of ontologies. We also expect to gather feedback from the use of the network in the FSDAS and its evaluation.

Based on the work presented here, we plan on expanding the coverage of the ontology on fish stock so as to accommodate data produced by an increasing number of fisheries bodies devoted to the management and protection of various fish stocks. Necessary steps to achieve this are: include in the network more divisions of water areas (since fisheries bodies usually adopt their own division of the water areas under their competence, or further divide existing divisions), include data about correspondences between water areas divisions, and include data about fisheries bodies. Finally, we plan on experimenting more on the reuse of the geopolitical ontology, especially for what concerns groups of countries. We also plan on modeling the concept of fishery, which is a very changeling task. In fact, although there is no unique definition of fishery agreed upon, this concept is fundamental to all work aimed at studying, managing and making decisions and policies in the area of fisheries. Therefore, we expect

[14] This point is related to the implementation of some ontology design patterns. A detailed explanation of this point is out of the scope of this paper: the interested reader should refer to [2].

that by modeling this concept and hooking it to the actual data collated by FAO (and other organizations) we would offer a very useful tool to domain experts and policy makers working (however, a tool does not replace or create neither awareness nor political will).

Acknowledgments. We thank all NeOn team members for useful comments, feedback and collaboration.

References

1. Caracciolo, C., et al.: D7.2.2 Revised/Enhanced Fisheries Ontologies, `http://www.neon-project.org/web-content/images/Publications/neon_2007_d7.2.2.pdf`
2. Caracciolo, C., et al.: D7.2.3 Initial Network of Fisheries Ontologies, `http://www.neon-project.org/web-content/images/Publications/neon_2009_d723.pdf`
3. Haase, P., et al.: D1.1.5 Updated Version of the Networked Ontology Model. NeOn project deliverable, `http://www.neon-project.org/web-content/images/Publications/neon_2009_d115.pdf`
4. Suárez-Figueroa, M.C., et al.: D5.4.1 NeOn Methodology for Building Contextualized Ontology Networks, `http://www.neon-project.org/web-content/images/Publications/neon_2008_d5.4.1.pdf`
5. Suárez-Figueroa, M.C., et al.: D5.4.2 Revision and Extension of the NeOn Methodology for Building Contextualized Ontology Networks, `http://www.neon-project.org/web-content/images/Publications/neon_2009_d542.pdf`
6. Presutti, V., et al.: D2.5.1 A library of Ontology Design Patterns, `http://www.neon-project.org/web-content/images/Publications/neon_2008_d2.5.1.pdf`
7. Erdmann, M., et al.: D7.6.2 Second Prototype of the Fisheries Stock Depletion Assessment System (FSDAS), `http://www.neon-project.org/web-content/images/Publications/neon_2009_d762.pdf`

Improving Information Exchange in the Chicken Processing Sector Using Standardised Data Lists

Kathryn Anne-Marie Donnelly[1,*], Joop van der Roest[2], Stefán Torfi Höskuldsson[3], Petter Olsen[1], and Kine Mari Karlsen[1]

[1] Norwegian Institute of Food, Fisheries and Aquaculture Research (Nofima), Muninbakken 9-13, Breivika
Kathryn.donnelly@nofima.no
[2] RIKILT - Wageningen UR, RIKILT – Institute of Food Safety, Bornsesteeg 45, P.O. Box 30, 6700 AE Wageningen, The Netherlands
[3] Maritech ehf, Hlidasmari 14, 201 Kopavogur, Iceland

Abstract. Research has shown that to improve electronic communication between companies, universal standardised data lists are necessary. In food supply chains in particular there is an increased need to exchange data in the wake of food safety incidents. Food supply chain companies already record numerous measurements, properties and parameters. These records are necessary for legal reasons, labelling, traceability, profiling desirable characteristics, showing compliance and for meeting customer requirements. Universal standards for name and content of each of these data elements would improve information exchange between buyers, sellers, authorities, consumers and other interested parties. A case study, carried out for the chicken sector, attempted to identify the most relevant parameters including which of these were already communicated to external bodies.

Keywords: Standard, Chicken Processing, Data Element, Traceability, Ontology.

1 Introduction

The modern world is characterised by ever increasing global trade with a constant increase in the need to communicate information precisely, effectively and also electronically[1, 2]. To facilitate electronic interchange of such product information, international, non-proprietary standards are required such as the ones highlighted by Jansen-Vullers et al [3]. Currently there are a multitude of information technology systems, software and formats, and no standardized way of electronically coding and transmitting information. Individual companies have made great progress in proprietary technologies for automated data capture and electronic data coding. However the benefit of these is lost when the data element transmission is required for use outside the originating company as it is only effective when there is an identical software system at the receiving end [4].

* Corresponding author.

F. Sartori, M.Á. Sicilia, and N. Manouselis (Eds.): MTSR 2009, CCIS 46, pp. 312–321, 2009.

Food scandals and the increasing demands on food business organisations, from consumers, for information about the food they buy are bringing the issue of data exchange to the attention of the agricultural and fisheries industries. These 'food producing' industries are faced with particular challenges, such as the speed at which food must be produced, delivered and consumed. The heterogeneity of the different food sectors which often cross paths during the production of the more complex food stuffs.

To enable effective, electronic information exchange, work needs to be carried out on a sector-specific level. Analysis of the precise product information the particular food sector already records should be carried out together with a method and format for identifying this information in a standard form.

The need for such systems has already been identified throughout the food industry, but particularly in areas where the authenticity of a product is in question. The viability of such non-proprietary standards were shown in the TraceFish project [5-7] where both sector-specific standards (for captured fish and farmed fish) and generic standards (for electronic coding and request-response scheme) were developed. The TraceFish work established sector-specific data models that not only contain information about data elements (including the relationship between them) relevant for product information in one link of the supply chain, but also information for each link. For further details see [7]. Standardized lists for data elements which can be included in data models have been acknowledged as a key technology for resolving semantic heterogeneity and are important in knowledge management in large organisations [8-11].

In this paper, the authors present work, which has been inspired by both the TraceFish work and other industry orientated research. Similar research to that of the work presented in Donnelly et al. [4] has been carried out for the chicken sector. Recent food safety incidents such as dioxines in eggs in Germany, incidents of salmonella and campylobacter in various European egg and poultry products, adulteration of UK chicken with cheap beef waste and water [12] has increased awareness of the industry for effective information exchange. The important new data in this paper is the information that the chicken processing sector communicates to external bodies. Previously there was little quantitative evidence presented regarding the amount or the type of parameters which it is appropriate to include in a standard. There are currently no European sector specific standards regarding how chicken product and processing information should be stored and transmitted, nor is electronic information exchange across companies in the supply chain common. This paper also outlines how the previously developed methods were used and modified. The method section in this paper describes how and what data elements were collected. Their relevance and wider uses are considered in the discussion.

2 Method

The method used to develop the basis for this standardized parameter list was devised during the creation of the TraceFish standards [5-7], the mineral water initial standard [13]. A questionnaire was developed in order to gather information about which data elements the chicken processing industry recorded. A list of possible data elements was collected using inputs from the process mapping of a chicken processor [14] In addition to previous work a new question (number three in tab.1) was introduced which was thought be of particular significance. This additional question gives an indication of how much data is already communicated.

Initially a list of chicken processors, from across Europe, was compiled. These companies were then contacted by telephone with a follow up e-mail or only e-mail when telephone contact was not possible. The companies were asked to fill out the questionnaire provided and/or give an indication of the data elements they recorded during chicken processing.

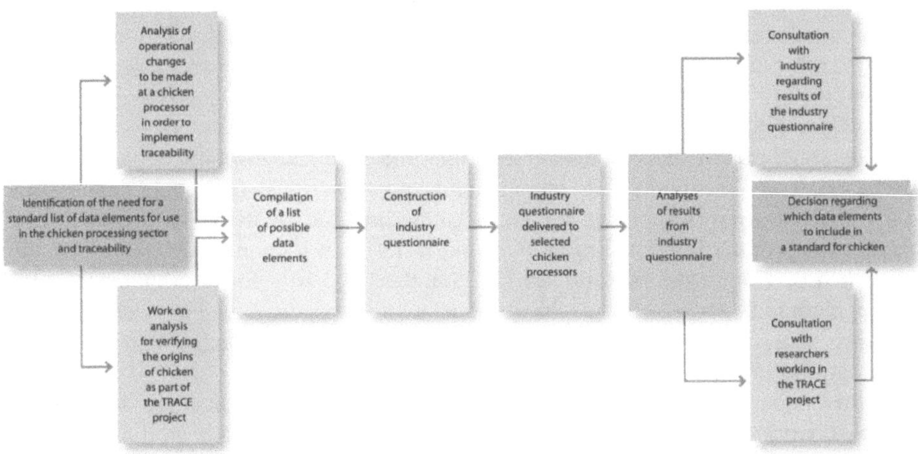

Fig. 1. Method modified from Donnelly et al. [4]

The questionnaire listed all known relevant data elements and for each asked a series of questions which could be answered simply by inserting a number or yes or no (Y/N). In addition there was also space for the company to fill in any data elements which they used, but which were not included in the questionnaire.

For each of the data elements the following questions were asked:

Table 1. The questions asked as part of the survey, modified from Donnelly et al (2008)

1. DO YOU RECORD THIS INFORMATION
2. HOW IMPORTANT IS THIS INFORMATION
3. DO YOU COMMUNICATE THIS INFORMATION WITH ANYONE OUTSIDE YOUR COMPANY
4. WHERE IS THE QUALITY CONTROL CHECK CARRIED OUT? (INTERNALLY OR EXTERNALLY)
5. WHICH METHOD FOR THE MEASUREMENT IS USED? (WHICH ANALYTICAL METHOD IS USED IF ANY?)

3 Results

Responses were received from the following countries; Belgium, Norway and The Netherlands. The response of 5 was not as high as has been reported in similar studies by Donnelly et al. [4]. The greatest weakness in the data is the poor geographic spread of responses.

Fig. 2. Shows the percentage of the parameters that were communicated to bodies outside of the company. The number of parameters communicated externally is almost equal to those communicated only internally emphasising the need for standardised data lists to ensure clear exchange of information.

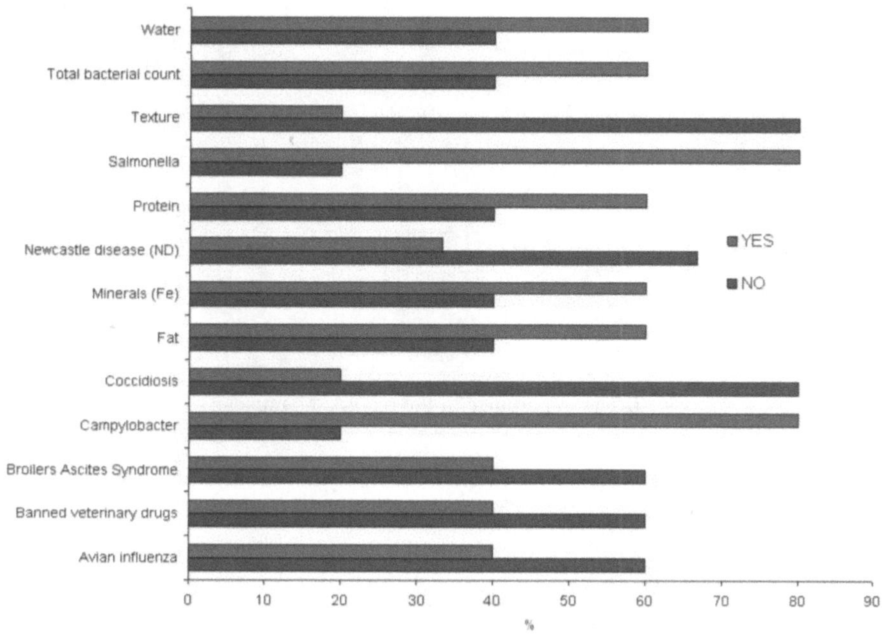

Fig. 2. Total percentage of companies communicating information about recorded parameters

Fig. 3 shows the importance the respondents allocated to the parameters which they recorded. From this small sample it was not possible to distinguish any specific pattern of importance that the respondents allocated to the different parameters.

As explained in detail in previous publications the data elements from tab. 2 were taken and included in table which can be found in the report Traceability of chicken – Specifications of the information to be recorded at chicken slaughter/processing establishments and other links in chicken distribution chains [15]. The classification of the data elements used is similar to that suggested by Folinas et al. [16] whereby the data elements are classified as 'mandatory data' or 'optional traceability data.' The difference between Folinas et al.'s [16] two classifications and the three classifications presented in this paper and also used by Denton [7] is that the data elements are categorized as 'shall' 'should' or 'may', these terms are explained in Denton [7].

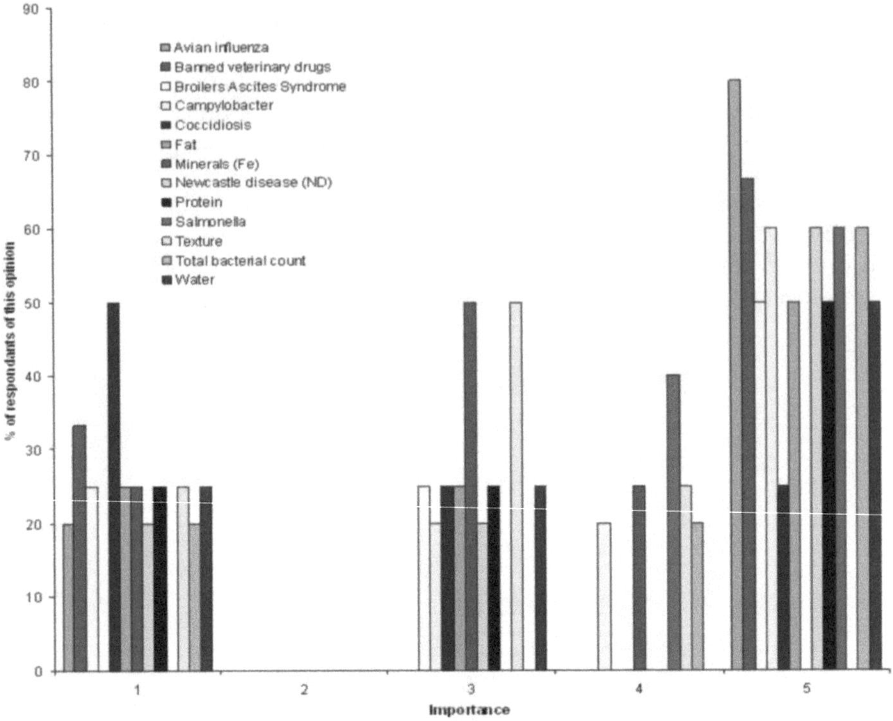

Fig. 3. The importance of the recorded parameters

Table 2. Shows the parameters which were recorded by one or more of the respondents

Water
Protein
Fat
Minerals (Fe)
Texture
Newcastle disease (ND)
Coccidiosis
Broilers Ascites Syndrome
Avian influenza
Total bacterial count
Salmonella
Campylobacter
Banned veterinary drugs

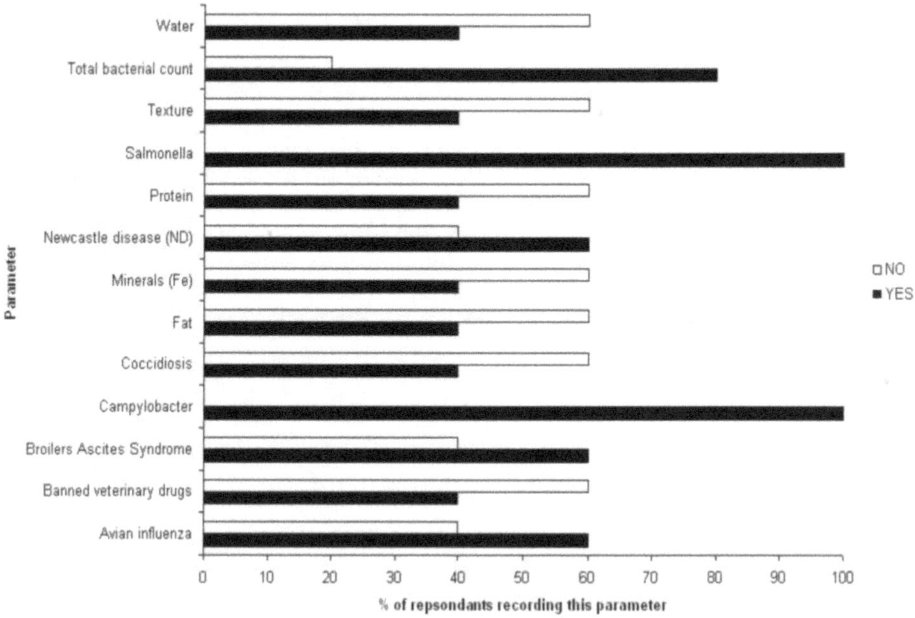

Fig. 4. Number of respondents to the survey who record this parameter

The identified data elements were then integrated into a sector specific vocabulary of the TraceCore Abstract model (TCA). TCA is a generic abstract model based on the TraceFish electronic standard that describes and specifies all the fundamental traceability elements which need to be in place to achieve both internal and external traceability. In addition to the TCA there is one core vocabulary, extensions or underlying vocabularies, and also various sector specific vocabularies. This can be seen in fig.5. Furthermore the TCA documentation describes how the model is mapped to other industry standards like Electronic Product Code Information Services (EPCIS) from Electronic Product Code Global (EPCglobal) and Universal Business Language (UBL) from OASIS.

4 Discussion

In food supply chains there are many data elements that must already be recorded for a variety of reasons. The results presented here are a contribution towards meeting the traceability needs of the chicken processing sector by the provision of a list of standard data elements.

The data elements presented in the results have been standardized by industry consultation and consultation with researchers. The standardization is limited as the research was devised for one specific purpose, that being to enable effective communication [16, 17]. In the European Union (EU) directive regarding nutritional labelling for foodstuffs [18] it is in fact stipulated that this labelling should take place in a standardized fashion. Since such work is becoming increasingly important an

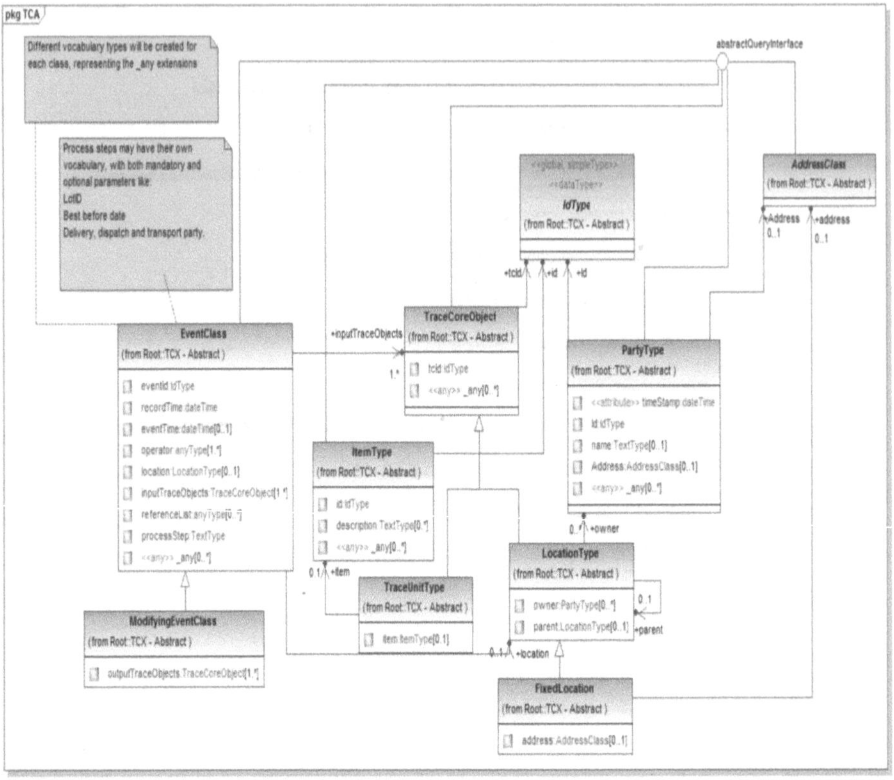

Fig. 5. TraceCore Abstract Model

extension of this work including multiple links in the chicken supply chains should be carried out at a European and also at a global level as import is often of major motivating factor with regards to cross border cross country information exchange.

One of the most interesting findings here is presented in fig.2. The companies clearly communicated over half of the parameters which they record, thus emphasising the need for standards. The corollary to this however is that the remainder of the parameters were not uniform, indicating the need for any standards that are adopted to be universal. Fig. 6 shows the difference between information exchange before and after standardisation which will enable effective communication between Food Business Organisations [16, 17].

The data presented in fig. 2 is particularly interesting as this an area which has until now been only rarely investigated [19]. Honey producers, in previous studies, showed a strong set of opinions regarding the importance of the different parameters, for instance classifying contaminants as extremely important and nutritional values as less important [4]. No such pattern can be seen in the results from the chicken processors. This may reflect a general characteristic in the chicken processing industry whereby no data elements are considered more important that others. The results may also be a reflection of the small number of respondents. The authors would credit the lack of response to the reluctance of chicken processors to share any information with bodies outside their own companies. This is understandable in light of the serious food incidents which have involved chicken in recent years, noted in earlier in this [12]. This once again emphasises the need for industry wide consultation and cooperation in appropriate ways to exchange information. The issue of confidentiality and the ownership of data in global food supply chains is an important area for further research. It brings up again the debate regarding what information should be shared and with whom, the so called 'confidentiality' debate. This is an area clearly highlighted for further research and which will be of the greatest importance with regards to international trade and future traceability initiatives.

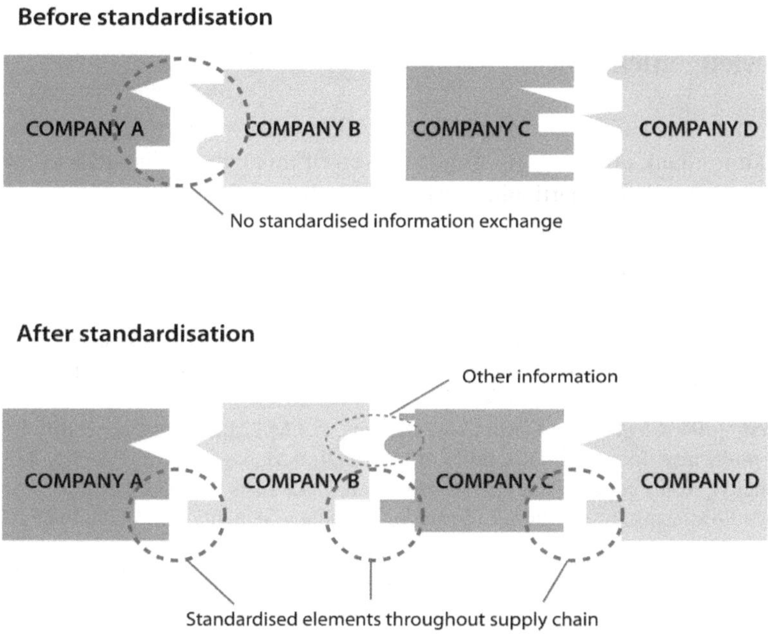

Fig. 6. Illustration of information exchange before and after standardisation

The standards, which the authors have attempted to create or 'initiate' and also to be seen in Donnelly et al. [4] are intended for use in a chicken supply chain to enable the effective–exchange of information across multiple software solutions. A final decision was taken to include all possible parameters in order to give companies the greatest possible choices for communications. The authors are interested to see whether this large choice of information causes the preliminary standard to be too large to be useful. The information gathered in such surveys could form the basis for standardized electronic interchange in the supply chain, for instance as an extension of the Universal Business Language (UBL); UBL is a library of standard electronic Extensible Markup Language (XML) business documents such as purchase orders and invoices developed and supported by Organization for the Advancement of Structured Information Standards(OASIS) and already supported by many national governments; in particular by Denmark and Iceland. It would be appropriate to extend this work and carry out a similar survey on all links in the chicken supply chain in order to test the use of such standards along supply chains similar to that carried out in the TraceFish work [5, 6]. A further appropriate extension of this work could be global consultation with organizations in the chicken supply chain which may result in the identification of new and useful data elements for inclusion in a universal standard. This work contributes to a larger aim of international standards which are necessary so that food business organisations and regulatory bodies in the international food market no longer have to handle multiple identification systems for data. It will be interesting to learn what effect new standards have on regulatory authorities and the chicken sector with regards to issues of confidentiality and data sharing.

Acknowledgements

This research was carried out as part of the EU-funded project TRACE. The authors would like to thank all involved in the creation of this paper. The authors would like to thank Oddvar Dahl in particular for help with the graphics.

References

1. FSA, Traceability in the Food Chian- A Preliminary Study, FSA, Editor (2002)
2. Moe, T.: Perspectives on traceability in food manufacture. Trends in Food Science & Technology 9(5), 211–214 (1998)
3. Jansen-Vullers, M.H., van Dorp, C.A., Beulens, A.J.M.: Managing traceability information in manufacture. International Journal of Information Management 23(5), 395–413 (2003)
4. Donnelly, K.A., et al.: Creating Standardized Data Lists for Traceability – A Study of Honey Processing. International Journal of Metadata, Semantics and Ontologies 3(4), 283–291 (2008)
5. CEN14659, CEN Workshop Agreement. Traceability of Fishery products. Specification of the information to be recorded in caught fish distribution chians. European Committee for Standardization (2003)
6. CEN14660, CEN Workshop Agreement. Traceability of Fishery products. Specification of the information to be recorded in farmed fish distribution chians. European Committee for standardization (2003)
7. Denton, W.: TraceFish: The development of a traceability scheme for the fish industry. In: Luten, J.O., Olafsdottir, G. (eds.) Quality of fish from catch to consumer, pp. 75–91. Wageningen Academic Publishers, Wageningen (2003)
8. FAO, AGROVOC (2006), `http://www.fao.or/aims/ag_intro.htm`
9. Haverkort, A.: The Canon of Potato Science: 36. Potato Ontology. Potato Research 50(3), 357–361 (2007)
10. Haverkort, A., Top, J., Verdenius, F.: Organizing Data in Arable Farming: Towards an Ontology of Processing Potato. Potato Research 49(3), 177–201 (2006)
11. Stuckenschmidt, H.: Ontology based information in dynamic environments. In: Twelfth IEEE Internation Worskshops on Enabling Technologies: Infrastructures for Collaborative Enterprises(WETICE 2003). IEEE, Linz (2003)

12. RASFF, Rapid Alert System for Food and Feed (2008), http://ec.europa.eu/food/food/rapidalert/reports/week44-2008_en.pdf, EC, Editor
13. Karlsen, K.M., Van der Roest, J., Olsen, P.: Traceability of Mineral Water – Specification of the Information to be Recorded in Mineral Water Distribution Chains, in Nofima Reports Nofima, Editor (2008)
14. Olsen, P., Forås, E.: Analysis of traceability in the chicken processor Beijing Dafa Chia Tai Co, in Nofima Report, Nofima, Editor (2008)
15. Donnelly, K.A.-M., et al.: Traceability of chicken – Specifications of the information to be recorded at chicken slaughter/processing establishments and other links in chicken distribution chains, in Nofima Report, Nofima, Editor (2008)
16. Folinas, D., Manikas, I., Manos, B.: Traceability data management for food chains. British Food Journal 108(8), 622–633 (2006)
17. Dreyer, H.C., et al.: Traceability standards and supply chain relationsships. In: Proceedings of NOFOMA (the Nordic Logistics Research Network Conference), Linkoping Sweden (2004)
18. EC, Guidance on the implementation of articles 11,12,16,17,18,19 and 20 of regulation (EC) NO178/2002 on General Food Law. Conclusions of the standing committee on the food chain and animal health, EC, Editor, EC (2004)
19. Pálsson, P.G., et al.: Traceability and electronic transmissions of qualitative data for fish products, in Status report no. 3. Danish Institute for Fisheries Research, Lyngby, Denmark, Department of Seafood Research (June 2000)

Navigation as a New Form of Search for Agricultural Learning Resources in Semantic Repositories

Ramiro Cano, Alberto Abián, and Elena Mena

Information Engineering Research Unit, University of Alcalá, 28871, Alcalá de Henares, Spain
{ramiro.cano,alberto.abian,elena.mena}@uah.es

Abstract. Education is essential when it comes to raise public awareness on the environmental and economic benefits of organic agriculture and agroecology (OA & AE). Organic.Edunet, an EU funded project, aims at providing a freely-available portal where learning contents on OA & AE can be published and accessed through specialized technologies. This paper describes a novel mechanism for providing semantic capabilities (such as semantic navigational queries) to an arbitrary set of agricultural learning resources, in the context of the Organic.Edunet initiative.

Keywords: Navigational search, Learning resources, Agriculture, Repository, Semantic Web.

1 Introduction

In the last few years, the emerging "organic agriculture" philosophy is becoming more important for our society as a mean to achieve the sustainability in the agricultural processes. In order to spread the knowledge related to organic agriculture and agroecology (OA&AE), education arises as one of the key elements, both in the teaching and researching contextes. This circumstance has led to a higher availability of digital learning resources [1] related the mentioned disciplines.

Learning resources are usually annotated with metadata to provide them with additional qualitative information, often making use of specific standard schemas such as IEEE LOM [2]. It is a common practice to store such metadata in specialized collections often called *metadata repositories*. These metadata repositories do not, in most cases, contain the resources themselves, but only their metadata according to given metadata schemas.

Semantic learning resources repositories [3], a certain type of repositories that make use of ontologies [4], are an efficient way to meaningfully manage these metadata. This ontology-based solution allows a knowledge-based representation of metadata which provides enhanced semantic capabilities, such as the ability to perform automatic reasoning on the data stored. An example of these semantic repositories for learning resources is the so-called LOMR (Learning Object Metadata Repository) [5] architecture, developed as a part of the EU-funded project LUISA (http://www.luisa-project.eu/).

F. Sartori, M.Á. Sicilia, and N. Manouselis (Eds.): MTSR 2009, CCIS 46, pp. 322–327, 2009.

Since it can be considered that the value of the learning resources stored in any repository is less significant without any interaction with final users or external repositories, searching capabilities arise as one of the key features for this kind of ontology-based systems. In the particular case of the Ont-Space framework, the semantic search approach is carried out by the so-called *navigational queries mechanism*, as it will be later described.

The rest of this paper is structured as follows. Section 2 provides general information on the technical foundations of this work. Section 3 details the algorithms responsible for the navigational semantic search. Finally, conclusions and future work are outlined in section 4.

2 Technical Foundations

2.1 Overall Architecture of Organic.Edunet

One of the main objectives of the Organic.Edunet Project is to integrate and specialize state-of-art technologies of the World Wide Web in order to provide end-users with a single European reference point (the Organic.Edunet Web portal [6]) that will offer advanced services such as ontology-based searching and social recommendation and will facilitate search, retrieval and use of the collected content. The global architecture of Organic.Edunet is thus composed of two major subsystems (as seen in Fig. 1).

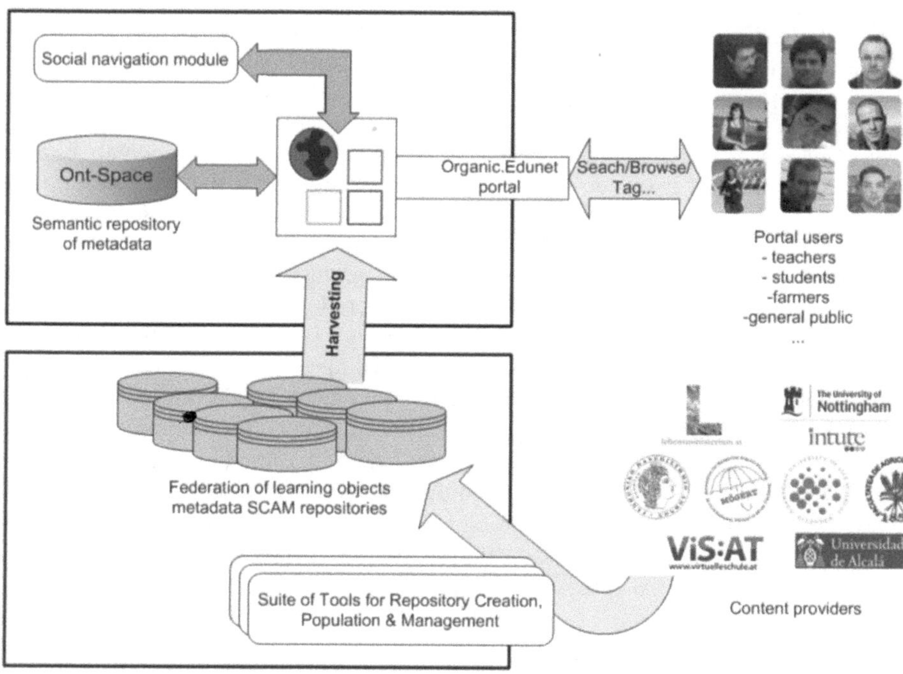

Fig. 1. The Organic.Edunet project architecture

The federation of repositories subsystem includes the repositories containing either the learning resources uploaded by the content providers of the Organic.Edunet project, or their metadata or both. As Organic.Edunet aims to spread high quality learning resources related to OA & AE (dealing with topics such as animal welfare, food safety, water quality, soil issues, etc.) the above-mentioned federation of repositories is only fed with resources certified by a quality process carried out by experts from european associations and universities.

The Organic.Edunet Web portal is the second subsystem, and provides general services to portal users (such as learners, teachers, farmers, etc.), each type defining a type of access to the learning resources cataloged. The information about the resources available is automatically fed into this subsystem from the federation of repositories [7]. Since the main scope of the Web portal is allowing the final users the utilization of the semantic search capabilities of the underlying repository, it is necessary to define the research technologies which sustain the foundations of such repository.

2.2 Semantic Technologies

The Ont-Space [8] semantic framework is responsible for providing to the Organic.Edunet Web Portal with the infrastructure for the meaningful management of learning resources metadata in a semantic format, using the Ontology Web Language (OWL) [9].

Built upon the Jena [10] Semantic Web framework, the Ont-Space repository makes use of these experimental technologies, often used only in research environments, and transfers them into a production environment aimed for end user interaction.

Using the OWL language, any specific knowledge domain (and particularly the OA&AE knowledge domain [11], for the purposes of the Organic.Edunet project) can eventually be represented as an ontology, and so it can be stored in such a semantic repository, and used in order to perform the semantic capabilities associated to it.

Ontology-based systems allow performing automatic reasoning, which can be achieved by the use of reasoners to infer new information from the data in the ontological model through the application of logical rules. In particular, Ont-Space uses the lom2owl ontology, a mapping of IEEE LOM to the OWL ontology language, to store learning resources' metadata in a semantic form, thus facilitating the application of reasoning capabilities based on the information stored in learning objects metadata.

In the above-mentioned lom2owl ontology, learning objects are represented as instances of the concept `learningObject`. Regarding its metadata, certain categories from the IEEE LOM standard are first level concepts in the ontology, such as IEEE LOM categories *7.Relation* (`lomRelation` concept), *8.Annotation* (`lomAnnotation` concept) or *9.Classification* (`lomClassification` concept); whilst other IEEE LOM elements are represented through several ontology concepts or attributes, such as category *3.Meta-Metadata*, represented as several attributes in the `learningObject` concept, while its sub-section *3.2.Contribute* is represented as a first level concept (`contribution` concept).

3 Semantic Search

The semantic navigation algorithm is based upon the existing relations between the ontology elements (namely, `OntResource` in the Jena API) for a given set of ontologies stored in the semantic repository. These ontology elements can be either a class in the ontology (`OntClass` in the Jena API) or an individual created from the instantiation of a given class (`Individual` in the Jena API). So, we can define the *sub-elements* for a given ontology class as both the subclasses of such class, and the individuals created from its instantiation.

From the above-mentioned definitions, the ontology elements altogether can be seen as a tree, being the inner nodes the ontology classes, and the leaf nodes the individuals, as they have no sub-elements so far. With this data structure, the ontology can be traversed through the navigation of its elements.

We also define a set of *related elements* for an arbitrary ontology element from its attributes. In the case of an ontology class, a given attribute will have an associated data type. This data type can be either a primitive data type or any ontology class. If the data type of the attribute is an ontology class, we call it a *related element* for the parent ontology class which has such attribute. In the case of an individual created from such an ontology class, the attribute value would be an individual, which would be instantiated from the ontology class that is the data type for the attribute of the parent ontology class. We call this individual a *related element* for the parent individual.

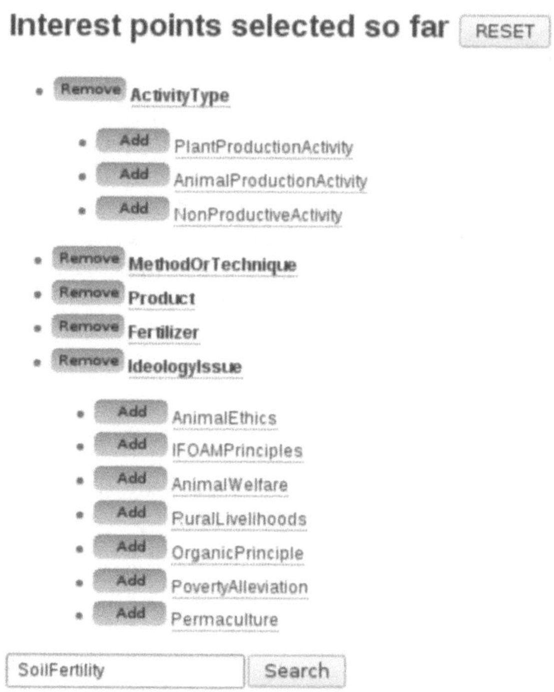

Fig. 2. A prototype on the semantic navigation interface

The semantic navigation is carried out through the navigation of both the sub-elements and the related elements of the interest points selected by the user. In fact, category *9.Classification* of the IEEE Learning Objects Metadata (LOM) standard has been specifically targeted for semantic search, even though other categories such as *5.2.Educational Resource Type* or *1.3.Language* can be also used as a basis for semantic searches by nation / language. The purpose for this category is '*describing where an object is classified in a given classification system*'. Each classification (being *none*, *one* or *many*) assigned to the object is described by four information elements (*purpose*, *taxon path*, *description* and *keyword*). According to this schema, each classification may have associated a given purpose. This attribute indicates the purpose aimed in the classification of this particular learning object, and so it allows, through the search by the classification, filtering the contents in the ontology so that we can obtain every learning object which have a given *purpose*.

The results from this search allow the categorization and classification of the learning objects, so we can group learning objects that share the same purpose. For example, if we navigate on a learning object whose *educational objective* takes the value *Fertilizer*, through the search by classification we will be able to navigate to other learning objects in the same classification schema, i.e. classified according to elements in the same ontology related to the term *Fertilizer*. This is particularly useful when different ontologies are in use, so that users performing semantic navigation do seamlessly change from one ontology to another in a transparent manner.

4 Conclusions

The presented navigational process is the basis for the semantic search mechanism embedded in the Web Portal of the Organic.Edunet Project. It is built upon the Ont-Space semantic framework technology used in the semantic repository.

As a main objective, this new navigational search mechanism aims at obtaining of meaningful results for the final users by mixing up the presented research technologies.

Future work will be focused on the improvement of the usability issues related to the user interface, so that the navigational search interface may be compliant with the most extended usability guidelines. Also, the improvement of the automatic reasoning capabilities of our framework is a very important concern regarding our work.

Acknowledgments

The work presented in this paper has been funded by the European Commission, project Organic.Edunet (ECP-2006-EDU-410012).

The authors would like to express their gratitude to Dr. Salvador Sánchez-Alonso, for his continued help and support.

References

1. Polsani, P.R.: Use and Abuse of Reusable Learning Objects. Journal of Digital Information 3(4), `http://jodi.tamu.edu/Articles/v03/i04/Polsani/`
2. IEEE LTSC, Learning Technology Standards Committee: IEEE 1484.12.1-2002 Draft Standard for Learning Object Metadata (2002)
3. Soto, J., García, E., Sánchez-Alonso, S.: Semantic learning object repositories. International Journal of Continuing Engineering Education and Life-Long Learning 17(6), 432–446 (2007)

4. Gruber, T.R.: Toward principles for the design of ontologies used for knowledge sharing. International Journal Human-Computer Studies 43(5-6), 907–928 (1995)
5. Sicilia, M.A., Sánchez-Alonso, S., Arroyo, S., Martín-Cantero, S.: LOMR overall architecture. Deliverable D4.1, LUISA Project (IST- FP6 – 027149)
6. Sánchez-Alonso, S., Mena, E., García-Barriocanal, E., Cano, R., Abián, A., Manouselis, N., Kastradas, K., Ebner, H., Rodríguez, D.: Design and Specification of the Organic.Edunet Web Portal. Deliverable D5.3.1, Organic.Edunet Project (ECP-2006-EDU-410012)
7. Abián, A., Cano, R.: Harvesting Learning Resources on Organic Agriculture and Agroecology to Semantic Repositories. In: Proceedings of HAICTA 2008, 4th International Conference on Information and Communication Technologies in Bio and Earth Sciences, pp. 533–537 (2008)
8. Ont-Space: Java-based software framework providing the services of a semantic metadata repository, `http://sourceforge.net/projects/ont-space/`
9. OWL Web Ontology Language Overview, `http://www.w3.org/TR/owl-features/`
10. Jena is a Java framework for building Semantic Web applications, `http://jena.sourceforge.net/`
11. Organic.Edunet Consortium: D2.2.3 OA and Agroecology Domain Model Representation (2008)

Author Index

Kolibri 360 GmbH
Ferdinand-Jühlke-Straße 7
99095 Erfurt, Deutschland
produktsicherheit@kolibri360.de